NUCLEAR ENERGY
IN THE GULF

Nuclear Energy in the Gulf

The Emirates Center for Strategic Studies and Research

THE EMIRATES CENTER FOR STRATEGIC STUDIES AND RESEARCH

The Emirates Center for Strategic Studies and Research (ECSSR) is an independent research institution dedicated to the promotion of professional studies and educational excellence in the UAE, the Gulf and the Arab world. Since its establishment in Abu Dhabi in 1994, the ECSSR has served as a focal point for scholarship on political, economic and social matters. Indeed, the ECSSR is at the forefront of analysis and commentary on Arab affairs.

The Center seeks to provide a forum for the scholarly exchange of ideas by hosting conferences and symposia, organizing workshops, sponsoring a lecture series and publishing original and translated books and research papers. The ECSSR also has an active fellowship and grant program for the writing of scholarly books and for the translation into Arabic of work relevant to the Center's mission. Moreover, the ECSSR has a large library including rare and specialized holdings, and a state-of-the-art technology center, which has developed an award-winning website that is a unique and comprehensive source of information on the Gulf.

Through these and other activities, the ECSSR aspires to engage in mutually beneficial professional endeavors with comparable institutions worldwide, and to contribute to the general educational and academic development of the UAE.

The views expressed in this book do not necessarily reflect those of the ECSSR.

First published in 2009 by
The Emirates Center for Strategic Studies and Research
PO Box 4567, Abu Dhabi, United Arab Emirates

E-mail: pubdis@ecssr.ae
Website: http://www.ecssr.ae

ISBN: 978-9948-14-116-7 hardback edition
ISBN: 978-9948-14-117-4 paperback edition
ISBN: 978-9948-14-118-1 electronic edition

CONTENTS

FIGURES AND TABLES

FIGURES

TABLES

ABBREVIATIONS AND ACRONYMS

ABWR	advanced boiling water reactor
ACNS	Advisory Committee on Nuclear Safety
ACPSR	Advisory Committees for Project Safety Review
ADFEC	Abu Dhabi Future Energy Company
ADS	accelerator-driven subcritical system
AEA	Atomic Energy Authority (Egypt)
AECC	Angarsk Electrolysis Chemical Complex (Russia)
AECL	Atomic Energy of Canada Ltd.
AEDB	Alternative Energy Development Board (Pakistan)
AEET	Atomic Energy Establishment Trombay
AERB	Atomic Energy Regulatory Board
AERI	Atomic Energy Research Institute (Saudi Arabia)
AFFF	advanced fuel fabrication facility
AGR	advanced gas-cooled reactor
AHWR	advanced heavy water reactor
AIP	air-independent propulsion
ANC	African National Congress
APWR	advanced pressurized water reactor
BARC	Bhabha Atomic Research Centre
BBC	British Broadcasting Corporation
BMR	balancing, modernization and rehabilitation
BOO	build–own–operate
BOP	balance of plant
BRIT	Board of Radiation and Isotope Technology
BSE	bovine spongiform encephalopathy (mad cow disease)
BWC	Biological Weapons Convention
BWR	boiling water reactor
CANDU	Canada–Deuterium–Uranium (reactor)
CASTOR	cask for storage and transport of radioactive material

CCGT	combined cycle gas turbine
CCS	carbon dioxide capture and storage
CDU	Christian Democratic Union Party (Germany)
CERN	European Organization of Nuclear Research
CFR	Code of Federal Regulations
CGPA	cumulative grade point average
CHASCENT	CHASNUPP Centre of Nuclear Training
CHASNUPP	Chashma Nuclear Power Plant
CHTR	compact high temperature reactor
Ci/km^2	curies per square kilometer
CIA	Central Intelligence Agency
CNNC	China National Nuclear Corporation
CNSTN	National Center for Nuclear Science and Technology (Tunisia)
COG	CANDU Owners Group
CPPNM	Convention on the Physical Protection of Nuclear Material
CRS	Congressional Research Service
Cs-137	caesium-137
CSU	Christian Social Union Party (Germany)
CTC	Computer Training Centre (Pakistan)
CWC	Chemical Weapons Convention
DAD	decide, announce, defend
DAE	Department of Atomic Energy (India)
DBT	design-basis-threat
DGFS	DAE Graduate Fellowship Scheme
DNSRP	Directorate of Nuclear Safety and Radiation Protection (Pakistan)
DOE	US Department of Energy
ECCS	emergency core cooling system
EDF	*Èlectricité de France*
EEG	Renewable Energy Sources Act (Germany)

EEWärmeG	Renewable Energies Heat Act (Germany)
EIA	Energy Information Administration
EIA	environmental impact assessment
ENEC	Emirates Nuclear Energy Corporation
EnEV	Energy Saving Ordinance (Germany)
EPA	Environmental Protection Agency
EPACT	Energy Policy Act
EREF	European Renewable Energies Federation
ESBWR	economic simplified boiling water reactor
ESP	Energy Security Plan
ETP	effluent treatment plant
EURATOM	European Atomic Energy Community
FBR	fast breeder reactor
FBTR	fast breeder test reactor
FDP	Liberal German Party
FMCT	Fissile Material Cutoff Treaty
FNCA	Forum for Nuclear Cooperation in Asia
FRAMATOME	*Franco-Américaine de Constructions Atomiques*
FRG	Federal Republic of Germany
GCC	Gulf Cooperation Council
GCR	gas-cooled reactor
GDA	generic design assessment
GDP	gross domestic product
GeV	giga-electron volt
GMO	genetically modified organism
GNEP	Global Nuclear Energy Partnership
GNPI	Global Nuclear Power Infrastructure
GW	gigawatt
GWd/t	gigawatt days per ton
GWe	gigawatt (electrical)
GWth	gigawatt (thermal)
HAW	high-level radioactive waste

HDI	Human Development Index
HEER	highly efficient environmentally-friendly reactor
HESCO	Himalayan Environment Studies and Conservation Organization
HEU	highly enriched uranium
HLW	high-level waste
HM	heavy metal
HTES	high temperature electrolysis of steam
HTR	high temperature reactor
IAEA	International Atomic Energy Agency
IGCAR	Indira Gandhi Center for Atomic Research
IIT	Indian Institute of Technologies
ILW	intermediate level waste
IMRSS	international monitored retrievable storage system
INFCC	international nuclear fuel cycle center
INFCE	international fuel cycle evaluation
INPRO	International Project on Innovative Nuclear Reactors and Fuel Cycles (IAEA)
INSC	International Nuclear Safety Center
IPCC	UN Inter-governmental Panel on Climate Change
IRBM	intermediate range ballistic missile
IREL	Indian Rare Earths Ltd.
ISF	Improving Safety Features (of KANUPP)
ITER	International Thermonuclear Experimental Reactor
IUEC	International Uranium Enrichment Center
JAEA	Japan Atomic Energy Agency
KAMINI	Kalpakkam MINI
KANUPP	Karachi Nuclear Power Plant
KAPS	Kakrapar Atomic Power Station
KASCT	King Abdulaziz City for Science and Technology
KESC	Karachi Electric Supply Company
kgHM	kilogram of heavy metal

KINPOE	The KANUPP Institute of Nuclear Power Engineering
KRUSHAK	Krushi Utpadan Sanrakshan Kendra
kWth	kilowatts (thermal)
LEU	low enriched uranium
LRBM	long-range ballistic missile
LWGR	light water-cooled graphite-moderated reactor
LWR	light water reactor
mA	milliampere
MAED	model for analysis of energy demand
MAPS	Madras Atomic Power Station
MCi	mega curies
MED	multi-effect distillation
MENWFZ	Middle East Nuclear Weapons-Free Zone
MeV	mega-electron volt
MHI	Mitsubishi Heavy Industries Ltd.
MHT	main heat transport
MINATOM	Ministry for Atomic Energy (Russia)
MITEI	MIT Energy Initiative
mmcfd	million cubic feet per day
MMR	measles–mumps–rubella
MoU	memorandum of understanding
MOX	mixed oxide
MSF	multistage flash
MT	metric tons
MTHM	metric tons of heavy metal
MW	megawatt
MWe	megawatt (electrical)
MWth	megawatt (thermal)
NAPS	Narora Atomic Power Station
NATEC	National Atomic Energy Commission (Yemen)
NATO	North Atlantic Treaty Organization

NCNDT	The National Center for Non-Destructive Testing (Pakistan)
NFC	nuclear fuel complex
NGO	non-governmental organization
NII	Nuclear Installations Inspectorate
NNSA	National Nuclear Safety Authority (China)
NNWS	non-nuclear weapon states
NOC	non-objection certificate
NPCIL	Nuclear Power Corporation of India
NPP	nuclear power plant
NPR	US Nuclear Posture Review
NPT	Non-Proliferation Trust
NPT	Nuclear Non-Proliferation Treaty
NRB	Nuclear Regulatory Body
NRC	Nuclear Regulatory Commission
NSG	Nuclear Suppliers Group
NSSS	nuclear steam supply system
NTDC	National Transmission and Dispatch Company (Pakistan)
NTI	Nuclear Threat Initiative
NUMEC	Nuclear Materials and Equipment Corporation
NWS	nuclear weapon states
PA	public acceptance (of nuclear energy/Japan)
PACT	Program of Action for Cancer Therapy
PAEC	Pakistan Atomic Energy Commission
PCEC	per capita electricity consumption
PET	positron emission tomography
PFBR	prototype fast breeder reactor
PHWR	pressurized heavy water reactor
PIEAS	Pakistan Institute of Engineering and Applied Sciences
PINSTECH	Pakistan Institute of Nuclear Science and Technology
PLO	Palestine Liberation Organization

PNRA	Pakistan Nuclear Regulatory Authority
PPC	public participation conference
PPED	Power Projects Engineering Division
PRC	People's Republic of China
PSA	probabilistic safety assessment
PTR	pressure tube reactor
PURNIMA	plutonium reactor for neutron investigations in multiplying assemblies
PVO	Pohjolan Voima Oy
PWR	pressurized water reactor
QA	quality assurance
QM	quality management
R&D	research and development
RAFAEL	Israeli Armaments Development Authority
RANF	Reliable Access to Nuclear Fuel
RAPS	Rajasthan Atomic Power Station Units
RES	renewable energy sources
RIA	radioimmunoassay
RO	reverse osmosis
RuTAG	Rural Technology Action Group
SARCAR	Safety Review Committee for Applications of Radiation
SARCOP	Safety Review Committee for Operating Plants
SFZ	Special Free Zone (Masdar)
SIPRI	Stockholm International Peace Research Institute
SNERDI	Shanghai Nuclear Engineering Research and Design Institute
SOK	safe operation of KANUPP
SPD	Social Democratic Party (Germany)
SRBM	short-range ballistic missile
SSR	super-heated steam-cooled reactor
SSSF	solid storage and surveillance facility

SWD	solid waste disposal
Synroc	synthetic rock
TAEK	Turkish Atomic Energy Authority
TAPS	Tarapur Atomic Power Station
TDP	Technology and Development Program (Masdar)
TEPCO	Tokyo Electric Power Company
THOREX	thorium uranium extraction
TNRC	Tajoura Nuclear Research Centre
TOE	tons of oil equivalent
TUP	technological up-gradation Project
TVO	Teollisuuden Voima Oy
TWh	terawatt hours
UCI	Uranium Corporation of India Ltd.
UCTE	Union for Coordination of Transmission of Electricity
UIL	university–industry linkage
UMP	uranium metal plant
UNSCR	UN Security Council Resolution
VVER	*Vodo-Vodyanoi Energetichesky Reactor* (Russian PWR)
WANO	World Association of Nuclear Operators
WAPDA	Water and Power Development Authority (Pakistan)
WEU	Western European Union
WiN	Women in Nuclear
WIP	waste immobilization plant
WMD	weapons of mass destruction
WMDFZ	WMD-free zone
WNA	World Nuclear Association
ZERLINA	zero energy reactor for lattice investigations and new assemblies

FOREWORD

The future heralds a potential crisis in power supply in view of growing global and domestic demand for fossil fuel energy, particularly in supplying power for seawater desalination in the Arabian Gulf region. The possibility of fossil fuel depletion on the one hand, and the increasing rate of development in the GCC countries on the other, leaves us to consider the development of peaceful nuclear energy as an alternative source of power for the Gulf region.

There are currently some 436 commercial nuclear power reactors operating in 30 countries worldwide, providing fifteen percent of the world's electricity needs. Nuclear energy is a clean and stable source of power, and is considered a viable strategic and economic option for the countries of the GCC states in light of the increasing demand for energy derived from hydrocarbons.

In order to ensure the success of peaceful nuclear energy programs in the Gulf, support in the form of international experts, technicians and specialists will be required. Furthermore, the establishment of adequate security infrastructure will be vital in assuring transparency and bolstering both national and international security.

In order to provide a realistic strategic view of this topic, including various strategies with which to guarantee its achievement and benefit from the experience of other countries, the Emirates Center for Strategic Studies and Research (ECSSR) convened its 14th Annual Energy Conference under the title "Nuclear Energy in the Gulf," on November 24–26, 2008 in Abu Dhabi, hosting a group of distinguished energy experts from various academic, professional and technical backgrounds.

This book comprises a valuable collection of the papers presented at the conference. It identifies the motives behind the development of peaceful nuclear energy programs in the Gulf and the local and regional implications of such programs, and examines the role of nuclear energy in the context of Gulf energy security and climate change. The inherent risks and opportunities of managing nuclear power are explored through a discussion of security infrastructure, non-proliferation standards and the role of international cooperation in the field of nuclear technology and long-term development.

The book also analyzes factors pertaining to the economic viability of the nuclear energy option and its political and social implications, drawing on international experiences by reviewing models of existing civilian nuclear programs in other parts of the world. Examples presented in this volume include case studies of Japan and Germany, and a comparative outlook assessing the respective programs of Iran, Israel, Pakistan and India.

It is fitting at this juncture to express my gratitude to all the speakers for their participation in the ECSSR 14th Annual Energy Conference. Their informative presentations compiled in this volume offer sound insight and informed perspectives on the theme of the conference. I would also like to express my appreciation to the distinguished academics who served on the referee panel, reviewing the conference papers prior to publication and offering their critical assessment.

Finally, thanks are due to ECSSR Editor Francis Field for coordinating the publication of this book, as well as to the other members of the Department of Publications who assisted during the course of the project.

Jamal S. Al-Suwaidi, Ph.D.
Director General
ECSSR

INTRODUCTION

Nuclear Power in the Gulf Energy Mix

The world is currently seeking new sources of energy and research is underway to find viable alternatives that are less harmful to the environment than the current reliance on fossil fuels. In light of this, nuclear energy will become one of the most important sources of global energy in the coming decades owing to diminishing reserves of fossil fuels, continued population growth, and ongoing economic and social development. Nuclear energy is considered the least harmful to the environment of all viable energy sources, but there is an obvious need to devise and enforce strict regulations for the organization and supervision of all phases of the nuclear fuel cycle. This, of course, requires a strengthening of regional security.

Despite the perceived risks of using nuclear energy, there are around 436 nuclear power reactors worldwide providing fifteen percent of global electricity needs. Nuclear energy is a clean and stable source of power, and is considered a viable strategic and economic option for the countries of the Gulf Cooperation Council (GCC) in light of the growing demand for energy derived from fossil fuels. Utilizing nuclear energy for peaceful purposes, it is suggested, would push ahead comprehensive national development plans in these countries.

However, the need for peaceful nuclear energy in the Gulf will be satisfied only with the support of relevant experts, technicians and specialists, the establishment of the necessary infrastructure, the implementation of adequate nuclear safety regimes, the adoption of measures to assure transparency and bolster both national and international security, and the formulation of strategies and policies that guarantee continuity of operation based on a national labor force. In this regard, the UAE prepared and formally declared its Policy on the Evaluation and Potential Implementation of Nuclear Energy in April 2008.

In his Keynote Address, Hans Blix stresses the importance of non-proliferation measures, as well as protecting nuclear installations from sabotage. He insists that states can and must take effective measures to ensure that all nuclear material within and crossing their borders remains in authorized hands to prevent non-state actors acquiring such materials. He highlights Iran's uranium enrichment program, which has engendered considerable tension in the region, and suggests that to avoid exacerbating this situation it would be desirable that any state embarking on nuclear power in the Middle East should declare – as Abu Dhabi has done – that it will not build facilities for the enrichment of uranium or production of plutonium. Such declarations, writes Blix, will enhance confidence and should be accompanied by all parties to the Non-Proliferation Treaty (NPT) – including those in the Gulf – accepting the Additional Protocol of the International Atomic Energy Agency (IAEA), strengthening its inspection powers and thereby ensuring no clandestine nuclear activities are taking place.

Jungmin Kang draws a similar conclusion, arguing that for their nuclear power projects to succeed, and in order to ease related international concerns, Middle Eastern countries must strongly commit to abide by the global nuclear non-proliferation regime; specifically, he states, they should accept both the full IAEA safeguards and the Additional Protocol to ensure their peaceful use of nuclear energy and high standards of nuclear safety. He also points out that it is important for these states to cooperate closely with international societies in the general areas of nuclear fuel supply assurance, spent fuel management, nuclear non-proliferation, and nuclear safety.

Charles Ferguson also supports this approach, adding that each state should be expected to sign the Additional Protocol as a condition of sale under the Nuclear Suppliers Group's guidelines. In his detailed overview of the requirements and risks associated with nuclear power programs he examines the security infrastructure required for states to establish nuclear programs, as well as the necessity of design-basis-threat analyses, defense-in-depth safety infrastructure, and the variety of potential modes of attack that might be employed by terrorists, saboteurs, or hostile military forces against nuclear facilities. He also examines whether new nuclear power plants can include design features to make them more impervious to attacks or sabotage and discusses the emerging international nuclear security regime, highlighting areas where further improvements are needed. Concerning non-proliferation standards, he underscores the fact that states have both rights and responsibilities when it comes to nuclear technology. All states have the right of access to peaceful nuclear technologies and technical assistance from the IAEA or developed nuclear power states, he explains, but these rights are contingent on a state assuming its responsibilities to maintain strong safeguards as well as transparency regarding its nuclear intentions and capabilities.

On the topic of nuclear transparency, John Steinbach discusses the status of nuclear technology in Israel, focusing principally on its nuclear weapons program. He argues that when nuclear proliferation is discussed in international forums, the issue of Israel's substantial nuclear arsenal is too often ignored or downplayed. Despite maintaining a policy of neither confirming nor denying possession of nuclear weapons for over 40 years, he writes, Israel is now the world's fifth largest nuclear power with some 200 nuclear warheads in its arsenal. Steinbach charts the assistance received from France, South Africa, the United States and others, that enabled Israel to clandestinely develop a sophisticated nuclear weapons arsenal including miniaturized fusion-boosted fission bombs, neutron bombs and possibly even thermonuclear bombs which, he states, places the entire Middle East, Europe and much of Asia and Africa within its range. Ultimately, he warns that Israel's refusal to sign the NPT, and its policy of nuclear opacity, reinforce the prospect that future conflicts in the Middle East could escalate into a regional or global nuclear confrontation.

The Arabian Gulf enjoys the advantage of access to the substantial financial resources required to establish and operate nuclear power programs. Nonetheless, in his review of the economic consequences of the nuclear power option, Steve Thomas warns that 2009 will not be a opportune time to embark on such programs, citing a combination of drawbacks in the form of loosely-defined construction prices, uncertain economic gains, the current shortage of expertise worldwide, the untested nature of modern reactor designs and the possibility that, should nuclear programs in the West falter, the Gulf states might be left without adequate support in maintaining new reactor designs. However, he suggests that while questions concerning the new generation of nuclear designs are unresolved, a comprehensive program employing energy efficiency measures and renewable energy technologies should be pursued by the Gulf states. This would lead to substantial gains in energy efficiency, while renewables would conserve fossil fuel supplies for export and offer opportunities for the Gulf states to become world leaders in a technology that will inevitably become increasingly important in the future.

In this regard, and with a particular focus on energy diversification and nuclear training in the Gulf, Youssef Shatilla and Mujid Kazimi provide an overview of the advances being made by the Masdar Institute of Science and Technology in Abu Dhabi. In particular, they focus on the importance of the role of technical training to the development of an indigenous nuclear program, even though they admit that the Gulf's first nuclear infrastructure development will almost certainly comprise turn-key projects from foreign vendors. Adequate specialized training is needed for all related personnel, they state, including engineers and scientists, managers of power plants, power sector lawyers, knowledgeable government officials (regulatory, atomic commission experts) and procurement officials, etc.

[6]

Shatilla and Kazimi also discuss the development of highly efficient and environmentally friendly reactors (HEERs), which they envisage will have high power-conversion efficiency with minimum production of spent fuel and waste. In their analysis, Abu Dhabi's economic interests are clearly served by investigation of HEERs for local power needs, the production of hydrogen for local refinery and transportation use, and provision of heat for seawater desalination facilities.

Aside from the technical considerations of adopting nuclear power in the Arabian Gulf region, Malcolm Grimston discusses the role of political and social factors in nuclear development. He concludes that while proposals to build new nuclear facilities in many developed countries – especially on 'greenfield' sites – provoke considerable opposition, in developing countries populations tend to be more accepting of nuclear power projects, partly as a result of the major employment opportunities represented by large nuclear construction projects in regions of considerable poverty. He recommends that in order for governments to both encourage support for nuclear projects among their populations and improve the quality of their decisions when it comes to nuclear development, more transparent decision-making approaches are needed that involve a wider range of parties. In this regard he proposes that the Gulf countries should commit to ensuring an open and honest relationship between leaders and the public from the outset.

Of equal importance to the issue of public support is the need for international partnerships to assist the Arabian Gulf countries in the construction and operation of nuclear reactors and to provide the necessary nuclear fuel. In this respect, the UAE leadership has established cooperative ties with major nuclear countries, thereby allowing for the potential procurement of nuclear technologies that are more efficient and less costly. Such partnerships will also bring the added benefits of cooperation with states that have long experience of operating nuclear programs. For this purpose, Junko Ogawa provides a comprehensive overview of Japan's peaceful nuclear energy program and highlights the fact that Japan is a country with extremely few natural resources, where energy continues to occupy a position of great strategic importance. She states that the stable supply of energy plays a significant role in maintaining peace, hence why Japan drew a line under its previous disastrous experiences and started to develop nuclear technology for peaceful uses. Because of the Japanese experience of having suffered a nuclear attack, most Japanese people, writes Ogawa, would not hesitate to show their disapproval of nuclear technology for military use. This places Japan in a unique position to be able to champion the peaceful use of nuclear power and act as a role model for the UAE and the world in general.

A second example of a state with long nuclear experience from which the UAE might benefit is Germany—the fourth largest nuclear energy producer in the world. Kirsten Westphal discusses at length the planned phase-out of nuclear energy in Germany, providing both an historical perspective on the use of nuclear power and a broader picture of the structures of the current German electricity market. The issue of nuclear energy in Germany, she writes, is intertwined with the expansion of renewable energy in electricity generation and its transmission. Westphal argues that the nuclear phase-out is very much an issue that has to be seen through the lens of the development and transformation of the energy system as a whole. She explains that the debate about the pros and cons of extended use of nuclear energy in Germany proffers strong opinion and is often biased due to diverse interest groups.

Presenting the Indian nuclear development experience, Rajagopala Chidambaram and Kumar Sinha explain that the Indian nuclear power program aims to achieve adequate economic and societal development in the country rapidly, while maximizing the energy potential of available uranium and thorium resources. They point out that for nuclear energy to be a sustainable mitigating technology, it is necessary to close the nuclear fuel cycle. In India, an extensive research and development program and a large human resource development effort back the nuclear program, they write, which also aims to provide societal benefits through applications of nuclear energy in the fields of agriculture, medicine, industry, food preservation, desalination and water resource management. Chidambaram and Sinha conclude that India, as a developing country but with advanced capability in the nuclear energy field, will be more than happy to share its experience with – and provide its expertise to – countries such as the UAE that are now seeking to enter the nuclear energy field.

Finally, Ansar Parvez and G.R. Athar provide a brief account of the progress made by India's neighbor, detailing the contribution of the Pakistan Atomic Energy Commission towards the development of a national nuclear power program. According to Parvez and Athar, the first initiative in the area of nuclear power in Pakistan was taken as early as the mid-1960s when the 137 MW Karachi Nuclear Power Plant (KANUPP) was contracted with Canada. The plant was connected to the national grid in 1971 but had to be operated without any vendor support owing to embargoes on the transfer of nuclear technology to Pakistan. This, the authors note, helped to spur the development of a systematic self-reliance program to ensure safe operation of the plant over its design life of 30 years, and to make safety retrofits to allow it to be re-licensed beyond its design-life, thus making Pakistan a perfect case to study the current and future role of nuclear power in a developing country with a small economy base and a medium-sized grid.

After reviewing these and other nuclear success stories from around the world, and in light of the projected supply–demand balance in electricity provision in the Gulf, Abdulghani Melaibari concludes that the sustainable nuclear vision being advanced in the GCC countries must be part of an overarching strategy of energy resource management, and be based on the rational use of energy, giving added priority to renewable sources and solar energy. He argues that the GCC must adopt a program of balanced expansion incorporating a variety of different sources of energy – encompassing nuclear, renewables and fossil fuels – while reducing negative impacts on the environment and taking into consideration the needs of coming generations.

Melaibari suggests that the option of incorporating nuclear energy into the energy mix in the GCC countries is promising for several reasons, including: reducing dependence on fossil fuels; reducing greenhouse gas (GHG) emissions that contribute to global warming; expanding the energy mix; and developing and upgrading industrial infrastructure—especially if the plan is to pursue gradual participation in terms of producing parts for the nuclear sector, or even entering into the nuclear fuel cycle. Similarly, he states that solar energy must play a part in the future energy mix of the GCC countries, and that this will require investment in research and development (R&D) in this field, as well as in the companies and factories involved in producing its associated components.

Ultimately, according to Mahmoud Nasreddine, the adoption of the nuclear energy option will essentially depend on both the internal environment in the Gulf states and the global, external climate. Internally, he cites the availability of sources of energy; environmental concerns and the position of civil society regarding nuclear energy; adequate investment climate; and technological progress and development of nuclear technologies and infrastructure as all being vital to the future of nuclear power in the region. At the same time, the extent of growth in global energy demand and reduced availability of fossil fuels in the light of international tension and security events will also have major influence on the viability of nuclear power. Nasreddine also highlights the importance of the availability of nuclear fuel, which depends on current uranium stocks as well as new discoveries or uranium extraction from phosphate (which is found in abundance in a number of Arab countries) or even from sea water by new technologies that keep prices economically competitive. Ultimately, he posits, issues such as the management of nuclear waste, safeguards on nuclear safety, the availability of nuclear expertise, and favorable public opinion concerning the introduction of nuclear programs will determine the success of what has come to be known as the 'nuclear renaissance.' To ensure this success, however, Nasreddine claims countries that possess nuclear capabilities will be required to contribute to the development of safer, less expensive reactors that are more attractive to commercial investors.

It is quite possible that in the next few decades the world will witness the beginning of the end of the age of oil and the prospect of a new age where peaceful nuclear energy and its related technologies will become the world's main source of electricity supply. Today, 30 countries benefit from nuclear energy from over 400 nuclear plants, with another 44 power plants under construction in 13 countries.

Considering this current resurgence of interest in nuclear energy worldwide, as well as the growing regional demand for electricity and concerns surrounding resource conservation, the states of the Arabian Gulf have made a clear commitment to exploring the possibility of employing nuclear power for electricity generation. Not only is it highly efficient, nuclear electricity generation also has the added advantage of being among the least harmful sources of energy in terms if greenhouse gas emissions. With global warming becoming a matter of increasing concern across the world, this fact will be difficult to ignore when charting the future of the domestic energy sectors in the Gulf countries.

Perhaps more important is the prospect of producing one of the scarcest, but most vital resources in the Arabian Gulf—fresh water. Arguably this is of even greater interest to the states of the region. Nuclear desalination has been put forward as a 'silver bullet' with which to finally eradicate the region's growing scarcity of potable water. As development in the Gulf states continues, even greater demand will be placed on the area's water and electricity resources, making it an absolute priority that a suitable solution to both problems is found.

Thus, it would appear that nuclear technology offers the most appropriate source of energy to satisfy the future needs of the Arabian Gulf states. However, proliferation concerns will complicate the spread of nuclear technology throughout the region. In view of the ongoing

[11]

controversy surrounding Iran's uranium enrichment program, continuing concern among Arab states regarding Israel's undeclared nuclear arsenal, and numerous announcements from Arab states of their intention to launch their own nuclear programs, some have warned of the possibility of a regional nuclear arms race, despite the repeated proclamations of peaceful intent on the part of the states of the region. Considering these concerns, major emphasis will need to be placed on the application of non-proliferation structures to bolster international confidence and acceptance of nuclear programs in the region, as is repeatedly highlighted by the authors of this volume.

KEYNOTE ADDRESS

Nuclear Energy in the Gulf

Hans Blix

This conference has been convened to discuss the prospect of using nuclear power in some of the world's most oil- and gas-rich countries. Twenty years ago no one would have dreamt of arranging such a conference; much has changed.

At that time the Chernobyl accident was fresh in the world's memory. Even though it occurred in a very specific type of reactor found only in the Soviet Union, it increased many people's doubts about the safety of nuclear power.

During the 1990s, new orders for nuclear plants practically disappeared in Western industrialized countries. One reason was less demand for electricity, another was that in these countries nuclear plants were not economically competitive when compared to combined cycle gas-fuelled plants that were faster to build and could adapt their output of electricity to changing demand. New nuclear construction occurred mainly in Eastern Europe and Asia where a steadily rising base load could be foreseen.

Regulators, nuclear industry and utilities made good use of the 1990s, however, to build on their considerable collective experience to improve regulatory processes, and strengthen safety and reliability, readying themselves for the day when demand would return. That day has come.

In 2008, the media are full of reports about decisions or plans for new nuclear plants. Construction is accelerating in China and India and continues elsewhere in Asia and Eastern Europe. In Western Europe, Finland is the pioneer, now building the first of a new generation of 1,600

MW plants, and a second plant is being built in France. The Finns, I should add, are also ahead of all other countries in building a facility for permanent disposal of high level nuclear waste. The project shows that with openness and a good technical concept it is possible to foster public acceptance of such facilities.

Several European countries are now resuming dormant nuclear programs. In the United Kingdom the decision has the support of both the Labour Government and the opposition Conservatives. Italy, which closed its few nuclear plants after the Chernobyl accident and has imported a lot of electricity from France, will now restart its nuclear program. Switzerland will expand the program that was frozen after Chernobyl.

I should also note, however, that some other European countries that have no nuclear power show no sign of changing course. This is true for Austria, Denmark and Ireland. On the other hand, in Germany, which still has a policy of phasing out nuclear power, the situation could change if the Christian Democratic and Liberal parties were to win the election in 2009.

The United States has about 100 nuclear power reactors but after the Three Mile Island accident in 1979 orders for new ones stopped. The current economic recession might affect calculations as to how much more electricity may be needed in the next few years. Nevertheless, with a need to reduce CO_2 emissions, a nuclear revival seems certain.

Before the recent financial turmoil and recession it was clear that important utilities in the US were determined to build new nuclear plants and the Bush administration encouraged this line. In the recent election campaign both Mr. Obama and Mr. McCain showed that they were fully aware of the risk of global warming and the relevance of energy policies. Obama expressed strong support for new technology, energy saving and renewable sources of energy. While not pushing for nuclear power – as Mr. McCain did – he stated his support for "safe nuclear power." I think it can be safely assumed that nuclear power construction will soon be resumed in the United States; indeed, there are over twenty requests on the table in the country to construct nuclear plants.

Elsewhere in the world many states have expressed intentions of varying degrees of firmness to embark on the road to nuclear power. This is true as regards practically all the states in North Africa, where Algeria and Egypt have long had well developed nuclear research programs and Egypt had a power program that it shelved after the Chernobyl accident.

Libya tried clandestinely to develop uranium enrichment and nuclear weapons programs. However, the preparations were revealed and Libya agreed to terminate the program. As a result, trade and industrial restrictions were lifted and there seems now to be an intention to acquire civilian nuclear power. Turkey has discussed the nuclear option several times in the past and the intention to start a program may now be firm.

For Israel, nuclear power plants would provide an indigenous source of energy based on uranium fuel that would be easily available. However, as Israel has nuclear weapons and is not a party to the Non-Proliferation Treaty (NPT) the state members of the Nuclear Suppliers Group (NSG) will not currently export nuclear equipment and material to the country. The Group recently reluctantly gave a green light to such exports to India but it seems unlikely that it would be ready to do the same for Israel. I think only peace in the Middle East and agreement on a zone free of weapons of mass destruction, enrichment and reprocessing would open the path to nuclear power for Israel.

When states in the oil-rich Gulf region began to announce that they intended to use nuclear power some brows were raised. Some voiced the suspicion that Iran's development of a capability to enrich uranium – opening the option to develop nuclear weapons – might have been the incentive for other states in the region to develop a nuclear power capability.

However, to my knowledge no governments have expressed concern about Gulf plans for the construction of nuclear power plants. The argument seems accepted that it may be economically advantageous for countries rich in oil and gas to use nuclear power to generate the increasing amounts of domestic electricity needed and to export the oil and gas that they would otherwise burn to make electricity. This reasoning is also accepted as valid for Iran. The states negotiating with Iran have explicitly declared that they are ready even to assist Iran in its program to expand nuclear power construction; it is the building of facilities to enrich uranium or separate plutonium that raises concern.

The Nuclear Spring is Justified

In my view, the current nuclear spring is justified and welcome after a long winter. A major driving force is the fact that nuclear power can provide huge amounts of energy without emitting CO_2 and contributing to global warming. Some other factors also support increased use of nuclear power in the global energy mix.

A starting point is the certainty that the world's – and especially the developing countries' – use of energy is going to increase. People strive for a higher standard of living and high energy use is seen as a means of attaining it. Some figures on energy use are telling: the United States uses 7.84 tons of oil equivalent per capita; the OECD average is 4.67; the world average, 1.69; China, 1.10; Nigeria, 0.72; and Bangladesh, 0.16.

The famous Indian nuclear scientist Homi Bhaba once said that "no energy is more expensive than no energy." It is easy to understand him, when you see pictures of women – always women – in Asia or Africa carrying heavy jugs of water on their heads or bundles of firewood on their backs.

How will the World Meet the Rising Energy Demand?

There is no doubt that we could squeeze out more energy from the sources we now use, and that we should reduce the wastage that takes place. However, where people use little energy there is little that can be saved.

Clearly our lives improve with access to more energy. From the earliest period of our existence humans have sought to reduce the burden on their shoulders and the sweat on their brows. For example: to move on water men first used their muscles to paddle or row; then they used sails to be moved by the wind; then they burnt coal in steam engines that moved ships regardless of winds; finally, the steam engines were replaced by more effective diesel engines.

The use of nuclear propulsion in commercial shipping has been prevented by anxiety about possible accidents in ports and territorial waters despite the fact that nuclear propulsion is used in huge aircraft carriers and submarines that operate for long periods over enormous distances without re-fuelling.

Fossil fuels – coal, oil and gas – have been terrific and their use has skyrocketed. These sources now provide the world with around 85 percent of its commercial fuel, whilst hydropower and nuclear power each provide 4–5 percent.

Oil, being less bulky than coal, has proved particularly useful in the transport sector. i.e., cars, trucks and aircraft.

While the coal resources of the world are very large, we now realize that in the short span of a couple of hundred years we have burnt much of the oil and gas that it took the earth millions of years to create. This has had some dramatic consequences:

Environmental Threats

The burning of fossil fuels has released so much CO_2 that when developing countries now want to increase their use of these fuels, industrialized countries tell them: "please do not put more CO_2 into the atmosphere. We have already filled it!"

Scientists are not fully agreed but a strong majority is convinced that we are destabilizing the climate of the earth with potentially disastrous consequences for human civilization. Glaciers and polar icecaps are melting. Africa's highest mountain, Kilimanjaro, is no longer snow clad. Other areas of the world may be affected by droughts.

More and more people and political leaders agree that we cannot wait to act. I confess that I am as scared about drastic global climate change in the long-term as I am worried about the threat of nuclear and other weapons of mass destruction in the short-term.

There is a near consensus that we must reduce greenhouse gas emissions, notably CO_2. The Kyoto Protocol – to which neither the United States, nor China or India are parties – is now seen as too timid.

How can we reduce the environmental threat from the global use of CO_2-emitting fossil fuels? One evident answer is: "by a reduction in the burning of these fuels." This, of course, is easier said than done.

It will be difficult to agree on the distribution of reductions among states. It is easy, however, to see that we should avoid wasting energy and should try to switch to energies that emit little or no CO_2—when this can be done without prohibitive cost. When the number of cars doubles it would be desirable that they use half as much gasoline per kilometer or run on bio-fuels or electricity. When more light is used it would be desirable to use lamps that require less electricity, etc.

Connecting electricity grids gives greater security of supply as well as energy savings, as peak load occurs at different hours in different time zones.

Innovation must be encouraged; for example, fuel cell engines using hydrogen would produce water rather than CO_2 as waste. However, these engines are not around the corner. Batteries that would allow us to run cars, trucks and buses with electric engines – charging the batteries at night and driving during the day – would be a revolution and increase demand for electricity.

An innovation currently undergoing its first large-scale tests is carbon capture—the sequestration of CO_2 from power plants fired by coal and the injection of it into underground cavities. A reasonably economic way of taking care of the CO_2 that is formed in the burning of fossil fuels would be of tremendous interest. However, thus far we are only talking about pilot schemes.

Switching from coal to oil and gas, and from gas to nuclear or renewable sources are also helpful ways of reducing CO_2 emissions, as the burning of gas results in about half as much CO_2 per unit of electric energy as does coal, and nuclear and renewable sources produce little or no CO_2.

The Economic Risks of Current and Rising Use of Oil and Gas

The ongoing global economic downturn has sent the price of oil and gas down, but in the long-term the increasing demand from China, India and other countries will push prices up. Sarah Palin's advice was: "Drill, baby drill!" Others will recommend energy saving and new sources of energy, like nuclear power and renewable sources.

In oil-rich countries we should perhaps remind ourselves that whatever volumes of oil and gas are not pumped out are not lost; they remain in the ground. Indeed, perhaps the long-term safety and dividends of oil and gas kept in the ground may be better than money in banks or equity.

Renewable Sources of Energy

By "renewables" we usually mean wind power, solar power and biomass, although statistically hydropower is the largest component.

Solar rays, wind and biomass are obtained at little or no cost but they are dispersed, it is costly to harvest them and they have low energy density. Much experience has been gained in decades of experimentation with renewable sources of energy and important improvements have been made. Yet, it is realized that while welcome as a part of the energy mix renewable energies (other than hydro) can only satisfy a small part of the greatly increasing energy demand that the world will face. Shanghai or Calcutta will not meet their demand for electricity by burning biomass, or building wind turbines or fields of photovoltaic cells.

Nuclear Power

I learnt long ago that one kilogram of firewood can generate about 1 kilowatt-hour of electricity (kWh). The values for the other fuels and for nuclear fuel are: 1 kg of coal—3 kWh; 1 kg of oil—4 kWh; 1 kg of uranium—50,000 kWh.

Nuclear fuel thus has an enormous energy density. This has an impact on the functioning and economy of nuclear power. The supply of the fuel has been steady and it can be transported relatively easily. In Russia, fuel has been flown in to some power reactors in the Arctic region. Nuclear fuel is relatively cheap and constitutes a small part of the cost of the electricity generated. On the other hand, the plants using the fuel are expensive and take quite a number of years to build.

The Potential Role of Nuclear Power

As of today, nuclear power is a technologically mature source of energy that can expand in a big way – but not overnight – giving the world CO_2-free energy.

Many governments, utilities and private industries around the world now favor the nuclear power option. The International Energy Agency (IEA) recently recommended an 80 percent increase in nuclear power worldwide up to 2030. One reason for the new support is that the economics of nuclear-generated electricity now seem viable. Improvements in efficiency and safety in the operation of nuclear power plants have gone hand in hand and have improved their economy.

It may well be that only prolonged trouble-free operation of nuclear power will overcome fears that may be as deep-rooted and common as people's fear of flying. The conclusion is that the safety culture of nuclear power must be higher than that of the aviation industry. However, confidence that nuclear power operates with high safety is growing, and with good reason.

The number of unplanned stoppages in nuclear plants has dropped and the availability of plants has gone up. The calculated technical life span of plants has gone up from some 25 years in the 1970s to some 50–60 years now. Thus, we now get more megawatts from the plants with the same safety and for a longer time.

It is sometimes argued that nuclear power is not a viable long-term energy source as the uranium resources of the world are finite. They are, but they will last a long time and suffice for a much bigger nuclear power plant park than the world has today. Furthermore, when uranium resources become scarcer and more expensive, breeder reactor technology is available. It is well tested and operative—but hardly economic at today's uranium prices. Using breeder reactors would improve the use of the energy contents of the uranium fuel by something like 80 times and make the uranium of the earth last hundreds of years longer, even with a much larger park of reactors.

I might also mention that nuclear power based on a thorium cycle is technically feasible and would, likewise, allow for reliance on nuclear power very far into the future. The more I learn about the thorium option the more interesting it seems to me. Thorium fuel would generate less waste than uranium fuel and would raise less proliferation risk. There is thought to be about three times as much thorium as uranium in the crust of the earth.

Many people feel concerned about the extremely long-lived, high-level radioactive waste. The concern is understandable and is taken seriously. Current plans call for deposits in deep geological formations and have been thoroughly examined. While one cannot perhaps be certain that no gram of plutonium could leak from a deposit within the next thousand years, one ought to remember that the 'alternative waste' from fossil fuels would most likely have been sent into the earth's atmosphere. We should also be aware that ongoing scientific work may one day lead to methods that reduce both the volume and longevity of nuclear wastes.

[23]

Challenges for the Nuclear Option: Safety, Security and Proliferation

In terms of the safe operation of nuclear installations there will be some challenges ahead. Very competent personnel will be needed both for the building and operation of nuclear plants and in supervising regulatory authorities. Such staff will now be sought after worldwide and the training of staff should start without delay.

Protecting nuclear installations from sabotage has been and will remain a high priority. The nuclear industry may be better prepared for this task than many other sensitive industries. Gas pipe lines, oil refineries and chemical industries have similar problems and may well need to catch up with the standards of the nuclear industry.

The Question of Proliferation

It is true that if a country masters the technique of reprocessing spent nuclear fuel it can produce plutonium for nuclear weapons – as North Korea has done – and if a country has mastered the technique of enriching uranium to somewhere near four percent on an industrial scale – as Iran is doing – it can enrich to 90 percent and above, which would be weapons grade. The North Korean and Iranian issues need to be handled urgently through negotiations. Failure in either case could lead to dangerous domino effects.

Today, there is some concern that the increased production of and trade in enriched uranium that will accompany greater reliance on nuclear power might increase the risk of production for weapons and diversion of uranium. Can we eliminate or reduce such risks?

States can and must take effective measures to ensure that all nuclear material within and crossing their borders remains in authorized hands to prevent non-state actors acquiring it to make dirty bombs, for instance.

The risk of states embarking on nuclear weapons programs is mostly – but not invariably – linked to a perception on their part that they require such weapons for compelling security reasons: India after its armed conflict with China; Pakistan after it was certain that India was developing nuclear weapons. Possession of nuclear weapons may also be perceived as a means to gain prestige, a right to be heard. We should not strengthen such perceptions by our policies. Isolating states is not a good idea. As to status, we may point to Germany and Japan as great powers without nuclear weapons. To reduce the incentives of states to embark on weapons programs it will be of great importance to strengthen regional security.

A peaceful settlement in the Middle East is desirable from all viewpoints, not least the view that it would reduce tensions and any perceived need to keep or acquire nuclear weapons. Such a settlement should make the Middle East – including Iran and Israel – a zone free of weapons of mass destruction and free – at least for a rather long period of time – of facilities for the enrichment of uranium and production of plutonium.

Iran's program for the enrichment of uranium, as we know, has engendered considerable tension. To avoid exacerbating that tension it would be desirable that even before a peaceful settlement is attained any state embarking on nuclear power programs in the Middle East should declare – as Abu Dhabi has done – that it will not build facilities for the enrichment of uranium or production of plutonium.

Such declarations will enhance confidence. Developing facilities to produce fuel for a limited number of reactors would not be economic and would raise concerns. To be without nuclear enrichment facilities is no more remarkable than to be without oil refineries. Naturally, such a policy might be coupled with arrangements for the long-term assurance of supply of fuel and even for the taking back of spent fuel.

Again, to avoid creating concerns and suspicions it would be desirable that all parties to the NPT – including those in the Gulf – accept the Additional Protocol of the IAEA, strengthening the Agency's inspection powers and thereby creating confidence that no clandestine nuclear activities are taking place. If inspections of the kind foreseen in the Additional Protocol had taken place in Iraq and Libya, the risk of detection might have deterred these countries from pursuing their clandestine programs for the enrichment of uranium. In any case, the programs would most likely have been discovered.

One last comment: while the big powers tell us much about the risk of destabilizing proliferation, we hear rather little about the risks posed by the some 25,000 nuclear warheads in their own hands and in India, Pakistan and Israel. Yet, perhaps preventing the spread of nuclear weapons to further states will be less difficult if those that have these weapons take the lead in marching out of the nuclear weapons era and seek disarmament. Preaching to others to stay away from the nuclear weapons that you have yourself is a little like smoking a cigar whilst asking your children not to smoke. With new leaders in several states – the United States, Russia, France, the UK – there is some hope for a revival of disarmament. It would be most fitting to match a revival of nuclear power with a dismantling of nuclear weapons; to concentrate on megawatts instead of megatons.

In 2007, the military expenses of the world amounted to about 1,300 billion US dollars—half of it falling on US tax payers. Yet, using armed force has proved disastrous in Iraq and Lebanon and will – for fear of further disasters – in all likelihood be avoided in the cases of Iran and the DPRK. Aircraft carriers and nuclear weapons are not needed when fighting terrorists. How about using at least half of the 1,300 billion dollars now budgeted for defense to defend our planet against global warming?

ROADMAP FOR A PEACEFUL NUCLEAR CAPABILITY

1

The Role of Nuclear Power in Achieving Energy Security and Countering Climate Change

Mahmoud Nasreddine

C ircumstances dictated that nuclear energy be harnessed for military purposes in the wake of its discovery in the 1940s. This affected public opinion in that nuclear energy was thereafter irrevocably linked to nuclear weapons. However, after the end of World War II, the world began to appreciate the potential for peaceful applications of nuclear fission, specifically in generating electrical energy. The first civilian nuclear power station was completed in Obnisk in Russia (at that time the USSR) in 1954. The United Kingdom followed by building the Calder Hall plant in 1956, whilst the Americans completed their Shippingport reactor in the United States in 1957. These countries continued to develop nuclear reactors and used them increasingly in electricity generation. Over time, other countries followed suit, building their own nuclear power reactors. The world presently produces electricity from nuclear energy in amounts equivalent to the total electricity generated in 1960 from all sources combined. Current global electricity generation by nuclear energy represents more than fifteen percent of total production, generated by 436 power reactors in 30 countries. There are about fifteen countries that depend on nuclear energy in generating more than one quarter of their electricity needs, and France generates around 75 percent of its requirements using nuclear energy. Nuclear power for generating electricity is considered the fastest growing major energy source, in spite of the reluctance that arose after the Chernobyl accident in the Soviet Union. At present there are about 44 power reactors under construction, as well as plans to build a further 110 reactors, and the efficiency and safety of modern nuclear technologies are increasing day by day.

There continues to be heated debate between those who advocate nuclear power and their opponents who have reservations relating to the management of nuclear waste, the economics of nuclear electricity generation, nuclear safety in comparison with other primary sources, and the potential for proliferation of nuclear technology. From the outset, this discussion favors the advocates of nuclear power, owing to the tremendous efforts that have been made to devise innovative solutions to these problems; it is safe to say that nuclear power is the most dependable source of energy for the future.

With the increasing need for electricity and fresh water in the Arab world and mounting shortages in the oil and gas reserves of some Arab countries, considering the option of resorting to nuclear energy as a source of electricity generation and desalination becomes a strategic necessity in the medium- to long-term.

Growing energy demand, particularly in countries with large populations and sizable economic growth such as China and India, makes the search for alternative sources of energy – and non-reliance on a single source that might be affected by economic, political and internal or external security factors – an inevitable objective. China and India have recognized this fact and have thus considered nuclear energy. In 2002 work began to build four nuclear plants in China, and India began building six new nuclear plants for electricity generation. It is worth mentioning that major nuclear countries have encouraged China and India to make use of nuclear energy in securing the energy resources they require so that oil demand is eased and prices lowered to help global economic growth.

Electricity generation from nuclear power is being considered by a number of countries worldwide; this is especially true of the Arab states, which are particularly interested in moving toward nuclear powered desalination. Heat can be used in the desalination of sea water either directly from a reactor (where the desalination plant is built on the same site as the reactor), or from electricity generated by a nuclear plant (with the plant built elsewhere).

Nuclear energy has been a bone of contention in Europe thanks to the presence of active opposition from environmental and political groups. This has led some governments, such as that of Germany, to announce the abandonment of nuclear power as a means of electricity generation. Nonetheless, in recent years there has been renewed interest in nuclear energy, particularly in the United States, Britain, France and other European and Asian countries. This has led to the development of a special European Union (EU) strategy for the utilization of nuclear energy to partly satisfy Europe's energy requirements by 2020.

Nuclear Electricity Generation

Electricity is generated by a nuclear plant using the same method as that employed by fossil fuel-based plants, but with one crucial difference—the source of heat. In nuclear plants, heat is produced by nuclear fission, which is accompanied by the release of an immense volume of thermal energy and neutrons.

The neutrons released by this reaction are moderated so that they lead to the fission of other uranium atoms, thus producing a recurring reaction process. This is called a chain reaction, and it is contained and controlled by the nuclear reactor.

The heat released from the core of the reactor boils water and converts it into steam that is transferred to turbines which transform mechanical energy into electricity, as is the case in fossil fuel-powered plants. This type of reactor is called a nuclear power reactor (NPR).

Conversely, nuclear research reactors produce an abundance (pack) of neutrons instead of thermal energy. These neutrons are then used in several applications such as producing the isotopes used in medicine and industry, the process of analyzing the elements of materials by means of neutron activation, or in processes such as the coloring of semi-precious stones like topaz, for instance.

Components of Nuclear Reactors

There are a number of known, common components of most reactor types. These are:

Nuclear Fuel

Nuclear fuel most often consists of pellets of uranium dioxide contained inside tubes to produce fuel rods. These rods are arranged in fuel groups inside the core of the reactor. Power reactors use uranium enriched by U-235 in percentages ranging from 0.7 percent (the percentage found in natural uranium) to 90 percent, but most modern reactors available for civil purposes use low-enriched uranium with a U-235 component of approximately three percent (the aim being to lessen the chances of the proliferation of highly-enriched uranium (HEU) which can be used for weapons applications).

Metallic uranium is not directly used as reactor fuel in view of the fact that it expands when exposed to irradiation by neutrons. This can destroy the fuel rods during reactor operation. Because uranium is chemically active and reacts strongly with water and air, uranium oxide is used instead of metallic uranium. However, in view of the fact that uranium oxide is a poor conductor of heat, it is necessary to use it in relatively small rods – 0.5–0.75 inches in diameter – or inside panels. The fuel itself is manufactured as tablets, rods, tubes, or panels from uranium. They are then placed inside tubes or panels of zirconium or steel that are rust-proofed. It has been agreed that enrichment should not exceed 20 percent U-235, so that HEU is not used in the production of nuclear weapons.

As a result of the fission process that takes place at the core of the reactor during operation, chemical changes occur in the nuclear fuel rods. To illustrate this, we can follow what happens in an amount of nuclear fuel in a power reactor where ordinary water is used as a coolant and moderator; the fuel is composed of U-238 which contains 3.3 percent U-235. A reactor producing 1 GWe (3 GWth) consumes about 7.3 kg of U-235 in one operational day. After a suitable period, Pu-239 – and later Pu-241 – which is formed by the absorption of U-238, begins to contribute to the fissile reaction. As a result, the consumption of U-235 diminishes gradually during reactor operation, and we can assume that about 40 percent of the total production of energy results from the fission of the plutonium isotopes generated in the fuel itself.

The products of fission that are found at the core of a reactor with a 1 GWe capacity are composed of a large group of isotopes whose atomic numbers range between 72 and 160 (i.e., about 80 types of isotopes). The results of detailed studies have shown that fission produces round 800 isotopes; some of these are radioactive and with various different half-lives, whilst some are stable and are not radioactive. If we exclude stable isotopes and those with half-lives of less than 26 minutes, we are left with 54 radioactive isotopes, the most important of which are: Co-58, Kr-85, Sr-89, Sr-90, Sr-91, Zr-95, Ru-103, Ru-106, Te-127, I-131, Cs-134, Cm-141, and Cm-144. The different isotopes of Plutonium, which form in large numbers, have a long half-life and constitute a burden on the environment if released from the reactor in the event of accidents. There are many studies directed towards studying the effect of products of fission in reactors on public health and the environment were they to leak out from the containment building. The most important of these studies are those on the effects of the Chernobyl accident on the environment, which persisted for a whole decade and necessitated changes in a number of prevailing concepts in the industry.

Reflector, Moderator and Coolant in Reactors

Fission chain reactions lead to a doubling of neutrons during successive fissions. For this to occur, the neutrons must be effectively contained and prevented from leaking out from the core of the reactor and thus altering the concentration of neutrons vital to the continuation of the reaction. Also, a certain proportion of neutrons is lost when they undergo lateral, non-fissile reactions with the components of the core of the reactor. Hence, to make available the greatest number of neutrons to contribute to the continuity of the ongoing fissile reaction at the core of the reactor, a material that reflects neutrons is placed around the core to return scattered neutrons heading towards the walls of the reactor. Naturally, the materials found at the core of a reactor are chosen for their particular characteristics: low neutron absorption and their non-reactivity with neutrons being the most important. This way the components of the reactor themselves will not interfere with the reactor's operation.

For the chain reaction to be sustained at the core of the reactor, it is necessary that one neutron remains after each fission state (usually several neutrons are produced in this state) to enter into a new process of fission. The remaining neutrons are lost as a result of leakage from the core, or owing to reaction with the components of the core or walls.

The neutrons produced by the fissile reaction have high energy and therefore it is necessary to moderate their energy to less than 1.0 electron-volt to improve their ability to enter into the next reaction. This takes place through moving fuel rods away from one another so that the moderating liquid passes through the vacuums that separate them. Usually the moderator is made of a low atomic number material that is efficient in moderating fast neutrons. Reactors that operate using this method (moderating neutrons) are called thermal reactors to distinguish them from fast reactors, where fast neutrons are used (whose energy reaches about 100 kiloelectron-volts or more) to form the next generation of fissions.

[34]

Table 1.1

Efficiency of Various Moderators

Moderator	Number of Required Collisions
Water	19
Heavy water	36
Beryllium	84
Graphite	114

Table 1.1 shows the capability of various moderators to reduce the energy of neutrons emitted from nuclear fission to the level of thermal energy that allows them to continue to contribute to the chain reaction inside the reactor.

The cores of giant reactors require a super capability cooling system to cope with the immense heat resulting from the fissile reaction in the fuel rods. This system transfers the resulting heat from inside the rods to the outside in such a way that allows it to be used in other applications. The coolant is usually composed of appropriate liquids such as water (light water) or heavy water. Sometimes, as in the case of fast reactors, the coolant takes the shape of a fused metal such as fused sodium. Usually the coolant enters from one side of the reactor and moves to the bottom of the pressure vessel in the void between the walls of the vessels and the barrier of the core of the reactor (made up of fuel rods) in such a way that it penetrates the vacuums between them. The boiled water comes out from the upper part of the reactor at about 325°C and passes to the steam generator outside the reactor.

Control Rods

These are composed of elements that absorb neutrons such as Cadmium, Hafnium, or Boron. These are introduced and withdrawn from the core of the reactor in order to control the reaction, adjust fuel use, allow the operators to undertake necessary procedures to ensure the continuity of the work of the reactor, or to attain a higher degree of safety.

Pressure Vessel or Pressure Tubes

This is a steel vessel containing the core of the reactor and the relevant moderator, or a series of tubes containing fuel and conveying coolant.

Steam Generator

The steam generator is a part of the cooling system where heat from the core of the reactor is used to produce the steam required to turn turbines that generate electricity, or to produce steam in a secondary circuit if the coolant is heavy water or any material other than light water. The steam generator is found only in power reactors or those that are used to generate electricity. In research reactors this steam is not required and therefore is not developed.

Containment Vessel

This is the construction which contains the components of the reactor, protecting it from external effects, and prevents radiation from affecting the surrounding environment. Commonly it is one meter thick and made of concrete and steel. Further details of the vessel are provided later in this paper.

Reactor Types

The basic aim of building reactors is to benefit from them as sources of neutrons for irradiation purposes or to generate heat in the following ways:

Reactors are used as sources of neutrons for the purposes of irradiation to obtain products such as radioactive isotopes or plutonium. They can also be used in high neutron flow testing of building or industrial materials, metals, or components of facilities. This type of reactor is called a research reactor.

Reactors are also used as a source of intensified heat, generated at the core of the reactor as a result of the fission process. The coolant (water) can be used – after being heated at the core of the reactor and transformed to pressurized water with high temperature – in heating applications or in producing steam in a secondary circuit through a joint heat exchanger for use in the turbines of electricity generation plants, or for producing steam that is intensified later in salt water desalination. These types of reactor are usually called power reactors, which include light water reactors (LWR) and pressurized heavy water reactors (PHWR).

Light Water Rectors (LWRs)

These are reactors that use normal water as a coolant and usually operate using fuel enriched to three percent U-235. Most often they consist of two types: pressurized water reactors and boiling water reactors. Also, they are usually connected to methods of generating steam that is used in operating turbines to generate electricity, or in desalination plants to produce fresh water.

- PRESSURIZED WATER REACTOR (PWR): A pressurized water reactor is composed of a pressure vessel made of carbon steel 20 cm thick and about 20 m high with a 4 m diameter. The manufacture of this vessel requires very advanced technical capabilities. The interior surfaces of the pressure vessel are covered with stainless steel. The vessel can be opened during maintenance processes or for the purposes of changing fuel. Usually the pressure vessel is designed to bear pressures reaching seventeen megapascals (MPa). Fuel is provided in the form of aligned groups of rods at the center of the vessel on the bottom, surrounded by the core cylinder (core tank). Through the fuel rods pass control rods and devices used to monitor the core of the reactor.

  The cooling liquid (water) enters through a special inlet where it enters the core cylinder, which lies between the core (where fuel is put) and the wall of the vessel. It then runs to the bottom of the vessel and proceeds to rise through the fuel bundles in the core of the reactor and exit from an upper opening to head towards the steam generator.

The steam generator is, in fact, a heat exchanger which uses heat carried by the cooling liquid from the core of the reactor (what is called the primary circuit) to generate steam from ordinary water passing around the central tubes in the heat exchanger. Thus the water of the primary circuit is cooled and returns to the core of the reactor, whilst in the secondary cooling circuit the steam generated is used to produce electricity or desalinate sea water.

The fuel used in pressurized water reactors is composed of uranium oxide in the form of tablets or small cylinders contained in thin tubes (fuel rods) made of Zircaloy-4 (Zr-4; composed of Zirconium and aluminum). The fuel rods are gathered in square, 17 x 17 fuel bundles, and boron carbide control rods are passed through them.

A 1,000 megawatt (MW) pressurized water reactor contains about 200 fuel bundles made up of nearly 50,000 fuel rods with a total fuel amount of about 48 tons. The reactor includes several systems – both basic and subordinate – to control its operation. The basic systems control the chemistry and volume of the cooling liquid and remove heat generated from the decomposition of radioactive materials. The subordinate systems include special cooling systems used to control reactor efficiency, and those that ensure operational safety in emergency cases.

The reactor and the steam generator are found inside a concrete building with steel reinforcements called the container or the containment vessel, which is architecturally designed to bear any increase in pressure when an accident occurs involving a loss of coolant. The containment building is equipped with sprinkling and cooling systems to cleanse the container, cool the interior vacuum and reduce pressure to below the tolerance of the building itself in an accident.

This type of integrated reactor design places both the core of the reactor and the steam generator inside the pressure vessel in order to overcome the problems associated with huge connecting tubes between the reactor and the steam generator.

- BOILING WATER REACTOR (BWR): The pressure vessel in a boiling water reactor is 22 m in height and 6 m in diameter, and is made of 16 cm carbon steel. The lid is made of stainless steel and is 3.0 cm thick. The vessel bears a pressure in excess of 7 MPa.

  The pressure vessel houses the core of the reactor and other components such as the control rods, which are found at the bottom of the vessel—instead of the top, as is the case in pressurized water reactors.

  Water enters via the feed inlet, is conveyed to the base by side pumps and then turns upwards to pass through the fuel system. As the outside pressure in the pressure vessel is 7 MPa, water boils at 285°C and steam is generated which is contained above the core of the reactor. This steam exits through steam driers before leaving through the steam outlet.

  The fuel in the boiling water reactor is composed of uranium dioxide placed in a large number of 8 x 8 or 7 x 7 bundles of fuel rods. The packs feature a special coating that reduces the flow of water to the inside of the pack. They also contain rods through which water passes, working as a moderator.

  A large boiling water reactor (say 1,000 MW) contains nearly 160 metric tons of uranium dioxide. Out of each four fuel bundles, a control element passes; it is cruciform in shape and contains boron carbide.

  The rate of flow of water in the core of the reactor is 13 kg/s; the water temperature is 191°C on entering the core and 288°C when it exits.

  The steam resulting from the boiling water reactor is radioactive owing to the presence of nitrogen-16 (with a seven second half-life), which will consequently be converted into radioactive water after being used in cycling turbines and cooling.

  Boiling water reactors include several control and command, and safety and security systems, all of which are placed together with the reactor itself in an armored concrete building that guarantees the separation of the reactor and its associated components from the circumambient environment.

- ELECTRICITY GENERATION PLANTS USING LIGHT WATER REACTORS: These plants are based on a pressurized water reactor through which cooling water passes to absorb the heat generated in the fuel rods. The water then completes its cooling cycle in the primary circuit by passing through a steam generator to produce steam in the secondary circuit. The cooled water returns once more to the core of the reactor through a cycling pump in the primary circuit. Thus the heat resulting from uranium fission is transferred to the steam generator. The primary cooling circuit operates under pressures greater than 14 MPa and therefore the water does not boil. The steam produced in the secondary circuit passes through a steam drier (to separate any water droplets), and on to the turbine and generator. After operating the turbine and generator the steam is collected again and passes through a condenser and subsequently the resulting water is reused to feed the steam generator, and so on.

    The plant depends on using a boiling water reactor which is cooled through converting water inside it directly to water vapor. This system, albeit simpler in terms of engineering, requires multiple precautions to prevent pollution, as steam is produced directly inside the pressure vessel, making its radiation level high and thus increasing the chances of polluting the turbine. This explains the limited reliance on reactors of this type compared to PWRs that contain a closed primary cooling cycle.

Heavy Water Reactors

Canada has focused on the development of a new type of generator (CANDU), which operates on natural uranium fuel in the form of tablets of uranium oxide and uses heavy water as a coolant. Heavy water is used as a coolant because it absorbs fewer neutrons in comparison to ordinary water. Reactors moderated by heavy water can operate using natural uranium (which contains 0.7 percent U- 235).

[40]

The CANDU reactor does not feature a pressure vessel like those in PWRs and BWRs. This system has been substituted by what are called pressure tubes—a large group of tubes in which fuel bundles are placed and through which the coolant passes.

The basis of this reactor is a huge horizontal cylinder called a "calandria." About 240 Zircaloy fuel channels pass horizontally through the calandria, which contains heavy water, pressurized to one atmosphere, as a coolant. The moderator has its own cooling system, which is composed of pumps that circulate the moderator through heat converters to cool it before it is re-pumped to the calandria to maintain its temperature in the region of 70°C.

The coolant heats up as it passes through the fuel tubes and is collected at a chamber leading to the steam generator, where the heat is used to generate steam from ordinary water, which powers turbines for electricity generation. After exiting the steam generator, the heavy water remains relatively cool and is pumped back to the inlets of the tubes distributed throughout the calandria to repeat the cycle.

There is a thin coating of graphite on the interior surface of the Zircaloy fuel tubes to prevent any interaction between the fuel tablets and the coating material. These tubes are closed after being filled with gaseous Helium so that gas fills any vacuums in the tablets in order to facilitate the conveyance of heat from the tablets to the surface of the Zircaloy tubes.

The CANDU reactor includes nearly 4,500 fuel bundles containing about 100 tons of uranium oxide fuel. The bundles are replaced by individually changing fifteen bundles per day without stopping the reactor.

CANDU reactors have a proven track record in several countries and are distinguished by their great flexibility in fuel selection. Natural uranium and low-enriched uranium, as well as uranium recovered from processing the fuel of combined oxides, thorium, or waste fuel from LWRs can all be used.

One of the most important distinguishing characteristics of this reactor type is the ability to recharge them with new bundles of fuel without affecting their operation. This is a big advantage, as LWRs equipped with pressure vessels must be stopped for a month each year to be recharged.

Generating Electricity from Nuclear Energy

One of the most important uses of nuclear energy is the production of electrical energy. The 1970s saw great demand for the construction of nuclear electricity-generation plants, especially during the oil shock of 1973. From the beginning of the 1980s, however, this tide started to ebb, especially after the accident at Three Mile Island. Despite the fact that nuclear energy accounts for around twenty percent of US electricity production – and the United States produces more nuclear energy than France and Japan combined – as of 1978 the demand for nuclear energy has steadily decreased. Applications to build new plants dried up and old plants were closed before the end of their design-life periods. When it became clear that plant construction required 6–12 years, the costs would reach US$ 4 billion, and that the plants that had been built were not operating quite as well as expected, the positive view of nuclear plants diminished. Plant safety and the problem of nuclear waste disposal further damaged public acceptance of nuclear power. To confront these issues, the nuclear industry turned to continuous and persistent development to overcome all the problems raised and to develop highly secure, ultra-reliable plants of more efficient construction (as reflected in the reduction of the period and cost of plant construction).

The nuclear industry has developed advanced designs of various types of large reactors, as well as small- and mid-sized reactors, which have been radically simplified and tailored to the needs of developing countries.

In spite of the general pessimism surrounding the future of nuclear energy, the diligent efforts made to develop new designs have served to boost the image of nuclear technology. In an opinion poll conducted in the early 1990s, the American public expressed its belief that nuclear energy must play a role in satisfying future US energy needs. In the spring of 1992, 35 percent of the adult population who participated in the opinion poll said that this role should be a prominent one, and 38 percent of the participants pointed out that its role should be 'important to some degree.' These findings do not require further comment.

Figure 1.1 shows the contribution of nuclear energy and other sources to electricity production around the world.

Figure 1.1
Contribution of Different Energy Sources to
Global Electricity Generation

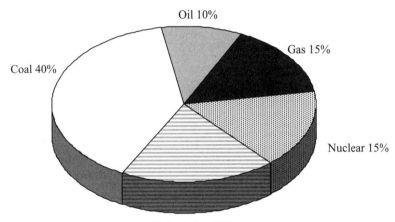

This figure illustrates the fact that the percentage of electricity generated by coal is the highest. Bearing in mind that coal is the most environment-polluting source of primary energy, the importance of shifting from coal to other sources of primary energy is clear. Foremost among these alternative sources are renewable and nuclear energy. The share of coal in generating electricity is greater than the contribution of oil and natural gas combined. The contribution of nuclear energy does not exceed sixteen percent and must be increased to meet the commitments of the Kyoto agreement relating to reducing the production of greenhouse gases causing global warming. However, this call for an increase in the use of nuclear energy in generating electricity will not block calls both globally and among the Arab countries to develop the use of renewable energies, especially solar and wind energy.

Figure 1.2

Percentage of Nuclear Electricity Generation by Country

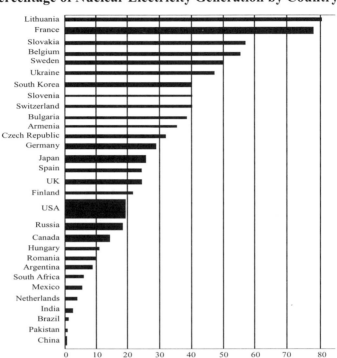

Figure 1.2 shows the contribution of nuclear energy to electricity production in countries that possess power generators.

It is quite possible that the world will find itself facing two options: either to expand the use of nuclear energy, or reduce energy-intensive standards of living and affluence; no doubt the world will head strongly towards nuclear energy eventually. For a breakdown of the present status of nuclear-generated electricity, Table 1.2 covers each country with a nuclear power generation capability as of the end of 2007, the percentage of the contribution of nuclear energy to electricity generation compared to other energy sources, and the situation until April 2009 in terms of operating reactors, reactors under construction, planned reactors, the capability of each in MWe, and their requirements of uranium.

Table 1.2
The Status of Nuclear Energy (April 2009)

COUNTRY	NUCLEAR ELECTRICITY GENERATION 2007		REACTORS OPERABLE 1 April 2009		REACTORS UNDER CONSTRUCTION 1 April 2009		REACTORS PLANNED April 2009		REACTORS PROPOSED April 2009		URANIUM REQUIRED 2009
	billion kWh	% e	No.	MWe	No.	MWe	No.	MWe	No.	MWe	Tons U
Argentina	6.7	6.2	2	935	1	692	1	740	1	740	122
Armenia	2.35	43.5	1	376	0	0	0	0	1	1000	51
Bangladesh	0	0	0	0	0	0	0	0	2	2000	0
Belarus	0	0	0	0	0	0	2	2000	2	2000	0
Belgium	46	54	7	5728	0	0	0	0	0	0	1002
Brazil	11.7	2.8	2	1901	0	0	1	1245	4	4000	308
Bulgaria	13.7	32	2	1906	0	0	2	1900	0	0	260
Canada	88.2	14.7	18	12652	2	1500	3	3300	6	6600	1670
China	59.3	1.9	11	8587	11	11000	26	27660	77	63400	2010
Czech Republic	24.6	30.3	6	3472	0	0	0	0	2	3400	610
Egypt	0	0	0	0	0	0	1	1000	1	1000	0
Finland	22.5	29	4	2696	1	1600	0	0	1	1000	446
France	420.1	77	59	63473	1	1630	1	1630	1	1630	10569
Germany	133.2	26	17	20339	0	0	0	0	0	0	3398
Hungary	13.9	37	4	1826	0	0	0	0	2	2000	274
India	15.8	2.5	17	3779	6	2976	10	9760	15	11200	961
Indonesia	0	0	0	0	0	0	2	2000	4	4000	0
Iran	0	0	0	0	1	915	2	1900	1	300	143
Israel	0	0	0	0	0	0	0	0	1	1200	0
Italy	0	0	0	0	0	0	0	0	10	17000	0
Japan	267	27.5	53	46236	2	2285	13	17915	1	1300	8388
Kazakhstan	0	0	0	0	0	0	2	600	2	600	0
Korea DPR (North)	0	0	0	0	0	0	1	950	0	0	0
Korea RO (South)	136.6	35.3	20	17716	5	5350	3	4050	2	2700	3444
Lithuania	9.1	64.4	1	1185	0	0	0	0	2	3400	0
Mexico	9.95	4.6	2	1310	0	0	0	0	2	2000	242
Netherlands	4.0	4.1	1	485	0	0	0	0	0	0	97
Pakistan	2.3	2.34	2	400	1	300	2	600	2	2000	65
Poland	0	0	0	0	0	0	0	0	5	10000	0
Romania	7.1	13	2	1310	0	0	2	1310	1	655	174
Russia	148	16	31	21743	8	5980	11	12870	25	22280	3537
Slovakia	14.2	54	4	1688	2	840	0	0	1	1200	251
Slovenia	5.4	42	1	696	0	0	0	0	1	1000	137
South Africa	12.6	5.5	2	1842	0	0	3	3565	24	4000	303
Spain	52.7	17.4	8	7448	0	0	0	0	0	0	1383
Sweden	64.3	46	10	9016	0	0	0	0	0	0	1395
Switzerland	26.5	43	5	3220	0	0	0	0	3	4000	531
Thailand	0	0	0	0	0	0	2	2000	4	4000	0

COUNTRY	NUCLEAR ELECTRICITY GENERATION 2007		REACTORS OPERABLE 1 April 2009		REACTORS UNDER CONSTRUCTION 1 April 2009		REACTORS PLANNED April 2009		REACTORS PROPOSED April 2009		URANIUM REQUIRED 2009
	billion kWh	% e	No.	MWe	No.	MWe	No.	MWe	No.	MWe	Tons U
Turkey	0	0	0	0	0	0	2	2400	1	1200	0
Ukraine	87.2	48	15	13168	0	0	2	1900	20	27000	1977
UAE	0	0	0	0	0	0	3	4500	11	15500	0
United Kingdom	57.5	15	19	11035	0	0	0	0	6	9600	2059
USA	806.6	19.4	104	101119	1	1180	11	13800	20	26000	18867
Vietnam	0	0	0	0	0	0	2	2000	8	8000	0
WORLD	2608	15	436	372,203	44	38,848	110	121,595	272	268,905	65,405

Source: World Nuclear Association (WNA), "World Nuclear Power Reactors 2007-09 and Uranium Requirements," April 1, 2009.

The above table shows that in April 2009 there were 436 power reactors working in 30 countries with a total electric capacity of 372,203 MWe. Presently, nuclear energy produces about fifteen percent of world electricity and this percentage is continuously increasing. As of April 2009 there were 44 reactors under construction, 110 reactors planned and 272 proposed. Perhaps the most interesting development is the rapid growth in the use of nuclear energy to generate electricity in both China and India—at a rate higher than the rest of the world.

Obviously, Arab countries do not feature in these statistics as none operate nuclear energy programs. However, they do need to develop this capability, not only to produce electricity but also fresh water. Some have expressed concern regarding the introduction of nuclear energy to the region for a number of reasons, including the question of nuclear safety and the long period of financial investment required before plant completion. Such concerns may seem reasonable, but one must bear in mind the fact that the acquisition and safe application of an industrial nuclear capability will play an essential role in securing energy supplies and satisfying the demand for both energy and water. In the light of the falling per capita surface water in the Arab world and the growing demand for energy as a determining factor of industrial and economic progress, and in order to spare coming generations devastating crises in both these areas, the Arab world must seek to familiarize itself with this technology with a view to developing the technical capabilities to build and manage industrial nuclear installations.

Both the Arab League (March 2007, March 2008) and the GCC (December 2006) have highlighted the necessity of developing peaceful atomic energy in the Arab countries, especially in terms of electricity production.

The Concept of Energy Security

Energy security does not merely comprise sustainability, efficiency and supply security; it is rather a multi-dimensional concept involving both internal and external factors as well as a number of economic, political and security considerations that must be applied collectively. Simply defined, energy security is credible supply at an acceptable price. If we examine the components of secure supply, we will realize the complexity of achieving energy security.

The triangle of supply security, sustainability and efficiency forms the foundation of the relationship between the various aspects of energy policies; but it remains insufficient as a framework for energy security. Likewise, an understanding of how to guarantee energy security by commercial methods is also insufficient. Furthermore, progressive dependence on foreign supply breeds suspicion and distorts energy security. Therefore, before politicians devise a conclusive answer to the increased threat to energy security, we must satisfactorily clarify the concept itself.

Energy security, or rather the steady and stable supply of energy, is synonymous with security *per se*. Any negative effect on the continuity of energy supplies gravely impairs national economy, political stability, and the social welfare of citizens. Energy affects the various different productive activities of society and those linked to improving standards of living. Therefore, sufficient energy supplies are a prerequisite for economic growth and confer legitimacy on the political system. There is also a broader and deeper definition of security in the international context, whereby security is the ability of countries and societies to preserve their identities, independence and performance. Security might be defined as the absence of any threat to gains; accordingly it comprises more than defensive policies and the preservation of material survival. Economic affluence, individual security, and the stability of the political system are viewed as intrinsic values in any society. Therefore, a lack of energy security constitutes a major threat to these values.

If we go beyond the traditional definition of energy security – i.e., sufficient supply of energy and the ability to procure or produce it – politicians and the private sector can add four different and overlapping dimensions to energy security:

- internal policy;
- economics;
- geopolitics; and
- security policy.

These dimensions are closely related to energy security; as such, security is considered a prerequisite for economic progress and political stability. Although energy touches every domain of life, these four dimensions are considered the key to guaranteeing the stability of energy supplies. Absolute security is beyond reach, but carefully formulated policies are able to reduce the negative effects of any crisis.

It is worth mentioning that international and regional cooperation in this field are necessary, as an isolated state cannot face these challenges alone. This is where the importance of grid interconnection between Arab countries to bolster cooperation in the field of securing electricity becomes clear. In view of the expansion of interconnection projects with non-Arab countries, however, it is important that Arab states do not resort to borrowing electric energy from foreign countries and continue to pursue self-reliance in terms of domestic electricity supply to avoid undermining their energy security.

Options for Securing Sources of Energy

Energy use will double by 2050. If we take into consideration the introduction of essential efficiency improvements in the uses of fossil fuels, in addition to a continuous shift from coal to natural gas, it will remain necessary to reduce the contribution of fossil fuels to the energy mix to not more than 30 percent of total energy use in 2050 in order to meet emission reduction targets and stave off global warming. It is therefore necessary for decision-makers to review all possible approaches to energy production to ensure that increasing global demand for energy will be met in spite of the constraints imposed by international emissions reduction agreements. Some scholars are of the view that there are four basic options in this regard:

1. setting clear and firm criteria for energy demand in such a way that it is reduced, or at least maintained at its present level;

2. resorting to the use of nuclear plants to generate electricity and/or desalinate sea water for drinking, industry and agriculture;

3. developing the use of renewable energies, especially in homes, public infrastructure, agriculture and industries that do not require a great amount of energy; and

4. isolating carbon emissions resulting from fossil fuel use.

Although these options appear independent of one another and in some ways contradictory, they are all required in the pursuit of securing energy sources without impairing the planet.

Decision-making in the field of energy is a complex matter owing to the presence of several factors from energy source availability and national and commercial policies to natural disasters and political- or security-related events, etc. The recent three-fold increase of oil prices during the course of one year (and their collapse shortly thereafter) testifies to our inability to envisage all eventualities in the realm of energy demand and production. In spite of this it remains necessary to establish a clear mechanism to govern energy sources in 2050 in accordance with the options then available.

Making a decision about adopting a specific option for energy production will not be useful in the short-term if a long investment period is required (such as for nuclear reactor construction). This will run counter to the wishes of investors as regards quick returns and economic risk avoidance. This suggests that the time-scale is an essential matter and is especially important in the field of nuclear energy, owing to the fact that the average age of a power reactor is 40 years. Considering such a time-scale, more than half of the present nuclear capability will disappear by 2020.

If we accept that all current projects will be completed, they will not yield more than twenty percent of today's nuclear capability.

Compensating this deficit in total nuclear capability will not be wholly realized by nuclear energy, but by developing the exploitation of renewable energies and methods of using fossil fuels (with reassurances as regards GHG emissions) in addition to implementing measure to reduce energy demand.

Amid all these realities the only acceptable approach is to preserve as many energy options as possible. In view of the uncertainty of the future, the world is obliged to devise diverse strategies that consider all available energy options, and to develop a broad range of electricity-production technologies in order to exploit all opportunities and avoid undue wastage.

In this regard, it is difficult to oppose proposals that reserve the option of building nuclear plants. China has adopted such an option for the coming ten years to meet its growing energy requirements—even if it is building 80 reactors in order to secure only four percent of its energy needs. Other European, Asian, American and African countries have begun to take clear steps towards a nuclear renaissance. Arab oil and non-oil countries must also move forward in exploiting nuclear energy for peaceful purposes, especially in generating electricity. The first step involves developing human resources that are capable of operating peaceful nuclear facilities. The political decision has already been made by the Council of the Arab League; what remains is for each country to take the necessary steps to put this decision into effect.

The Nuclear Energy Option

Adopting the nuclear energy option in the future will essentially depend on two important factors: the environment inside and outside the state.

The internal environment will depend on several elements, including the following:

- The availability of sources of energy (oil, coal, natural gas, renewables).

- Environmental concerns and the position of civil society regarding nuclear energy.

- The investing body (state, private sector).

- Technological progress and development of nuclear technologies and infrastructure.

Meanwhile, the external environment will involve additional elements, including:

- The extent of growth in global energy demand, which is linked to population growth (especially in developing countries) and global economic activity.

- Devising long-term approaches to the system of energy production.

- The extent of the availability of fossil fuels in the light of international tension and security events.

- The environmental effects of fossil fuels in the light of the decisions of the Inter-governmental Panel on Climate Change (IPCC).

- The extent and speed of the development of renewable energy technologies and their economic returns.

- The extent of advancements in GHG reduction and carbon capture.

- Development of electricity storage technologies—a very important matter in the field of renewable energy.

- Economic factors, especially in the least developed countries. This necessitates development of reactors at a cost lower than at present.

- Availability of nuclear fuel, which depends on current uranium availability as well as new discoveries or uranium extraction from phosphate (which is found in abundance in a number of Arab countries) or from sea water by new technologies that keep prices economically competitive.

- Managing radioactive waste, especially in light of waning availability of areas prepared for storage, and the high cost of developing areas where this waste can be buried (deep), or stored at surface level.

- Nuclear safety, which has an impressive record in spite of the few tragic events that have occurred in the past: safety must be bolstered and improved in the case of a greater expansion in nuclear capability which will naturally increase the likelihood of accidents.

- Proliferation of nuclear weapons: fear of the proliferation of nuclear weapons contributes to the inability to adopt the nuclear option in some countries, especially those that must purchase reactors and nuclear fuel from abroad.

- Public opinion concerning nuclear energy, which differs from one country to another: the degree of acceptance among the public determines the adoption of nuclear energy and other technologies by decision-makers.

- Availability of expertise in the nuclear field: this is considered an important matter in light of the presence, or absence, of a decision to adopt the nuclear option. Developing national skills in any field, especially that of nuclear science, requires great effort and time. In spite of the availability of these skills in major countries, these countries have begun to worry about the diminishing numbers of students in this field and have designed international and national projects (in cooperation with the IAEA) to preserve knowledge bases. This necessitates additional efforts among other countries to provide indigenous cadres skilled in the field of peaceful nuclear use.

- Research and development (R&D): the existence of nuclear skills will lead to accelerated R&D in pursuit of the optimal exploitation of nuclear energy as a major future source of energy.

- The magnitude of the challenges associated with the question of sources of energy and their effects on the environment and public health will remain in the coming decades, as will their controversial nature. The countries that possess nuclear capabilities will be required to contribute to the development of safer, less expensive reactors that are more attractive to commercial investors. It is particularly important to target investors in the private sector, owing to the prevailing hesitance to invest heavily for long periods before receiving returns.

Nuclear Energy and Climate Change

The Kyoto Protocol obliges signatories within the Framework Convention on Climate Change sponsored by the UN in 1997 to pursue reductions in GHG emissions that cause global warming. In this respect, Kyoto urges countries to develop scientific research, increase the use of new and renewable forms of energy, and develop inventive and advanced technologies to conserve the environment. One of the important options in this regard is that of using carbon-emissions-free nuclear energy, which plays an essential role in curbing the phenomena associated with climate change.

As a source of electricity generation, nuclear energy presents great benefits to the environment. It does not contribute to raising the temperature of the earth via the emission of greenhouse gases such as carbon dioxide, nor does it produce any sulfur or nitrogen oxides or environment-polluting particles. When electricity is produced by nuclear energy heat is produced by fission, whereas the burning of fossil fuels emits huge amounts of environment-polluting gases (the burning of 12 g of carbon produces 48 g of carbon dioxide).

We can estimate that the amount of electricity produced globally by nuclear energy (approximately 15 percent of the total) saves the production of 1.8 billion tons of carbon dioxide. We can save more by building new nuclear plants, or by upgrading existing plants and prolonging their design-lives. In Europe alone, half a billion tons of carbon dioxide is saved (i.e. 75 percent of what is emitted from all the cars in Europe). As for the issue of nuclear waste, this is radioactive and solid, but is small in size in comparison to the electricity produced. This waste can be controlled, however, unlike that of fossil fuel plants which is simply released directly into the atmosphere. If fossil fuel plants are provided with measures to control pollution, the cost of building or managing them will increase, necessitating larger financial investments as well as an increase in the price paid by the consumer not only for electricity, but also for all other materials produced in factories, farms, or institutions using electricity. Higher electricity prices will also be reflected in the cost of services such as healthcare.

The basic concern regarding electricity generation is the emission of carbon dioxide (the basic element in the phenomenon of global warming) sulfur dioxide (which causes acid rain) and nitrogen oxides (which have a negative effect on the ozone layer). Nuclear energy provides great environmental benefits as it ultimately reduces atmospheric pollutants. It does not contribute to releasing these gases except in an indirect way (i.e., in manufacturing the components of the reactor, or when using machines operated by fossil fuels to transport these components). It is worth mentioning that renewable forms of energy also contribute indirectly to releasing greenhouse gases, and at a percentage higher than that of nuclear energy.

The Environmental Protection Agency (EPA) estimates the average emission of CO_2, sulfur dioxide and nitrogen oxides per kilowatt hour according to Table 1.3 below:

Table 1.3

Emissions per Kilowatt Hour (lb.)

Emissions	Natural Gas	Oil	Coal	Nuclear
Carbon Dioxide	1,135	1,672	2,249	0
Sulfur Dioxide	0.1	12	13	0
Nitrogen Oxides	1.7	4	6	0

There is no doubt that substituting nuclear energy gradually for other polluting sources of energy – especially coal – will substantially reduce global emissions of greenhouse gases and pollutants. Nonetheless, developing technology with which to better control pollution from fossil fuel plants might also be an option.

Carbon dioxide emissions from fossil fuel plants are estimated at 25 billion tons annually, nearly 40 percent of which is emitted by coal alone.

Figure 1.3 illustrates the emissions of greenhouse gases from different sources of energy; nuclear energy leads in terms of the trivial comparative volume of its emissions, and highlights the urgent need to switch to nuclear energy use. One can go further and say that the world has only one, single option in avoiding the catastrophic effects of GHG emissions; for every 22 tons of uranium used as fuel, one million tons of carbon dioxide are prevented from entering the atmosphere.

Figure 1.3

Direct and Indirect Emissions of Energy Generation

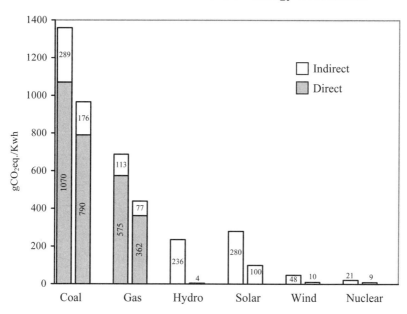

Notes: measured in grams per CO_2 equivalent per kilowatt hour; bars represent lowest and highest limits.

Conclusion

The magnitude of the challenges associated with the question of energy sources and their effect on the environment and public health will not diminish in the coming decades. Countries that possess the relevant expertise must contribute to the development of safer, more economical nuclear reactor designs to attract investment. Renewable energy technologies must also be furthered through appropriate R&D programs in order to contribute to the satisfaction of the world's energy requirements.

It is hoped that international efforts will be fruitful in the areas of nuclear integration and waste disposal. Irrespective of the degree of success of these efforts, however, it will nonetheless be a matter of decades before major nuclear electricity-generation capacity will truly come to fruition.

2

Economic Consequences of the Nuclear Power Option

Steve Thomas

B y the time of the Three Mile Island accident in 1979, the fortunes of the nuclear industry were already in decline. The huge programs of nuclear orders that many governments had foreseen in the wake of the first 'oil shock' were collapsing and existing orders in the United States were being cancelled. The main reason for this was not public opposition, but the fact that the economics of nuclear power were increasingly revealed to be poor, and demand for electricity was growing much less than had been forecast. Since then, there have been numerous predictions of nuclear revivals, often justified as being necessary to meet important strategic objectives such as diversifying energy supplies, combating acid rain and conserving fossil fuels, but all have come to nothing.

Since 2002, a new revival, or so-called 'nuclear renaissance,' has been forecast. This has been based on the availability of a new generation of reactor designs, described by the nuclear industry as Generation III+, but also backed by apparently much stronger commitments from governments such as those of the United States and the United Kingdom. While the economics of this new generation of plants is said to be superior to that of their predecessors, one of the main claimed benefits of the 'renaissance' is that it is a cost-effective way to reduce greenhouse gas emissions. Despite this powerful governmental support and these apparently compelling strategic justifications, orders have been slow to materialize. In the West,

[57]

only two units have been ordered—one for Finland and one for France, countries where the climate for nuclear orders has remained relatively healthy throughout the past 30 years. No orders have been placed in countries such as the United States, the UK and Italy, where orders would herald a major improvement in the prospects for nuclear power.

Finance

An important issue that has emerged as a potential barrier to a nuclear renaissance is that of obtaining affordable finance. As nuclear power plants are by far the most capital-intensive forms of electricity generation, economic viability depends heavily on the availability of low-cost finance.

The nuclear industry's record of building plants on time and to cost is poor – much worse than for other generating technologies – and many plants have far over-run their forecast construction time and cost. In the past, this economic risk did not have an impact on the cost of finance. Whilst electricity-production was a monopoly industry, electric utilities worked on a 'cost-plus' basis, passing on whatever costs were incurred to consumers. This meant that consumers bore all the risk of loans to electric utilities, which resulted in interest rates that reflected this low level of risk. However, as nuclear plants in the United States began to come on-line – far over budget – in the late 1970s, electricity tariffs for consumers were increased dramatically to recover the high costs incurred. Economic regulators therefore began to demand that utilities bear at least some of this over-spend from their profits rather than recovering it from consumers. Investing in nuclear energy thus became economically risky for utilities and their financiers because additional costs were likely to reduce their profits and could even lead to bankruptcy, resulting in the financiers not being repaid. Ordering, therefore, quickly stopped in the United States and no order that has not subsequently been cancelled has been placed in the country since 1974.

In the 1990s in Europe, the European Union's policy of opening up electricity generation to competition also made investing in nuclear power risky. In a competitive electricity market, companies whose electricity was expensive would quickly lose out and go bankrupt. This risk was graphically demonstrated in Britain in 2002. In that year, forty percent of Britain's generating capacity – including all its nuclear capacity – was owned by insolvent companies. A consortium of banks that had repossessed the fossil fuel plants from the companies they had lent to became the second largest owner of generating capacity in Britain. The bankruptcy of the nuclear companies also illustrated another risk of competitive electricity markets—at times of over-capacity, particularly when fossil fuel prices are low, wholesale prices can fall below the marginal price of production of even a nuclear plant.

The current 'credit crunch' will make banks even more cautious about lending money for risky investments such as nuclear power plants. Finance will only be available if electric utilities can find ways of at least partially insulating banks from risk. There are three possible ways this could be achieved: reverting to the old system where consumers take the risk; providing government-backed credit guarantees that ensure banks will be repaid by taxpayers in the event that electric utilities fail; or turnkey contracts that include a guarantee from the company selling the plant regarding the price electric utilities will have to pay.

From the point of view of risk reduction, the first option is the most effective of the three because both the bank and the utility are protected. However, it does assume a permissive regulatory body that can be guaranteed to permit full cost pass-through. Credit guarantees provide full assurance of cost recovery for the banks but they are of little comfort to the utility, which could still be bankrupted. In December 2008, the South African state-owned electric utility Eskom was forced to abandon its nuclear power program for at least the next few years – despite being offered export credit guarantees by the French government – because of the impact a nuclear program would have had on its credit rating. *Engineering News* reported:

In fact, ratings agency Standard & Poor's said on Thursday that South Africa's National Treasury needed to extend "unconditional, timely guarantees" across all Eskom's debt stock if it hoped to sustain the utility's current BBB+ investment-grade credit rating. The National Treasury was still to announce the details of the package. The Eskom board had, as a result, decided to terminate the commercial procurement process to select the preferred bidder for the construction of the Nuclear-1 project.[1]

Fixed price 'turnkey' contracts do protect the utility from additional construction costs but they generally offer little or no protection against time over-runs, and if things go wrong, they can lead to disputes between the utility and the vendor—as has become clear in Finland (see below). Turnkey contracts offer some protection to banks and consumers because the price paid for the plant is capped but the utility is not well protected against bankruptcy.

The issue for the 'renaissance' is whether consumers, taxpayers or vendors are willing to shoulder the economic risk of building nuclear plants. If they are not, financing new nuclear plants will be highly problematic.

New Reactor Designs

The nuclear industry has categorized reactor designs into four generations, with the first generation comprising prototypes and early commercial designs, and the second the majority of plants ordered from the mid-1960s onwards. The third generation represents designs made available from about 1980 onwards, while a fourth generation is not expected to be commercially available for another twenty years or more. Within Generation III, there is now a Generation III+ category, which has been available since about 2000; the Generation III+ designs are those upon which the hopes of the 'nuclear renaissance' depend. There are no hard and fast definitions about what criteria should be used to define the category of a particular design, but Generation III+ is distinguished from Generation III by the greater use of 'passive safety' systems; in other words, the safety of the design relies more on the inherent characteristics of the plant than on active engineered systems, such as emergency core cooling.

Five designs are being reviewed by the US Nuclear Regulatory Commission (NRC). Which of these should be regarded as Generation III and which as III+ is not clear. The Russian nuclear company Atomstroyexport is offering what it claims to be Generation III+ designs, but Russian designs are not under active consideration in any Western markets.

Table 2.1
US Nuclear Power Plants Announced Since 2006

Plant	Owner	NRC Status	Loan Guarantee	Design
Calvert Cliffs 3	Unistar	COL application submitted 3/08	Applied	EPR
South Texas 3, 4	NRG (Exelon)	COL application submitted 9/07	Applied	ABWR
Bellefonte 3, 4	TVA	COL application submitted 10/07	Not eligible	AP-1000
North Anna 3	Dominion	COL application submitted 11/07	Applied	ESBWR
Lee 1, 2	Duke	COL application submitted 12/07	Applied	AP-1000
Harris 2, 3	Progress	COL application submitted 2/08	Not applied	AP-1000
Grand Gulf 3	Entergy	COL application submitted 2/08	Applied	ESBWR
Vogtle 3, 4	Southern	COL application submitted 3/08	Applied	AP-1000
Summer 2, 3	SCANA	COL application submitted 3/08	Applied	AP-1000
Callaway 2	AmerenUE	COL application submitted 7/08	Applied	EPR
Levy 1, 2	Progress	COL application submitted 7/08	Applied	AP-1000
Victoria 1, 2	Exelon	COL application submitted 9/08	Applied	ESBWR
Fermi 3	DTE Energy	COL application submitted 9/08	Not applied	ESBWR
Comanche 3, 4	TXU	COL application submitted 9/08	Applied	APWR
Nine Mile Point 3	Unistar	COL application submitted 10/08	Applied	EPR
Bell Bend	PPL	COL application submitted 10/08	Applied	EPR
Amarillo 1, 2	Amarillo	–	–	EPR
River Bend	Entergy	COL application submitted 9/08	Applied	ESBWR
Elmore	Unistar	–	–	EPR
Turkey Point 6, 7	FPL	COL application planned 3/09	–	AP-1000

Notes: COL = Combined Construction & Operating License.

Source: Author's research.

AP-1000

The AP-1000 (Advanced Passive) was developed by the US-based company Westinghouse, which has been owned by the Japanese company Toshiba since 2006. It was developed from the AP-600 pressurized water reactor (PWR) design, a small reactor of 600 MW with more passive safety systems than in previous designs. When the AP-600 was designed, Westinghouse claimed that scale economies did not exist, so the large plants previously ordered offered no economic advantage over smaller ones. The AP-600 design received regulatory approval from the Nuclear Regulatory Commission (NRC) in 1997 but was judged to be uneconomic and was therefore never ordered. The design was scaled up to about 1,200 MW – ironically in pursuit of scale economies – and received regulatory approval from the NRC in 2006. Four orders have been placed, all for China, but construction work was not expected to start on these until end-2008. The design is also being evaluated in the UK. Of the 28 units that have had applications for construction licenses submitted to the US regulator (see Table 2.1), fourteen are based on the AP-1000 design.

ESBWR

The ESBWR (economic simplified boiling water reactor) is a boiling water reactor (BWR) generating about 1,550 MW, supplied by the US–Japanese consortium GE-Hitachi, formed in January 2007. No orders have been placed and the design is not expected to pass scrutiny by the NRC until 2009/10. It was also being considered by the UK regulatory authorities but in 2008 GE-Hitachi withdrew it from the procedures. Of the 28 units that have had applications for construction licenses submitted to the US regulator (see Table 2.1), six are based on the ESBWR design. However, the utility proposing two of these (Exelon's Victoria site) said in November 2008 that it was looking for a more proven design. Given these problems in the UK and USA, the prospects for this design are looking poorer and it may be that the more proven, but less advanced ABWR will replace the ESBWR in the USA.

EPR

The European pressurized reactor (EPR) is a 1,600 MW PWR supplied by the Franco-German company Areva NP (66 percent Areva, 34 percent Siemens). Two orders are under construction, one in Finland and one in France and two are on order for China but construction has not yet begun. It has completed regulatory approval in France but is still being evaluated in the UK and USA with final approval not expected before 2012. Of the 28 units that have had applications for construction licenses submitted to the US regulator (see Table 2.1), four are based on the EPR design.

APWR

First orders for the APWR (advanced pressurized water reactor), a 1,700 MW PWR to be supplied by Mitsubishi, are expected in Japan within the next couple of years but this has long been delayed. Regulatory approval in the USA is not expected before 2012. Of the 28 units that have had applications for construction licenses submitted to the US regulator (see Table 2.1), only two are based on the APWR design.

ABWR

The advanced boiling water reactor (ABWR) is a 1,350 MW BWR that was developed by GE and its Japanese licensees, Hitachi and Toshiba. Four units are in operation in Japan, with one more under construction, and two are being built in Taiwan. It is being offered in the USA by GE-Hitachi and, independently, by Toshiba. It was first ordered in 1989 and received NRC approval in 1997. That certification was for fifteen years and GE-Hitachi notified the NRC in December 2008 that it intends to apply for a renewal of the certification in mid-2010.[2] Toshiba must replace some elements of the design specific to GE-Hitachi and will also have to renew NRC certification. Of the 28 units that have construction licenses submitted to the US regulator (Table 2.1), two are based on the Toshiba ABWR design. If there is resistance to the GE-Hitachi ESBWR, which is unproven, the ESBWR orders could be replaced by ABWRs.

[63]

Construction Costs

When Generation III+ designs were first mooted in the late 1990s, the nuclear industry claimed they could be constructed for US$ 1,000/kW (kilowatt) so that a 1,000 MW plant would cost US$1 billion. This figure was regarded as optimistic by some observers and by around 2002 – when a number of nuclear economics studies were published, for example by the Massachusetts Institute of Technology (MIT) – the assumed construction cost was around US$ 2,000/kW.[3] In 2004, an order for Olkiluoto was placed at a cost of about US$ 3,000/kW (see below). This figure was widely regarded as a 'loss-leader,' somewhat below the economic cost, but given that it was a turnkey contract and the vendor, Areva NP, would have to pay for any cost overruns, it was assumed to be reasonably realistic. The problems at Olkiluoto mean the actual final cost may be at least US$ 5,000/kW. Since the Olkiluoto order, estimated costs have escalated even further. For example in 2007/08, the following costs have been estimated:

- Keystone: US$ 3,600–4,000/kW (June 2007);[4]
- S&P: US$ 4,500/kW (May 2007);[5]
- Moody's: US$ 5,000–6,000/kW (October 2007);[6]
- FP&L: US$ 5,700–8,020/kW (Fall 2007);[7]
- E.ON (UK): US$ 6,000/kW (May 2008);[8]
- Duke Power: US$ 4,700/kW (Oct 2008);[9]
- Progress Energy: US$ 4,000/kW (Oct 2008).[10]

There is necessarily a wide range of estimates because these projects are at an early stage, but US$ 6,000/kW now seems a reasonable central figure. This represents a six-fold increase in cost in the decade up to 2009 and a doubling since the Olkiluoto order was placed.[11] There are a number of factors that seem likely to have contributed to this cost escalation, but as yet no analytical work to apportion this cost increase between these factors. These factors are:

Rapidly Rising Commodity Prices

Since 2003, global commodity prices have escalated at an unprecedented rate. In the period 2003–2007, nickel and copper increased in price by over sixty percent per year, cement by more than ten percent and steel by nearly twenty percent.[12] These price rises have increased the construction cost of all generation options but because nuclear plants are physically larger than other options, the impact on nuclear is much greater. In the second half of 2008, however, as the credit crunch began to lead to recession, commodity prices began to fall steeply.

Lack of Component Production Facilities

The low number of nuclear orders in the past twenty years has meant that many component manufacturing facilities have closed down and there are now only one or two certified suppliers of key components. For example, the ultra heavy forgings needed to fabricate the pressure vessels are only produced in one factory (in Japan). This will tend to raise prices, although if there are large numbers of new nuclear orders placed, component production facilities will be built. However, building and certifying such facilities will take some time and until ordering on a large scale is re-established, building such facilities will carry significant risk.

Shortages of the Necessary Nuclear Skills

As with components, the lack of recent nuclear orders and the ageing of the existing workforce have led to a serious shortage of qualified personnel. For example, Standard & Poor's states:[13]

> We expect that the first few new nuclear units in the US will rely somewhat on project management experience from countries such as France and Japan where construction of nuclear units has continued relatively unabated since the US program's demise. Specifically, we expect companies like Electricité de France S.A. (EDF) and Tokyo Electric Power Co. Inc. (TEPCO) to provide operational expertise.

In the UK, the British government has been unable to recruit sufficient safety inspectors to meet targets regarding reviews of reactor designs. *Nucleonics Week* reported:[14]

> NII [the Nuclear Installations Inspectorate] has been chronically understaffed, a situation that threatens the timing of the GDA [generic design assessment] process. NII spokesman Mark Wheeler said in September that the agency needs 40 inspectors to complete the GDA on time, in addition to 20 now working on design assessments. Excluding the GDA work, the agency is about 22 inspectors short of the 192 needed, he said. NII Chief Inspector Mike Weightman has said, however, that he believes about 232 inspectors are needed to do all the regulatory work required by existing and new nuclear installations.

Weakness of the US Dollar

The increase in costs measured in US dollars may in part have been due to the weakness of the US dollar since the end of 2005. This has meant that costs measured in Euros, for example, will have escalated somewhat less than dollar prices. Weakness of the dollar may also have contributed to commodity price increases measured in dollars. The value of the dollar fell from November 2005 when the Euro was worth US$1.17 to July 2008, when it was worth US$1.57. By November 2008 the value of the dollar had increased sharply so that the Euro was worth US$1.27, yet by December 2008 it had fallen again to US$1.40.

Greater Caution among Utilities

As noted above, utilities can no longer assume that they will be allowed to pass on whatever costs they incur when building a power plant. Where there are competitive markets, if costs are too high, utilities risk bankruptcy, as happened in 2002 to the privatized British nuclear generator, British Energy.[15] Where tariffs are still regulated, utilities will rely on electricity regulators to allow them to recover costs. This will force utilities to be more conservative in their cost estimations to reduce the risk that the actual costs will exceed those forecast.

Olkiluoto

The Finnish electricity industry, which had been attempting to obtain Parliamentary approval for a fifth nuclear unit since 1992, finally succeeded in 2002. The Olkiluoto order placed in 2004 was a huge boost for the nuclear industry in general and for the vendor, Areva NP, in particular. It represented the first order outside Japan for a Generation III design, which, when complete, will provide a demonstration and reference for other prospective buyers of the EPR. Equally significant was the fact that it seemed to show that nuclear power orders were possible in competitive electricity markets. Many commentators had suggested that new nuclear orders were incompatible with competitive electricity markets for the reasons detailed above. However, Finland is part of what is generally seen as the most competitive electricity market in the world— the Nordic market covering Norway, Sweden, Finland and Denmark. Finland also has a good reputation for the operation of the four units that it owns, so there were high hopes that this would answer many of the questions concerning the 'nuclear renaissance.'

The Contract Terms

Closer examination of the deal, however, reveals some very special features that raise questions as to how representative this deal is of conditions in other markets. The details of how the plant is being financed were not published, but the European Renewable Energies Federation (EREF) and Greenpeace separately made complaints to the European Commission in December 2004 that the financing arrangements contravened European State Aid regulations. The Commission did not begin to investigate the complaints until October 2006 and in September 2007 the Competition Commission dropped the case. According to the EREF, the Bayerische Landesbank (BLB, Barvarian-owned by the state of Bavaria) led the syndicate (with Handelsbanken, Nordea, BNP Paribas and J.P. Morgan) that provided the majority of the finance. It provided a loan of €1.95 billion – about 60 percent of the total cost – at a remarkably low interest rate of 2.6 percent.

Two export credit institutions are also involved: France's Coface, with a €610 million export credit guarantee covering Areva supplies, and the Swedish Export Agency SEK for €110 million. Again, this is a surprising feature as export credit guarantees are usually offered only for exports to developing countries with unstable economies, not a category that Finland falls into.

The buyer, Teollisuuden Voima Oy (TVO), is an organization unique to Finland. PVO (Pohjolan Voima Oy), the largest shareholder, holds 60 percent of TVO's shares. PVO is a not-for-profit company owned by Finnish electric-intensive industry that generates about fifteen percent of Finland's electricity. Its shareholders are entitled to purchase electricity at cost in proportion to the size of their equity stakes. In return, they are obliged to pay fixed costs according to the percentage of their stakes and variable costs in proportion to the volume of electricity they consume. The other main shareholder in TVO is the largest Finnish electricity company, Fortum, with twenty-five percent of the shares. The majority of shares in Fortum are owned by the Finnish Government. This arrangement is effectively a life-of-plant contract for the output of Olkiluoto 3 at prices set to fully cover costs.

Overall, the economic risks of building this plant are borne by consumers, through the cost-plus terms of the contract, French (and Swedish) taxpayers through the credit guarantees, and the vendor, Areva NP (majority owned by the French public), through the turn-key contract. Whether the loan is economic to the banks is a moot point. So, far from surviving in the market through its competitiveness, the plant has been very fully and deliberately insulated from the market.

Experience at Olkiluoto

From the very beginning, the construction process was delayed by major setbacks, to the extent that after three years of construction – in November

2008 – the plant was three years behind schedule and the vendor, Areva, was suffering severe losses.[16] This was not the result of any one particular problem but of a range of failures, including welding issues, design delays, and problems with concrete and the quality of some equipment. More generally, it seemed that none of the parties involved, including the vendor, the customer, or the safety regulator, had a clear enough understanding of the requirements of building a nuclear plant.

In December 2006, the French Ministry of Industry (the French government owns more than ninety percent of Areva) said that the losses to Areva had reached €700 million on a contract fixed at €3 billion. The turnkey contract should ensure that this cost escalation is not passed on to the customer, although the deal appeared to be under strain. Philippe Knoche, an Areva representative stated, "Areva-Siemens cannot accept 100 percent compensation responsibility, because the project is one of vast co-operation. The building site is joint so we absolutely deny 100 percent compensation principle."[17] TVO did not accept this interpretation and the TVO project manager, Martin Landtman, when asked about Knoche's statement said, "I don't believe that Areva says this. The site is in the contractor's hands at the moment. Of course, in the end, TVO is responsible [for] what happens at the site. But the realization of the project is Areva's responsibility." It may well be that this dispute will have to be settled in court.

On the finance side, the leader of the syndicate of banks, BLB, had to be rescued by the German government. In November 2008, it received the first installment of guarantees from the German government for €15 billion.[18]

Compensation for delays has already reached the limit of €300 million that would be payable for a delay of eighteen months. The buyer will not receive compensation for further delays beyond those already incurred by September 2006. The possibility that the cost of buying the power Olkiluoto was expected to produce during the years 2009–12 from the Nordic market might be high enough to cause TVO to default can no longer be ignored. TVO will have to buy power from the Nordic electricity wholesale market for at least three years. Generating capacity is getting tight in the Nordic market and because the system is dependent on hydro-electricity, it is vulnerable to dry winters, which in the past few years have led to wholesale prices increasing by up to five times. For the energy-intensive consumers that make up PVO's customer base, high and volatile electricity prices are intolerable.

The question that is so far unanswered from the Olkiluoto experience is whether this is an isolated failure or whether it is a demonstration that the old problems of unpredictable construction costs and times have not been solved. It is worth noting that the second EPR order – for Flamanville in France, construction of which began in December 2007 – has already encountered problems. In May 2008, the French safety regulatory authorities temporarily halted construction because of quality issues pertaining to pouring the concrete base mat.[19] Delays had led the vendor, Areva NP, to forecast the plant would not be completed until 2013 – a year late – but in November 2008, EDF claimed the delays could be made up.[20] EDF did acknowledge that the expected construction costs for Flamanville had increased from €3.3 billion to €4 billion.[21]

The United States' Nuclear 2010

From 1972 to 1990, US utilities cancelled orders for 117 plants and no order placed after 1974 has survived.[22] In February 2002, the Bush administration announced a program aimed at re-starting nuclear ordering in the USA. The rationale was that new nuclear designs would be economically competitive but that financial and regulatory hurdles would prevent ordering. To overcome these barriers regulatory processes had to be streamlined, approval assured for a number of new designs and subsidies provided for units at up to three sites (perhaps six units). The objective was:

> ... to complete the first-of-a-kind Generation III+ reactor technology development and to demonstrate the untested Federal regulatory and licensing processes for the siting, construction, and operation of new nuclear plants.[23]

The program was unrealistically optimistic in terms of its time-scale, and was based on bringing a new nuclear unit on-line by 2010. Reference was made to 'cost sharing' but the type and extent of subsidies was not made explicit. Two main forms of subsidy were proposed:

- **Production tax credits**: in order to make electricity generated from new nuclear power plants competitive with other sources of energy, an US$ 18/MWh tax credit would be paid for the first eight years of operation. According to the Energy Information Administration (EIA), this subsidy would cost US taxpayers US$ 5.7 billion by 2025.[24]

- **Loan guarantees:** to ease the difficulty of financing new plants, loan guarantees were offered so that utilities could borrow at government treasury bond rates. The Congressional Budget Office concluded that the risk of loan default by the industry would be 'well above 50 percent.'[25] The Congressional Research Service (CRS) estimated that the taxpayer liability for loan guarantees covering up to 50 percent of the cost of building 6–8 new reactors would be US$ 14–16 billion.[26]

The Energy Policy Act (EPACT) 2005 also offered up to US$ 500 million in risk insurance for units 1–2 and US$ 250 million for units 3–6. This insurance would be paid if delays not attributable to the licensee slowed licensing of the plant. It also offered support for R&D funding worth US$ 850 million and help with historic decommissioning costs worth US$ 1.3 billion.

It soon became clear that not only were the loan guarantees the key element of the package, but also that the extent of coverage offered was insufficient to allow utilities to place orders. Federal loan guarantees would cover up to 80 percent of the debt involved in the project and if debt accounted for about 60 percent of the cost of building the plant (the rest from equity), this would mean about half the cost of the plant would be covered. Utilities successfully lobbied for 100 percent coverage of debt up to 80 percent of project cost.

The scope of the subsidies also grew from covering just three sites to loan guarantees for up to three units of each 'innovative' design and by 2008 five qualifying designs were being processed by the NRC.

The extent of the loan guarantees required has therefore grown alarmingly. In 2002, when the program was launched, construction costs of US$ 1,000/kW were still expected and the guarantees needed for five units of about 1,400 MW, each covering 50 percent of the total cost, would have been about US$ 3.5 billion. By 2005, when the number of units to be covered had risen to 6–8 and the estimated construction cost had reached US$ 3,000/kW, guarantees worth US$ 15 billion would have been needed. But in 2008, if we assume fifteen units will be eligible and will be covered up to 80 percent of their total cost of US$ 6,000/kW, guarantees worth in excess of US$ 100 billion would be required.

The Energy Bill passed in 2007 gave the Department of Energy a budget of up to US$ 18.5 billion for 2008/09 for loan guarantees covering nuclear plants. One of the key pieces of unfinished business for the new Obama administration in 2009 will be to determine what the availability of loan guarantees for nuclear orders should be. How much assistance with loan guarantees the US administration will receive from the countries where the reactor suppliers are based remains to be seen. The French agency, Coface, which could support Areva orders, has some experience of guaranteeing nuclear orders, while the Japanese agency, Japan Finance Corporation, which might support orders for Toshiba, Mitsubishi, and Hitachi, was only formed in autumn 2008 and has no such experience.

So orders in the United States are heavily dependent on political backing and if this is not maintained, orders will not materialize. Even if subsidized orders are placed, it seems highly questionable whether orders without subsidies will be possible. If the US government is not prepared to make an open-ended commitment to grant subsidies for nuclear projects, a handful of heavily subsidized orders will not be the demonstration of nuclear power's viability that the industry wanted.

Nuclear in the UK: Back with a Vengeance

Contradictory Messages

The British government's position on nuclear power has been changing rapidly since 2005, with numerous reversals of policy positions. In 2003, the government published a White Paper (statement of government policy) which stated:

> Although nuclear power produces no carbon dioxide, its current economics make new nuclear build an unattractive option and there are important issues of nuclear waste to be resolved. Against this background, we conclude it is right to concentrate our efforts on energy efficiency and renewables. We do not, therefore, propose to support new nuclear build now. But we will keep the option open.[27]

Tony Blair subsequently acknowledged that the decision to 'keep the option open' for new nuclear investment was only at his insistence. Energy White Papers are generally separated by decades, yet just two years later the Blair government set in motion the process to produce a new White Paper. This was surprising given that in the two years since the previous White Paper, the government provided no evidence that the economics of energy efficiency and renewables had deteriorated, yet there was ample evidence that the economics of nuclear were deteriorating. There was also no evidence presented that the 'important issues of nuclear waste' had been resolved.

There were questions over how open-minded the process of generating the new White Paper would be when Blair pre-empted the results of the investigation in May 2006, two months before a Green Paper (a consultation paper that precedes a White Paper) on energy policy was published. He stated, "These facts [on CO_2 emissions and energy export dependence] put the replacement of nuclear power stations, a big push on renewables and a step change on energy efficiency, engaging both business and consumers, back on the agenda with a vengeance."[28] Greenpeace subsequently mounted a successful legal challenge that the information given to the consultations on nuclear power was not balanced and the courts demanded that the whole consultation process should be repeated. It was not until 2008 that the White Paper on nuclear power was finally published, stating that:

> Following the consultation we have concluded that, in summary, nuclear power is:
> - Low-carbon – helping to minimize damaging climate change.
> - Affordable – nuclear is currently one of the cheapest low-carbon electricity generation technologies, so could help us deliver our goals cost effectively.
> - Dependable – a proven technology with modern reactors capable of producing electricity reliably.
> - Safe – backed up by a highly effective regulatory framework.
> - Capable of increasing diversity and reducing our dependence on any one technology or country for our energy or fuel supplies.[29]

There were also conflicting indications of how much new nuclear capacity would be built and how much support the government would offer. In October 2006, Malcolm Wicks, the then energy minister, told the Parliamentary Trade and Industry Select Committee:

> Government will not be building nuclear reactors, will not say they want X number of nuclear reactors. I always thought myself that if at the moment one fifth of our electricity is from nuclear, if the market came forward with something to replicate that broadly in the future, from my own point of view it seems to me that would make a useful contribution to the mix. We are not going to do anything to facilitate that, nor this percentage nor that percentage.[30]

Yet less than two years later, the new Prime Minister, Gordon Brown said, "We will have to do more than simply replace existing nuclear capability in Britain. We will be more ambitious for our plans for nuclear."[31]

On whether the government would offer direct or indirect subsidies, Wicks told the Trade and Industry Select Committee in October 2006, "No cheques will be written, there will be no sweetheart deals,"[32] adding:

> No, there will not be any special fiscal arrangements for nuclear. It should not be a surprise, with respect, because we have said it very clearly in the Energy Review. You could pursue this if you wanted by saying that nuclear waste is quite a complex subject and we are going to look very carefully at that to make sure that the full costs of new nuclear waste are paid by the market.[33]

Yet in a consultation on nuclear waste, the government stated:

> The Government would expect to set a fixed unit price based on the operator's projected full share of waste disposal costs at the time when the approvals for the station are given, prior to construction of the station. Should the actual costs of providing the waste disposal service prove lower than expected, these lower costs will not be passed on to nuclear operators.[34]

Such a cost guarantee can only be viewed as a subsidy.

The Program

The UK government's program is based on very different underlying assumptions to that of the United States. The US government believed that nuclear power was competitive with fossil fuels but that initial barriers would have to be overcome for unsubsidized and sustainable ordering to take place. The UK government has never claimed that nuclear power is competitive with fossil fuels but that ordering would take place without subsidy provided a few non-financial enabling decisions were taken, particularly on planning processes and certification of designs.

The government's nuclear regulator, the Nuclear Installations Inspectorate (NII), started to examine four separate designs in 2007, the Westinghouse/Toshiba AP-1000, the Areva NP EPR, the GE-Hitachi ESBWR and a Canadian heavy water reactor design, the ACR-1000 (advanced CANDU reactor). The rationale was that three designs would be finally certificated, thus giving utilities a choice of designs. It was expected that the EPR and AP-1000 would be the final choice, and so it has proved. The ACR-1000 was quickly withdrawn and in late 2008, the ESBWR was also withdrawn.

[75]

The NII has experienced severe difficulties recruiting sufficient inspectors to carry out its tasks and in November 2008, it was still forty inspectors (about twenty percent) short of the required number. Some of the utilities operating in the UK, especially EDF, have said they expect to be able to order plants without subsidy. EDF took over the privatized UK nuclear generator, British Energy, in September 2008 for over £12 billion sterling, to gain access to sites where nuclear plants might be built and to draw on British Energy's skilled workforce.

However, realistically, orders cannot be placed for at least five years in order to allow regulatory approval for the chosen design and planning approval for a site. Until this point, the commitment of the utilities to build is limited and could easily be dropped—they will have spent in the order of a few *million* pounds so far, while an order would commit them to a few *billion* pounds. The five-year period until orders can be placed is ample time for the utilities to claim that circumstances have changed. The government would be in a very difficult position; for example, if in five years' time EDF claim they can build without credit guarantees, prevailing conditions might dictate that finance will become prohibitively expensive without such guarantees.

So, as in the United States, the UK program is very dependent on continued political backing and if that backing does not exist after 2011 when orders might be placed in the UK and the USA, nuclear ordering programs could easily come to nothing

Issues for the Gulf States

It seems clear that the Gulf States have the financial resources and the ability – at least in the long-term – to produce the human resources necessary to sustain a nuclear ordering program. However, there are a number of factors that suggest that 2009 is not a good time for the Gulf states to embark on a nuclear program:

- **Construction cost:** there is substantial volatility in estimates of construction costs that will only be resolved when there is real and verified experience of the actual construction cost of current designs. There is also a serious risk that the economics of nuclear power could be very much poorer than expected.

- **Skills:** there is a severe shortage of nuclear expertise worldwide that cannot be quickly remedied. Attracting the right number of well-qualified staff in the Gulf states would require paying high costs to outbid all the other countries competing for such expertise.

- **Unproven designs:** none of the available modern designs has any operating experience and cannot be regarded as proven until there is at least, say, three years operating experience; this will not happen before about 2015. Indeed, only one design, EPR, has any significant construction experience and this has been very poor, as shown above.

- **Potential exposure:** the nuclear programs in the United States and UK are still in their early stages and if very strong and costly government backing does not continue to be on offer, these programs could collapse. If the Gulf States have already placed orders, this could leave them in a very exposed position with little international back-up, i.e., of owning most of the examples of the new designs.

There is also the issue of opportunity costs. While the Gulf States are better able than most countries to finance nuclear plants, there is still an opportunity cost to this finance. This will be more important if the credit crunch leads to a severe global recession, which would result in low fossil fuel prices for the next five years. There is also the opportunity cost of skills, because a nuclear program would absorb a generation of the region's best scientific skills. If there is no realistic chance of the Gulf states becoming world leaders in nuclear technology, and given that there are many countries with a fifty-year head start in this regard, the chances do look slim. Is this the best use of these skills? Or are there technologies where the Gulf States have a more realistic chance of developing internationally competitive skills that would allow a smooth transition to the post-oil era?

These arguments seem to make a strong case that now is a bad time for the Gulf states to embark on a nuclear program. This is not to argue that they should do nothing, however. In the interim, while questions concerning the new generation of nuclear designs remain unresolved, a strong program involving energy efficiency measures and renewable generation technologies would have major attractions. Improving energy efficiency in households and industry, as well as preserving oil and gas supplies for lucrative exports, would leave Gulf consumers better able to cope with the higher energy prices that will be felt when the Gulf's reserves are further depleted. Renewables would conserve fossil fuel supplies for export, but would also offer the potential – particularly regarding solar – for the Gulf states to become world leaders in a technology that will inevitably become increasingly important in the future.

3

Nuclear Energy in the Gulf: The Role of Political and Social Factors

Malcolm Grimston

T he later years of the first decade of the 21st century have seen many
countries either considering the revival of their nuclear power
industries or employing the technology for the first time, driven by
growing concerns over the cost and availability of hydrocarbon fuels and
fears over climate change. However, the previous major wave of
investment in nuclear technology – in the 1970s and 1980s – ground to a
halt as fossil fuel prices fell and nuclear economics bred disappointment,
but also because of falling levels of public support.

There has been some agreement – at least in some developed countries
and until relatively recently – on nuclear power's particular unpopularity
as an energy source. It has thus been assumed that unless and until this
unpopularity is overcome, nuclear power will not flourish, even if a strong
case could be made for it on other grounds. For example, it has proved
particularly difficult in many countries to designate new sites for nuclear
facilities. In addition, there have been fears about nuclear installations in
neighboring countries (for example, Austrian concerns about the Temelín
plant in the Czech Republic, and Irish objections to Sellafield in the UK).

In its early days, nuclear power was broadly accepted, often with
enthusiasm, in many of those countries where it has recently been
unpopular. Yet even from those early days, there was a group of people –
including some scientists who had worked in the field – who expressed

considerable unease regarding nuclear technology in both its military and civil forms. The *Bulletin of the Atomic Scientists* was first published in December 1945, and in an early edition, Albert Einstein wrote: "The unleashed power of the atom has changed everything save our modes of thinking, and thus we drift toward unparalleled catastrophe."

By the late 1970s and early 1980s, those skeptical about both the technology and its practitioners were having a considerable effect on public opinion, aided by a number of factors involving the technology itself. The construction times and costs of many plants were far higher than projected. The performance of many plants was disappointing. The accidents at Three Mile Island and Chernobyl further served to exacerbate growing mistrust of the 'nuclear industry' and its supporters within governments. This mistrust had its origin – at least in part – in the arrogance and secretiveness of nuclear spokesmen in many countries. The suspicion that the industry and its supporters were able to put undue pressure on regulators further damaged their public credibility. Critics of the industry often had no apparent vested interest, while the industry's responses increasingly came to be discounted with phrases such as: "they would say that, wouldn't they?" The passion which has surrounded the nuclear debate in recent years is to a considerable degree a legacy of these factors.

At the same time, perceptions of the availability of alternatives to nuclear power were changing. When global fossil fuel supplies were under apparent threat (notably in the 1950s and again in the ten years from 1973 onwards), nuclear programs were instituted in a number of countries with relatively little objection—at least by today's standards. The discovery of vast reserves of gas, as well as oil, coupled with low prices and the development of the highly efficient Combined Cycle Gas Turbine by the mid-1980s, reduced the apparent need for nuclear power in many countries.

Without any pressing economic or supply security arguments apparent for nuclear power, it was perhaps natural that the negative aspects of the technology should come to the fore. Public opinion toward nuclear power through the 1990s and early 2000s seemed to be more positive in those countries with limited indigenous alternatives (such as Japan and France) and those with rapidly growing energy demand (such as China and India) than in those to whom alternative sources were available (like the USA, Germany and the UK).

The international picture was to change dramatically once again from the middle years of the first decade of the new century. Growing concerns about climate change, projections of hydrocarbon reserve shortages and, of course, the burgeoning price of oil served to bring nuclear energy back into favor among previously skeptical governments and citizens.

The challenge for the nuclear energy industry then, and indeed for those countries returning to nuclear energy or entering the field for the first time, is to learn the lessons of the failure to sustain the first major wave of nuclear investment and take suitable steps to prevent a repeat performance.

The 'Nuclear Debate'

The nuclear debate can be characterized as a battle between two diametrically opposed groups, who nonetheless seem to share a large number of common features.

The extremes of the debate are also characterized by unwillingness – or perhaps even inability – to engage in meaningful discussion with stakeholders. There have been many examples of nuclear advocates claiming, often with apparent frustration, that 'the public' simply does not understand how beneficial nuclear technology is, while making little apparent attempt to fully understand the concerns. A similar attitude – although opposite in argument – is to be found among the industry's adversaries.

In reality, most people holding a view on nuclear technology fall between the two extremes and are more prepared to accept both strengths and weaknesses in each side's view. Many of those who are not firmly committed to one side or the other show a willingness to alter their position, for example as new information becomes available.

[81]

Table 3.1
Nuclear Advocates and Opponents

The Advocates	The Opponents
Belief that major elements of the future are predictable; certainty regarding general projections of the availability of various energy sources. For example, renewables demonstrably have the potential to remain only relatively minor players in world energy supply.	Belief that major elements of the future are predictable; certainty regarding general projections of the availability of various energy sources. For example, renewables demonstrably have the potential to dominate world energy supply.
Absolutely certain about the future role of nuclear power (a major and important one), and about issues such as nuclear waste (not a difficult technical problem).	Absolutely certain about the future role of nuclear power (no role at all), and about issues such as nuclear waste (a technically insoluble problem).
Arrogance born out of belief in infallibility of own analysis.	Arrogance born out of belief in infallibility of own analysis.
Belief that the public is irrationally frightened of nuclear power. If only people could be properly educated they would become more pro-nuclear and support the nuclear industry.	Belief that the public is irrationally complacent about nuclear power. If only people could be properly educated they would become more anti-nuclear and support anti-nuclear campaigns.
Characterization of opponents as either fools or ill-intentioned.	Characterization of opponents as either fools or ill-intentioned.
Belief that government is not to be trusted to take wise decisions as it is under too much influence from anti-nuclear media and pressure groups.	Belief that government is not to be trusted to take wise decisions as it is under too much influence from the nuclear industry and its supporters.

There is evidence that public relations departments within the nuclear industry are learning from the past, and are moving from the 'one-way' model of communication that was prevalent in the 1950s towards a more balanced, 'two-way' approach, carrying with it a willingness to compromise where this presents a feasible way forward. Similarly, regulators and decision-makers are now well aware of the range of views that face them.

However, the level of emotion displayed by the die-hards on both sides of the debate is unlikely to aid the process of formulating careful and considered decisions about the future of nuclear power. Ultimately, decisions must be taken—and those decisions will not satisfy everyone.

Public Opinion

Considerable caution must be exercised when interpreting 'public opinion.' The very concept 'public' is of limited usefulness in a modern pluralistic society. The population is better viewed as an interlocking pattern of smaller 'publics'; any particular individual may moreover move from one public to another, if, for example, proposals are revealed to construct a major project near their home.

A useful distinction can be drawn between 'opinions,' 'attitudes' and 'values.' Worcester has offered an oceanographic analogy[1]:

- opinions – ripples on the surface;
- attitudes – the tides;
- values – the deep currents.

The results of 'opinion' polls, then, are notoriously dependent on the particular question asked. They can also be very volatile. However, several themes may be detected; whether a particular person or group of people tends to be pro- or anti-nuclear at a particular moment depends on a number of variables, including:

- perceptions of the 'need' for the technology—nuclear power tends to be more popular in countries with serious concerns about energy security than in countries with a wide range of energy options and relatively modest growth in energy demand;

- perceptions of risk—nuclear power tends to be less popular in the immediate aftermath of an accident, for example especially if the accident occurred locally or had local consequences, or after other major events which may be relevant to the safety of nuclear installations, such as the terrorist attacks of September 11 2001, while people who are more familiar with the technology, perhaps through having lived near an operating plant for some years, tend to be less worried; and

- social/political/psychological factors—political parties within a single country can hold radically different views on nuclear technology (e.g. the CDU and the SPD in Germany); individuals who are attracted to large economic or technological projects tend to be more pro-nuclear, or at least be less impressed with the arguments of anti-nuclear pressure groups than people who are suspicious of globalized markets and 'capitalism' in general; people whose jobs depend on a local nuclear facility naturally tend to be more pro-nuclear than those who do not; men tend to be more pro-nuclear than women.

It should be stressed, though, that these are at best tendencies—there is always a range of 'opinions' among people of the same nationality and from apparently very similar backgrounds.

A number of specific explanations have been suggested for the apparent special unease felt about nuclear power in many countries. They include:

- links to the military, both real (the development of shared facilities) and perceptual;

- secrecy, coupled sometimes with an apparent unwillingness to give 'straight answers' (in part, perhaps, because of links to military nuclear operations in some countries or because of commercial issues);

- the historical arrogance of many in the industry, dismissing opposition, however well-founded or sincerely held, as 'irrational';

- the apparent vested interest of many nuclear advocates, to be contrasted with the apparent altruism of opponents;

- the perceived potential for large and uncontainable accidents and other environmental and health effects, notably those associated with radioactive waste;

- the overselling of nuclear technology, especially in its early days and in particular with regard to economics, leading to a degree of disillusionment and distrust;

- a general disillusionment with science and technology and with the 'experts know best' attitude prevalent in the years immediately after the Second World War; and

- the wider decline of 'deference' towards 'authority' (including, for example, politicians and regulatory bodies).

It should be noted that many of these factors are relevant in other debates over public policy, such as the controversies over genetically modified organisms, road building or global trade, for example. This point will be developed later.

Perceptions of negative public opinion, whether justified or not, can be extremely costly for investors in nuclear power, and can even act as an absolute barrier. Opposition to the construction or operation of nuclear facilities could increase the costs of nuclear-generated electricity in a number of ways. There may be delays during construction or in achieving an initial operating license, or interruptions in operation. Extra physical or operational security measures might be demanded, for example in response to a potential terrorist situation, even if there is no direct evidence of a threat. Implementing such measures may be especially costly if they involve retrofitting new features to an existing plant or one under construction. The costs of site selection, evaluation and the licensing process itself can increase. The costs of transporting nuclear materials can escalate, because of increased requirements for security against protest or the need to find new routes. The economic risk associated with uncertainty results in demands for higher rates of return on investment, an especially serious issue for highly capital-intensive technologies such as nuclear power.

In the most extreme cases, fears of public reaction can lead to a fully completed plant being refused an operating license or a government taking steps to prevent nuclear construction or to close down existing facilities before the end of their technical lifetimes. Since 1978, for example, some 14 GW of nuclear power plants, and one MOx (mixed oxide) fuel production plant, have been closed or halted in advanced stages of construction for non-economic reasons in six OECD countries (Austria, Germany, Italy, Spain, Sweden and the USA), some as a direct result of referenda. Most of these closures were carried out in the years after the Chernobyl accident in 1986, although the 720 MW Tullnerfeld reactor in Austria was refused an operating license on completion after a referendum in 1978. Italy no longer operates nuclear power reactors, having closed three operating plants after a 1987 referendum. Germany, the Netherlands and Sweden adopted formal phase-out policies by law (subsequently rescinded in the Netherlands), Switzerland adopted a ten year moratorium on new construction in 1990 (again subsequently abandoned), and Belgium took a policy decision to phase out nuclear power. A number of countries which did not have operating nuclear power plants, such as Australia, Austria, Denmark, Greece, Ireland, Norway and Poland, put in place legal or policy obstacles to nuclear power.[2]

Whether or not the public really was deeply suspicious of nuclear technology in these countries, and whether or not public fears were justified, perceptions about the public mood has at various times had profound implications for nuclear power's development.

[86]

The Complexities of Public Opinion

It seems that, at least until recently, when asked whether new nuclear power stations should be built, people in most developed countries tend to say 'no.' However, the same is true of a number of other activities which appear to continue relatively unfettered. In the UK, for example, the 'chemical industry' has regularly appeared less popular than the nuclear power industry in opinion polls (when people are asked which industries they believe cause environmental damage, for example). However, with certain exceptions such as PVC, most chemical companies have not seen their business as severely curtailed as in the catalogue of nuclear rejection listed above.

The phenomenon of holding apparently contradictory attitudes is a common one in the field of public perception. A number of technologies, or indeed the concept of technology itself, can be simultaneously exciting and frightening. The wording of the question asked can be crucial in determining the response that is elicited, a fact well known to opinion pollsters. Questions about whether nuclear power is likely to be desirable 'as fossil fuels run out and concerns over global warming increase' are far more likely to receive positive responses than questions about whether nuclear power is likely to be desirable 'given the risks of a major accident like Chernobyl and the problem of nuclear waste.'

The rapidity with which public attitudes – or at least public responses to questionnaires – can change was demonstrated in California around the time of the power crisis of 2000–2001. As the power cuts bit, support for new nuclear stations rose significantly, falling away just as rapidly when power shortages eased after the cool summer.

'Public opinion' and 'public confidence,' then, would seem to be movable feasts, heavily influenced by the context in which questions are put to respondents. The degree to which support or opposition is expressed, however, is only one aspect of an opinion. The strength with which the opinion is held and the likelihood that the topic will be mentioned without specific prompting are also important.

[87]

Evidence from a number of developed countries would seem to suggest that the level of overt public concern among those not directly affected by nuclear projects is low. When people are asked, unprompted, about concerns regarding the environment, for example, nuclear power comes some way behind urban pollution from motor vehicles and climate change. Public concerns that are widespread but 'back of mind' are especially difficult to evaluate.

Perceptions of Public Perceptions

When it comes to considering the effect that public opinion has on decision-making, a further complication occurs. Decision-makers, naturally, will in part base their decisions on their perception of public opinion, in other words the perception of a perception. There is some evidence that these second-order perceptions may also be subject to some systematic errors.

Opinion polling carried out by MORI in the UK[3] suggested an interesting pattern of perceptions regarding public opinion held by opinion-formers and decision-makers.

Table 3.2
MPs Perceptions of Public Opinion and the Nuclear Issue

	Favorable towards nuclear energy industry	Unfavorable towards nuclear energy industry	Neither favorable nor unfavorable / don't know
Public opinion	28 %	25 %	47 %
All MPs	43 %	44 %	13 %
MPs' perception of national public opinion	2 %	84 %	14 %

A similar pattern has been observed in the USA. This data implies that, at least in some countries, the perception of public opinion among decision-makers may not be accurate, and therefore that the decisions being taken may be skewed by erroneous assumptions. Possible reasons include the attitude of certain elements of the popular media, and the greater effectiveness of anti-nuclear pressure groups in organizing letter-writing and other publicity campaigns.

Following public opinion is not always the motivation for opposing nuclear energy. Opinion polls in Sweden, for example, have shown a considerable majority in favor of the continued operation of existing nuclear power stations for some time[4], a view also taken by most of Swedish industry, yet the governing coalition forced the closure of two nuclear plants, in 1999 and 2005.[5] The internal politics of governing coalitions has also been a factor in determining policy in countries such as Germany.

Public Attitudes toward Particular Projects

Attitudes (and the values on which they are based) towards nuclear technology, and how those attitudes and values are perceived by decision-makers, are matters of considerable importance. For example, companies considering investment in new nuclear power stations may be more willing to do so if they consider 'the public' to be broadly supportive. However, if the public is clearly skeptical about nuclear technology, fears may arise about changes in government policy that could work against the industry—for example, the possibility of a considerable delay between the completion of a plant and the granting of an operating license or the imposition of stricter emission limits. Even if this did not deter investors entirely, it might well lead to higher rates of return being demanded to compensate for the higher risks involved.

However, even if there were little evidence of front-of-mind public opposition to nuclear power in principle in a particular country, proposals to build a particular facility at a particular site might provoke a considerable backlash from the local population and others. Widespread public action against waste transportation in Germany and shipments of plutonium to Japan have indicated that many people, who might regard themselves as unconcerned about nuclear power when not directly involved, can change their views and behavior considerably when confronted by a concrete proposal, especially if that proposal concerns a nearby area. One might expect that such concerns could be more easily addressed if some of the issues that underpin opposition were to be taken into account – for example, participation in decision-making and ensuring a fair distribution of risks and benefits – but this cannot be guaranteed.

A Brief History of Nuclear Perceptions

It seems an untenable assertion that, for some individuals, fears about nuclear power are more acute than fears about other societal risks which, based on past record, are more dangerous, e.g. use of coal for electricity or petrol for transport, each of which have been associated with numerous deaths, for example through direct accidents or the health effects of emissions. Furthermore, the debate about nuclear power has an emotional tone which is unusual even within the controversial and uncertain field of energy. The fascination of some elements of the mass media with matters connected with radiation may be part of the explanation.

Some commentators have observed that radiation and nuclear energy have a number of properties that connect with deep-seated, apparently universal human images and myths—both positive and negative in their emotional tone. Elements of this 'pre-fission nuclear imagery' include:

- 'Eden' myths—tales of a golden age before the fall of man was brought about by dabbling in knowledge which is forbidden, i.e., the 'small boy playing with fire';

- images of the destruction of mankind, often by fire—the concept of 'atomic bombs,' for example, was introduced in H.G. Wells' 1913 novel *The World Set Free*;

- the 'taming of nature' and unleashing of massive forces, for good or evil;

- the alchemists' dream—transmutation of elements, which occurs during radioactive processes, had held a fascination for many centuries; and

- the idea, found in many civilizations, of invisible rays, undetectable by unaided human senses, that could bring life or death or blight future generations.

Before the Second World War, debate about the use and effects of radioactive materials was not widespread. Before 1945, there did not seem to be any particularly unique fear of radiation. It had benefits (especially in medical treatments) and risks and many people seemed happy to believe that the former outweighed the latter. High levels of trust in the scientists involved may also have been important.

The atomic weapons dropped over Japan at the end of the Second World War had a profound effect in many ways. They revealed to the public at large that atomic energy was a reality, not merely a distant dream. The image of world destruction was all too believable in the face of photographs and reports from Hiroshima and Nagasaki. Post-war programs such as those concerned with civil defense seemed to cause greater fear among the population. Concerns both about the idea of nuclear technology, and about the practices of some of those in charge of it, were raised by some of the very scientists who had been involved in its development, perhaps the most notable being Albert Einstein, Robert Oppenheimer and Joseph Rotblat.

It became clear that there was a danger that nuclear fear in the population could result in a rejection of nuclear technology, and weapons in particular (by now regarded by the US government, and increasingly by other countries, as essential). In the immediate aftermath of the war scientists in many countries, notably the USA, spoke of their determination to turn the dreadful power of the atomic weapon into a positive force by using it as an alternative way of making electricity and providing propulsion. The US 'Atoms for Peace' program was launched by President Eisenhower in 1953. Under its auspices, the Atomic Energy

[91]

Commission carried out exhibitions throughout the 1950s in places such as Karachi, Tokyo, Cairo, São Paulo and Tehran displaying the benefits of atomic energy.

The Atoms for Peace program was officially promoted as a step to controlling weapons proliferation. Considerable help in developing nuclear technology was offered to any nation which would allow international inspection of facilities to demonstrate that materials were not being diverted for military use. Another purpose, however, was undoubtedly to reduce the level of potential public opposition to a continued US nuclear weapons program in the face of the Soviet nuclear threat.

In Europe and Japan the most easily-mined coal had been exhausted, the best hydropower sites were in use and almost all oil was being imported. The energy shortages of the war were followed by coal crises in the post-war years. In the extremely severe winter of 1946 much of Europe faced shortages of fuel to heat homes and lacked electricity to light them, while Japanese cities faced brownouts well into the 1950s.

Owing to the highly complex nature of the subject, even many politicians felt unable to engage meaningfully in the intricacies of the debate. Responsibility not only for implementing nuclear policy, but also for setting that policy, was to a considerable extent vested in bodies representing nuclear experts. A technocratic mode of decision-making became dominant, to the detriment of dialogue with – and control by – normal democratic structures. This was exacerbated by the secrecy associated with military uses of nuclear materials. It is perhaps unsurprising in such circumstances that the spokesmen for the nuclear industry became increasingly arrogant.

The ground was fertile, therefore, for nuclear power to be oversold as the new miracle fuel. The rate of publication of popular books and articles on nuclear power trebled after Eisenhower's Atoms for Peace speech. Prior to an international conference in Geneva in 1955, most nuclear scientists had warned that as an economic venture nuclear power was still decades away, but the 3,000 scientists and their followers attending the conference heard from representatives of several countries that nuclear power was close to being commercially profitable. The American people were told, famously, that nuclear power would become "too cheap to meter," an incomprehensible but highly influential statement.

The enthusiasm of many more countries, such as West Germany, for the new technology was fired, while the Suez crisis of 1956 added urgency to the perception that a new source of energy was required. By the end of 1957 the United States had signed bilateral agreements on nuclear technology with 49 countries, and American firms had exported 23 small research reactors. The Soviet Union had also invested enormous resources in the development of nuclear technology for export within the Communist bloc, both as a Cold War challenge to the USA and through a similar belief in a nuclear utopia.

However, enthusiasm for new technologies and other human endeavors is often exaggerated, and disillusionment sets in. The Atomic Scientists of Chicago (the site of the first controlled fission experiment in 1942) was formed in 1945 to sound warnings against believing all of the official pro-nuclear pronouncements. Over the next three decades further groups were founded who argued this case, perhaps the most notable being the Campaign for Nuclear Disarmament (CND, 1957), Greenpeace and Friends of the Earth (1971).

Attitudes and Feelings

It might appear obvious that our attitudes to issues such as nuclear power are built on our interpretation of the facts presented to us. The industry assumed that public fears about nuclear waste, for example, which arose because of ignorance or irrationality, could simply be countered by 'education.' The 1980s offered many examples of full-page advertisements about nuclear waste, reactor safety, the health effects of radiation and the like, designed to increase the public's level of knowledge, and hence level of comfort, regarding nuclear matters. Levels of public knowledge about nuclear power are low. A public survey carried out in the UK in the mid 1980s[6], for example, found that the best-known fact about nuclear power was that uranium was the fuel, and that this was known by 31 percent of respondents. Other surveys have shown that some 50 percent of people believe nuclear power to be responsible for acid rain and that only 13 percent of people are aware that radiation can come from both natural and man-made sources.

Nonetheless, it has become increasingly clear that the opinion that opposition to nuclear power is caused by low levels of knowledge is at best simplistic. One need only reflect that many of the 'professional' opponents of nuclear power are extremely well versed in the issues.

The gap between the 'rationality' of nuclear scientists and the 'rationality' of the public is perhaps exemplified in the debate about nuclear waste.

It is the settled view of most people in the nuclear industry that radioactive waste management and disposal offers no insuperable, or even very interesting, technical challenges. It has even been said that radioactive waste is the only long-lived waste stream for which a solution exists. Intermediate level waste (ILW) is described as a material which is about as dangerous as paint stripper or petrol. It could certainly cause problems if not treated carefully, but with proper safeguards it can be handled perfectly safely.

Yet at the same time as the scientists are explaining why ILW is no more dangerous than many other types of waste, and less dangerous than some, they are also explaining how this material requires burial 800 meters underground in an area with very little groundwater flow and without valuable minerals in case people dig it up by accident some centuries into the future.

These messages are clearly contradictory. No such measures are envisaged to deal with any other industrial waste streams. The 'rational' assumption is therefore that nuclear waste must be far more dangerous than any other type of waste, and furthermore that the nuclear scientists must be insulting the public's intelligence by pretending it is relatively benign.

The response of the industry has been to propose ever 'safer' solutions to radioactive waste management, on the assumption that people will be reassured when they see greater steps being taken to contain the waste. In reality, however, the 'rational' response on the part of the non-expert would be to assume that radioactive waste was even more dangerous than had been admitted before, especially if regulators produced progressively more stringent safety conditions. Not only would fear of radioactive waste be exacerbated but also mistrust of the scientists involved.

It seems, then, that the rationality of physical science and the rationality of everyday life can diverge quite radically. For example, people put considerable stress on issues such as the credibility of the source of information and whether the overall message makes common sense. Churchill once said that when deciding what weight to put on a statement, he would first consider who was saying it, then how they were saying it, and finally what they were saying. In everyday life, most of us seem consistently to make up our minds about many matters based on whether we like and, especially, trust the messenger, rather than on critically examining the message.

The ideology of science, however, is that when a piece of research has been published, after suitable peer review, it can be assumed to be objective, to be judged primarily on its consonance with other similar research. Only relatively recently has this ideology been seriously challenged, it being observed that scientists too have their vested and psychological interests. There seem to be plenty of cases, e.g. in the early research carried out into the health effects of tobacco, where skepticism about the objectivity of some of the scientists involved, or at least the extent to which they were prepared to make public statements detrimental to their paymasters, has proved well founded.

It is difficult to believe that the overall level of decision-making over the twentieth century would have been better had scientists been refused an input into those decisions. However, there have been several cases where accepted scientific understanding has been overturned by subsequent discoveries. 'Science' has often been perceived by many in the public, the media and politics as a source of 'right' or 'wrong,' but always 'absolute,' answers. When complex scientific debates result in a change in view from time to time, one result has therefore been disillusionment with the process of science itself. To improve the understanding of what science can and cannot offer the decision-making process would be of major benefit in many complex areas.

Factors affecting our Perceptions of Risk

Society's attitudes to risk sometimes appear paradoxical. While risk is clearly regarded as undesirable and to be avoided in most cases, there are many examples of some risks being valued highly for pleasure (e.g. 'extreme sports') and individuals who take major and apparently unnecessary risks are often admired.

There are several possible reasons why the historical level of risk, as calculated by risk professionals, may not be accepted as a firm basis for risk perception by 'the public.'

The first question, of course, is what is meant by 'risk.' To the safety professional, the term has a fairly clear technical meaning, associated with the amount of harm (measured in deaths, or sometimes injuries, for a given quantity of the activity in question, say 'per 1,000 participants per year' or 'per individual per 1,000 hours').

However, it is not clear that the public makes the same interpretation of the word 'risk.' Oughton[7] asked individuals to respond to two identical lists of activities, and to rank them with respect to two questions – "which of the following do you consider to be the most risky?" and "which of the following do you associate with the highest probability of premature death?" (see Table 3.3; bold text indicates hazards drawing a statistically higher response rate than the others).

These results seem to suggest that when we refer to an activity as 'risky,' we refer to more than simply its potential for causing death. There are two possible ways of explaining such observations. It may be that some aspect of particular risks leads people to feel more uncomfortable about them than others, given the same likelihood of causing harm. Alternatively, there may be something about the social context of certain risks – for example, a particular mistrust in the individuals associated with the activity – which may lead to people regarding the activity as 'risky,' even when they are aware that it is unlikely to be associated with direct harm.

Table 3.3

Risk Perception Survey

"Which of the following do you consider to be the most risky?"	"Which of the following do you associate with the highest probability of premature death?"
Nuclear power	**Smoking**
Genetically modified food	**Alcohol**
Food additives	**Car accidents**
Car accidents	**Air pollution**
Weapons	Bad diet/lack of exercise
Smoking	Weapons
Alcohol	Pesticides
Aircraft accidents	Aircraft accidents
Skiing	Nuclear power
Pesticides	Skiing
Radon	Food additives
Air pollution	Radon

A second point concerns uncertainty, especially when associated with new technologies or new circumstances relevant to established technologies. Unlike technologies such as air travel, where there exists a very large database concerning failure affecting particular aircraft designs which allows the likelihood of accidents to be estimated with some confidence, the precise likelihood of a major nuclear accident, like other very low probability/very high consequence events, cannot be calculated in such a way. The estimates have to be constructed from calculations of failure rates of individual components, and are inevitably less robust as a consequence. (It can be very difficult, for example, to be certain how the various components may interact in an emergency situation or how operators may behave.) Furthermore, the threat represented by a nuclear facility (or any other) does not involve only problems with the internal workings of the plant, but also the possibility of external threats. Many such attack scenarios are taken into account during the design of such facilities, but it is impossible to be certain that all options have been identified. The terrorist attacks on the USA in September 2001, for example, represented an apparently 'new' – and unquantifiable – threat within the nuclear safety equation. There are parallel uncertainties concerning what the precise effects of a major release of radioactivity would be; they would almost certainly depend on prevailing weather conditions at the time, for example.

Uncertainty inevitably leads to greater concern, even if the worst fears ultimately prove to be pessimistic.

Three particular groups of factors can be identified which might lead people to respond to calculated risks in different ways:

- those associated with the particular nature of the risks themselves;
- those associated with the social context in which the risks are communicated; and
- those associated with different types of personality.

The Nature of Different Types of Risk

Pioneering work in Oregon investigated the factors that act between calculated risk, as determined by a Probabilistic Safety Assessment (PSA), and public perceptions of risk.[8] (Probabilistic Safety Assessment, in effect, makes estimates of the risk associated with a particular activity by examining the dangers the activity has represented in the past. Results are often quoted in terms of 'deaths per mile' or 'injuries per hour,' etc.) They asked various groups of people – the (US) League of Women Voters, college students, 'experts' (who used PSA), etc. – to rank 30 potential risks in order of severity.

The correlation between public perception of risk and calculated risk was in general very good. However, in some cases there were considerable discrepancies. For example, college students ranked swimming as the least serious of the risks (below, for example, food colorants and hunting), while in reality over 100 swimming deaths occur each year in the USA. By contrast, all groups put nuclear power near the top except the PSA experts, who placed it twentieth.

Slovic concluded that there are three factors which act when we convert 'real' risk into 'perceived' risk. The first is associated with familiarity. If a risk is an old, well-established one, familiar to the individual and easily detectable by unaided human senses, it will tend to be underestimated compared to a risk of the same actual magnitude which is new, unfamiliar and difficult to detect.

The second involves controllability. If a risk is run voluntarily and is easy to control, it will be perceived as less serious than one which is imposed on people and is difficult to control. (Note that this is not a matter of the *acceptability* of a self-imposed risk against one imposed on us but of the perception of the actual *severity* of the risk itself.)

The third concerns the number of people affected by the risk. If a risk represents a small chance of damage to a large number of people (especially future generations), it will be perceived as more serious than a risk which has the same overall health effect, but where one can identify the likely victims.

Car travel, responsible for several thousand deaths each year in a country like the UK, lies near one extreme of all three factors. We are all familiar with car travel, having (by definition) taken part in this activity without suffering a fatal accident. We choose to indulge in it and we all know that we are better drivers than the majority and so are less likely to suffer an accident. Finally, although there are many deaths on the roads, it is always possible to identify the victims, and indeed relatively few such fatalities are reported in the mass media.

Nuclear power appears at the other extreme. It is perceived to be a new risk (although of course radiation is not a new phenomenon by any means). Radiation is not detectable by unaided human senses, and in a number of countries most people usually see images of nuclear stations in negative contexts, the mass media generally preferring 'bad news' stories to good ones. It seems to be 'imposed' on local communities and on society at large; few people choose, or would choose, to have a nuclear power station built nearby in the sense that they choose to go for a drive;[9] and although nuclear accidents are rare (and there has been only one with demonstrable off-site health consequences caused by radioactive releases), the impression is that a single extreme event could affect large numbers of people, perhaps everyone on earth and for many generations, maybe even interfering with the genetic stuff of life itself. Many individuals in northern Europe, for example, were aware that fallout from Chernobyl fell on their homes, especially during rainfall, during the week after the accident.

As a result, nuclear power tends to cause more unease than motor transport, even among some people who suspect they are being 'irrational,' in the same way that air travel tends to cause more anxiety than car travel, at least for many less frequent travelers.

In addition, perceptions of benefit are relevant to our perceptions of risk. In the case of car travel, the benefit – a convenient and private journey – is delivered immediately, in contrast to rail travel, for instance. There is therefore something of a psychological 'vested interest' in underestimating any associated risks. The perceptual link between nuclear power stations and the electricity coming from a wall socket is somewhat less obvious by comparison, as are the claimed benefits in terms of, say, reducing climate change.

The Social Context of Risk

The above analysis cannot fully explain the issue of how risk is perceived in everyday life. It is clear that other factors, such as the way in which 'information' is disseminated, and the apparent motivation of those giving the information, are also of significant importance.

It is undeniable that some of the technical issues in the nuclear field are so complex as to be inaccessible to non-specialists. Furthermore, 'experts' – people who clearly understand the field in great detail – disagree, publicly, about certain aspects of the technology.

When people seek to make up their minds about the desirability or otherwise of nuclear energy, the credibility that they attribute to the source of the information becomes a very important issue.

There is evidence that 'the public' has become considerably more skeptical of claims made by industry and government about complex technologies, a phenomenon which is often referred to in terms of a 'decline of deference' (see Figure 3.1). People are more likely to believe statements by scientists working for environmental organizations, for example, than scientists working within industry or government.

Figure 3.1

"Tell me how much you trust various institutions to do what's right ..."[10]

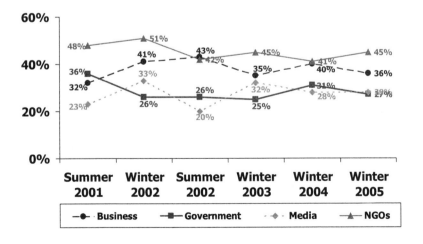

In its early days, the nuclear industry, as noted earlier, was heavily influenced by the views of technocrats, with relatively little scrutiny from the political establishment which often seemed to accept it was simply not equipped to frame suitable questions or to understand the answers. In such an atmosphere, 'communication,' when it was deemed appropriate at all, was a one-way process of informing the public of intentions.

The poor outcome of some of these decisions, especially in commercial terms, coupled with the industry's secrecy and intimate relationship with government, left a legacy of mistrust, as a result of which its later pronouncements were treated with some skepticism by significant sections of the population. This is of particular importance when it comes to evaluating the role of government as a regulator. There has been suspicion in some quarters that the apparently 'cozy' relationship between the industry and government makes proper vigor in regulatory processes difficult to guarantee.

In reality, it is much more difficult to identify large pro-nuclear factions within governments in many countries today than it might have been 30 years ago, and as noted earlier a number of governments have had explicit or implicit policies of nuclear phase-out.

There is evidence that communication will be most credible if its source is seen to be:

- open;
- accountable;
- inclusive;
- equitable.

The industry does appear to have recognized the importance of becoming a genuine partner in decision-making with other stakeholders and with society at large, and including in the consultative process representatives of a wider range of interests than was previously the case. Nonetheless, it may take some time before the change is accepted by elements of the public, the media and the political establishment.

'Personality' Issues

It should also be noted that people from apparently similar backgrounds, subject to the same information and within the same single country, can have radically different views on an issue such as nuclear power. This suggests that personality factors may be important to forming attitudes. Responses to the Three Mile Island accident are illustrative. Starting from the same event, the pro-nuclear community concluded that the incident confirmed the safety of nuclear power – 3,000 MW of heat energy going out of control without any significant radioactive releases – while of course for the anti-nuclear groups it is an icon of the dangers of the technology.

It would seem that simply challenging, or even changing, people's understanding of the facts will not (always) change the underlying attitudes, to the extent that the attitudes stem from more fundamental psychological factors. Studies tend to find relatively little difference in the level of factual knowledge among pro- and anti-nuclear members of the public. Providing technical information does not appear effective in building support. (Indeed, the emphasis on providing technical information may be counterproductive with respect to some sectors of the population. It may foster a sense that the industry has missed the point of people's real concerns, and therefore that, literally and metaphorically, it does not speak the same language as the public.) Of course, those who oppose nuclear power would argue that raising levels of knowledge among the population would cause people to turn against the technology.

Even where knowledge and favorability are correlated, it does not prove a simple causal relationship. People studying science in schools and universities tend to be more pro-nuclear than the population at large. However, it may be that such people tend to 'like' nuclear (and other) technology, and hence wish to study science courses in order to find out more about it.

Equally, it appears that some people who oppose nuclear technology will go to great lengths to find out about it, the better to be able to articulate an opposing view.

It is unlikely, even in principle, that it will ever prove possible to reconcile differences between the extremes of the pro-nuclear and anti-nuclear communities. In a Consensus Conference on radioactive waste held in the UK in 1999, for example, one witness argued against research on transmutation as a way of reducing the lifetimes of long-lived radioactive wastes on the grounds that solving the waste problem would make nuclear power more attractive and should therefore be resisted.

If attitudes towards nuclear power are more fundamental than the arguments used to support those attitudes, reconciliation of the entrenched pro- and anti- positions is unlikely to be achievable simply by examining 'facts.' Fortunately, most stakeholders do not operate from the extremes of the debate.

The focus in many countries has moved towards earlier and more thorough public participation, focusing on the expectations of various public groups. With respect to a particular major proposal, society can be regarded as being made up of three tiers:

- people who are *already* stakeholders—those who know they will have an interest in the decision in question, and will make their views known;

- people who will *become* stakeholders—unaware at present that the proposals will affect them 'directly,' and who will wish to make their views known when they become aware; and

- people who will *never* be stakeholders—and are therefore uninterested in the decision-making process as it refers to the proposal in question.

It can be argued that, at present, the problem with the traditional approach lies with the second group. Significant numbers of people become aware only quite late in the day that a particular proposal would affect them, and by then they appear largely excluded from the decision-making process. Often they can only make an input when details of a scheme have been concluded and its proponents are seeking only final permission. Public inquiries into proposals have therefore become more bitterly fought and more protracted, as individuals have perceived them as the only meaningful chance to express their concerns or opposition.

Volunteer Communities

The problem of site selection is an especially difficult one, for example in developing waste repositories. In many countries proposed sites have been rejected after political or legal campaigns. This has led to increasing attention being paid to the 'volunteer community' approach.

The development of a major facility such as a waste repository would bring both benefits and disadvantages to a local community. In addition to the natural advantages of improved infrastructure and employment opportunities, a program of 'planning gain' – a series of benefits to the community not directly connected to the development but designed to compensate for the disadvantages – could be offered to communities prepared to consider hosting such a facility. This approach has been followed with success in a number of countries, notably Finland, Sweden and France, and is now being pursued in the UK.

A long list of perhaps ten or fifteen geographical locations which are likely to satisfy technical requirements would be assembled. The communities in question would then be invited to register an interest in hosting the project.

Experience of this approach in a number of countries is mixed, but there are indications that success is more likely if local communities are guaranteed:

- the option of withdrawing from the siting process at any stage prior to the start of construction;

- that any final decision to proceed with repository development should be subjected to local referendum;

- to be empowered so that they can participate meaningfully in appraisal of site investigations, for example through the setting up and funding of a local advisory group to ensure that the concerns of the community are adequately addressed as investigations progress.

The Interface between Science and Society

Increasingly, problems which can be regarded as arising from the interface between scientific industries and the political establishment are being seen in other fields: BSE ('mad cow disease'), the MMR (measles–mumps–rubella) vaccine, mobile phone masts, genetically modified organisms, animal experimentation, etc. Paradoxically, outbreaks of profound public concern over activities that in all likelihood represent very small risks to human health are seen at a time when life, at least in the developed world, has never been safer. Disputes in these areas have come to follow a common path.

If the scientific and political establishments are to be reunited behind a common agenda, two related themes must be examined and resolved. First, the philosophical and practical mismatch between the political and technical mindsets. Secondly, the periodic changes in societal ethics between a basically utilitarian approach (often observed during times of societal stress) and a Kantian or rights-based approach (often more dominant during times of relative comfort) and how these must be managed if long-term solutions to technological problems are to be found and implemented.

It is difficult to identify solid themes in the evolving relationship between the scientific/technical industries and the political establishment in various settings. For example, it is in the nature of politics that values are debated, often passionately, between members of the same community. Prevailing political fashions, for example the neo-liberalism of the last decade of the twentieth century, however deeply they may seem to be ingrained, will always have dissenters. Even if we believe we can identify national political 'styles,' they can differ considerably from country to country and from time to time. If the politics of Scandinavia can be described as 'consensual,' that of the Anglo-Saxon countries as 'confrontational' and 'fragmented,' those of some of the Romance countries as 'centralist' (in which the State is expected to play a major role), these labels are useful only insofar as their limitations are recognized. Prevailing political cultures during wartime or other times of national stress are often very different from those which may pertain in prolonged periods of peace and economic prosperity.

Nonetheless, certain tentative themes can be recognized in this changing relationship.

Science and Politics: A Changing Relationship

When nuclear energy emerged in the post-war years it was the recipient of large amounts of support, both political and financial, from a wide range of governments. This was partly as a result of recognition of its potential as a new source of energy; partly perhaps as a political cover for nuclear weapons programs which might otherwise have attracted more opposition; and partly owing to the great faith that politicians put in scientists and the very concept of 'progress.' (On the scientists' part, developing nuclear energy represented a way to salve their conscience by turning the devastating destructive force of nuclear technology towards a more benign purpose.)

Nuclear energy at this stage mapped well onto the prevailing political and social fashion in most developed countries—one of considerable 'deference' towards experts and decision-makers, driven in part by a recognition that after the Second World War things would be difficult for a long time and that governments needed to take bold decisions, be they in

[107]

the realm of welfare reform or industrial policy. In effect, politics (as the arena for decision-making), science (as a source not only of technical input but also of legitimacy for those decisions) and the public (prepared to defer to both and to surrender some of their individual 'rights' as long as overall progress was maintained) were able to work in considerable harmony. The prevailing societal ethic was a utilitarian one; there was widespread assumption that policy was to be made on the basis of the greatest good for the greatest number, even if some individuals suffered disadvantage in the process.

As time passed, this consensus broke down, though to different extents in different countries. At first the fault lines appeared between politics and science on the one hand and the public – or growing numbers of members of the public – on the other. As post-War austerity was replaced by economic prosperity, so other profound social changes came to threaten the hegemony of the political/scientific establishment. People became more individualistic, traditional religion was replaced by cults, formal dress was replaced by informality among younger generations, drug taking became more common and young people were encouraged to 'turn on, tune in and drop out' by academic gurus like Timothy Leary and Allen Ginsberg.[11] It is of course bizarre to claim that all members of the population, or even more than a small minority, ever took up such activities, but a general decline in the awe with which royalty, politicians, clergymen and 'experts' were held was clear. The rights of individuals became more important than was typical in societies under external pressure, for example during wartime.

For a while the establishment could ride these changes – for example, large numbers of new nuclear reactors were ordered in the 1970s and 1980s – and indeed the economic decline of the 1970s, driven largely by the massive increase in oil prices in 1973, presaged a more subdued decade in which the strong political leadership (or out-of-touch, even uncaring, authoritarianism, depending on one's standpoint) of figures like Margaret Thatcher and Ronald Reagan could flourish in a way that might have been more difficult a decade earlier (or later). Science continued to play its role of not only supporting but to a large degree setting the direction of government policy, politicians by and large continuing to view the pronouncements of well-established scientific figures with close to blind faith. One can imagine that scientists were torn between a desire to make sure that politicians understood the inherent limitations of scientific inquiry (which cannot even in principle offer certain predictions of the future, only well-based guesswork) and enjoying the social cache of exerting such influence over the great and good.

Through the 1980s and 1990s (though it had been presaged by writers like C.P. Snow a quarter of a century earlier[12]) the commonality of interest and the close interaction between scientific and political cultures began to disintegrate. 'Big science,' as represented perhaps most spectacularly by the space program, fell out of favor and governments, with a few notable exceptions (Japan, France), drastically reduced expenditure on 'blue skies' research and development. In part, this schism came about through deliberate (and often successful) attempts by the opponents of technology, increasing numbers of whom were being elected into parliaments, to undermine the scientific basis of decision-making, including the assumption that scientists could be regarded as fonts of unbiased 'truths.' In this they were aided by the behavior of some scientists working for major corporations who seemed to be prepared to allow their views to be used (or covered up) in such a way as to promote the financial interests of their sponsors. (Research into the health effects of smoking tobacco became something of a *cause célèbre*, both because of the delay in the scientific community taking up the cause and because of the failure of some scientists to make their findings public.) It was also partly a result of a growing number of examples of 'failure' in the science on which decisions had been based—the nuclear accidents at Three Mile Island in 1979 and Chernobyl in 1986 were particularly influential, as were such events as the Bhopal chemical disaster in 1984 and the explosion of the Challenger space shuttle in 1986.

It was surprisingly easy to create a schism between science and politics because it seemed that the two never really understood each other. Indeed, the end of the twentieth century was a time of growing suspicion, scientists becoming frustrated that politicians were requiring 'right' answers to simplistic questions of the kind that science cannot answer and accusing them of only being interested in the next election, while politicians were becoming increasingly annoyed at the difficulty of getting a straight answer which could subsequently be relied upon and the scientists' inability to recognize the constraints of decision-making in a democratic context. So, for example, scientists during the BSE affair of the 1980s/1990s were:

> ... both deliberately and inadvertently utilized to provide spurious scientific legitimation for policy decisions which government officials believed ministers, other government departments, the meat industry and the general public might not otherwise accept.[13]

Science and Politics in Conflict

The similarities between the paths followed by controversies in widely ranging scientific fields are striking. Typically they involve issues in which an activity is alleged to have an effect on a small number of people, in circumstances where no clear causal relationship can be determined. At first politicians and (some) scientists dismiss fears as unfounded, an action which often exacerbates the initial concerns by adding suspicions of a 'cover-up.' The media identify and promote individuals who claim to have been affected by the activity in question and these individuals are often given equal (or greater) airtime and prominence to large-scale studies which imply a very small risk, if any. Politicians now set up a committee, drawn from 'experts' and non-experts, in the hope that it will provide a clean bill of health for the activity. This cannot happen—the scientists will always argue in effect that a negative cannot be proved, while the activists will stress that it has not 'yet' been proved that the activity is harmless and so it should be stopped. The outcome, in many cases, is regulatory action that puts barriers in the way of developing new technologies however beneficial they may be.

Tensions among the scientific and political establishments and the public have emerged at times of economic prosperity in many developed countries. As far as many members of the public are concerned there is no apparent need for radical political action to protect the fabric of our way of life (as there might have been, say, at times of war or prolonged industrial unrest). Society is not necessarily any happier – our anxieties simply get transferred onto other potential threats such as mobile phone microwaves in the environment, which by any reasonable standards are patently much less severe[14] – but there is less space for strong political decision-making in response to future threats where this may be seen to violate the rights of some individuals, however few.

Ironically, it seems that while 'strong/authoritarian' or 'consensual/weak' modes of leadership can be successful in delivering implementable policy in the realm of scientific controversy, to fall between these extremes may be less so. The search for a site for a radioactive waste repository seems to have been concluded in Finland and Sweden (where local people have been kept at the centre of the decision-making process throughout) and in the USA (where a firm federal decision to use Yucca Mountain in Nevada had been pursued in the face of widespread local opposition). By contrast, those countries which have made some attempt to accommodate public concerns, e.g. by including non-experts on advisory bodies, but which have continued to operate in conditions of considerable secrecy (it was only in 2005 that the list of sites considered for waste repositories in the UK in the 1980s was released to the public after an application under the then recent Freedom of Information Act[15]) have made relatively little progress. Undoubtedly national political characteristics and history play a part in these different approaches to decision-making—there can be no assumption that what has 'worked' in one country or region would be appropriate for another.

As a response to the growing public skepticism about the role of science in decision-making (and perhaps to the need to be seen to be 'doing something' without having to do something), politicians have attempted to rebuild a relationship of trust with the public by downgrading technical expertise or even writing it out of the loop. The panels charged with finding 'solutions' to matters such as radioactive waste management, the health effects of mobile phone masts, BSE, foot-and-mouth, etc., are increasingly populated by individuals with no technical knowledge of the topic in question and indeed often an antipathy towards such expertise.

The Proper Role of Science?

Yet, properly used, science has a vital role to play in decision-making. The scientific method, while not offering certain knowledge – especially when dealing with relatively uncommon potential health threats – is likely to provide advice which is closer to the truth, and therefore more useful, than that emerging from religion, gossip or ideology.[16] As long as a suitable attitude is taken to uncertainty this must lead to better decision-making. Even Hume, who takes the extreme skeptical position that simply because something has always happened in the past is no proof that it will happen in the future, argues that only a fool would live one's life on that basis.[17]

Such reflections are particularly vital when decisions taken or ducked today will have implications long after the end of the term of office of the politicians who take them. It is wrong to say that politicians are never motivated by long-term factors – leaders want to ensure their place in history – but how these desires can be integrated with the short-term stresses of the electoral timetable is not always clear. Nuclear energy is particularly (but not uniquely) vulnerable to any impression that politicians in the future may change the rules in a more or less capricious and unpredictable way. The initial investment costs of nuclear energy represent a higher proportion of total costs than is the case with most other ways of making electricity (notably Combined Cycle Gas Turbines [CCGTs]). To ensure a fair rate of return on the project it therefore requires a stable business environment for a rather longer period of time. If, say, it is perceived that a change in political control might bring with it more stringent regulations (or even a formal phase-out policy) then the economic risk associated with investment in nuclear plants becomes high and possibly unmanageable. In the most extreme case, the Shoreham nuclear station at Long

Island, New York, was closed in 1989 before commercial operation began because it was refused an operating license on the grounds that it could not comply with evacuation requirements introduced after construction had started. The Long Island Lighting Company was effectively bankrupted by the affair.

The challenge for politicians and the 'consumers' of their decisions, then, is twofold. First, how to reintegrate science into decision-making without making the mistakes of the past in which some scientists were given almost a free hand over policy development.[18] Second, how to take strong and possibly unpopular decisions before the impending crises of energy shortages and climate change become unmanageable (quite possibly after the politician in question has left office, thereby risking taking the short-term pain while receiving little of the long-term gain). Unless these are overcome, complex technologies like nuclear energy, even if they have a useful potential role to play, are likely to be excluded on grounds which, from an external viewpoint, will be regarded as irrational. (This is not, of course, to argue that there are no rational grounds for opposing any particular technology.)

There are certainly political risks associated with firm action over controversial issues, especially when society is not yet ready to acknowledge the need for such action. But so too are there political risks in ducking difficult questions—the risk of being seen to be weak during or after one's term of office. Politics is not merely a matter of getting through the next election—some decisions are inevitably longer-term and the success or otherwise of politicians in dealing with them will cast long shadows. The challenges which faced the world in 2008 – credit crunch, food shortages, spiraling oil prices – may well herald a return to public demands for 'strong,' decisive leadership and difficult decisions. However, the key issue for those investing in heavily capital-intensive technological industries is how to ensure that their investments remain viable should society turn again to a more complacent mindset.

[113]

Summary

There is little evidence that populations at large in most developed countries are committedly anti-nuclear. Despite the heated and acrimonious nature of the debate among those who devote their time to such things, most observers and stakeholders seem to take a more balanced view of the issues involved. It does seem, however, that politicians in a number of countries mistake the heat of the debate for major public disquiet. Polling of politicians in countries like the USA and the UK show that they overestimate public opposition by a large margin.

However, proposals to build new nuclear facilities in many developed countries, especially on 'greenfield' sites, provoke considerable opposition. This is true even in countries like Japan, which have a history of support for nuclear power and limited access to alternative sources of energy. This opposition, which seems to have grown steadily from the inception of nuclear power, may be ascribed to a number of factors, including:

- Disillusionment and mistrust of the use of science (especially science sponsored by commercial concerns) in policy-making. This has been caused in part by widely publicized and exaggerated claims on behalf of technology, notably nuclear technology, particularly in its early days, and examples of dishonesty and 'cover-up.' The decline of 'deference' is also relevant.

- Poor plant performance, during both construction and operational phases.

- The accidents at Three Mile Island and Chernobyl, and other more localized events.

- Campaigns against nuclear power from pressure groups and the media.

- The particular nature of nuclear risks—unfamiliar, involuntary and potentially affecting many people.

[114]

- Discovery of major resources of natural gas and oil, reducing the impression that nuclear power would be required in the short term.

- Failure of government and the industry to proceed with waste disposal facilities in most developed countries.

It appears that public opinion in some developing countries, though by no means all, may be more favorable, partly as a result of the major employment opportunities represented by large nuclear construction projects in regions of considerable poverty. However, one might speculate that the same forces that have affected public opinion in some developed countries might in due course come to bear in the developing world.

The traditional approach to taking decisions about major projects in the nuclear field has been for the industry and government to carry out early steps in private, with little or no public discussion, followed by an announcement of the result of the deliberations and a program of 'selling' the decision to the public, regulators and planning authorities (the decide, announce, defend – DAD – model). Quite apart from legitimate questions about whether such a decision-making process is equitable in a complex society, it has become increasingly clear that this approach no longer works in many developed countries. Decisions which are taken in secret and which do not reflect the needs of a wide range of stakeholders are increasingly being rejected, for example at the planning stage, or because of political lobbying, or because of direct action driving up costs.

This state of affairs benefits opponents of nuclear energy more than it does its advocates. Paralysis in decision-making about new power reactors, for example, would inevitably lead to a decline in nuclear output, and within a few years, given the age profile of nuclear stations in many developed countries (84 percent of installed global nuclear capacity is fifteen years old or more).

Figure 3.2

Global Nuclear Capacity Operating by Age

(MW/Years, September 2008)[19]

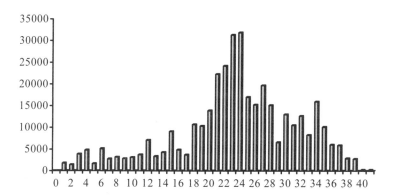

Several new decision-making approaches are therefore being tested in different countries. They share a number of features:

- involvement of a wider range of parties – both existing stakeholders, potential future stakeholders and members of the wider public – in the decision-making process;

- involvement of these parties at an earlier stage of the process;

- greater transparency over the reasons for proposals; and

- a greater say among possible host communities as to whether, and how, they wish to be involved in discussions.

All aim both to improve the quality of final decisions, and to increase confidence in those decisions by fostering a sense of trust among the stakeholders, and a wider sense of 'ownership' of the decisions.

Many of the innovative approaches are still at the stage of being evaluated. Although early experience does seem quite promising in some countries, there are major concerns about the time such procedures might involve in areas where delay is very expensive. However, recent stresses to the world economic, energy and food systems – to name but three –

may herald one of the periodic changes in societal ethics that are encountered in many countries. In this case the consensus view of the late 1990s, that the traditional centralized approach to decision-making was no longer an option and that new procedures of engaging the public were needed, may no longer hold. Should the public come to recognize that difficult decisions are needed they may well support such decisions against the protests of special interest groups.

The challenge for technical industries, though, is to ensure that if and when society once more becomes complacent and no longer sees the need for difficult decisions, the rug will not be pulled from under the investment in nuclear energy and industries with a similar profile. The nuclear industry in countries such as the Gulf states could do much to reduce the likelihood of such an outcome by ensuring an open and honest relationship with politicians and the public alike from day one.

4

Ensuring Domestic Capability: Education and Training at the Masdar Institute

Youssef Shatilla and *Mujid Kazimi*

Abu Dhabi has embarked on an historic initiative to use nuclear power to meet its future energy needs and to position itself as a sustainable energy hub of the Middle East with a technology-driven economy. Nuclear power can play a central role in the future energy mix of the UAE, as it has done for decades in many developed countries. Nuclear power is clean, affordable, expandable, sustainable, safe, and consistent with international policy.

The development of adequate local knowledge and expertise for the deployment of nuclear power is necessary to form part of the industry's basic infrastructure, and in the long run is the most economical option for the UAE. This involves building a suitable knowledge base in the areas affecting the implementation of nuclear projects and developing human resources and educational institutions for the training of personnel. Figure 4.1 illustrates the plant organization and technical qualifications of staff at a typical CANDU-6 700 MWe nuclear power plant. Using Figure 4.1 as a guide, the knowledge base required for the operation of a nuclear power program can be summarized as follows:

Knowledge Base

Government Policy

In the first stage of the development of nuclear policy, conceptual studies must be conducted. For this purpose, dedicated expertise in a number of areas should be employed by the relevant government implementation organization in order to carry out the task of determining national nuclear policy and priorities. The government should determine if a nuclear power program is beneficial to the country and, if so, develop a policy on the basis of which the private sector and/or government utilities can proceed to implement the nuclear power program.

Figure 4.1
Typical Plant Organization, Qualifications and Staff:
CANDU-6 700 MWe Unit

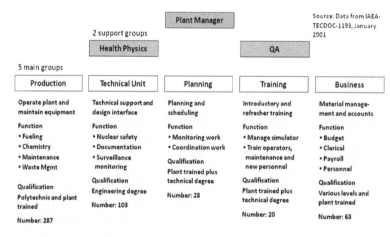

- Corporate support (6) including vice-president for nuclear operations
- Health physics (24) for radiation protection and emergency planning; university degree
- Quality Assurance, QA (8) for quality assurance and audits, etc; plant trained

Nuclear Regulatory Knowledge

A Nuclear Regulatory Body (NRB) is responsible for licensing of nuclear plants, timely issuance of the various permits needed, and conducting inspections. The laws and regulations governing the NRB must empower it with broad statutory authority and functional autonomy to carry out its functions independent of owners, operators, manufacturers and suppliers of nuclear power plants and other interested parties in both the public and private sectors. Qualified professionals are required to perform the major functions of regulation, assessment, enforcement, and public information dissemination and exchange. The scope of activity and the requirements of staff in each section of the regulatory body will vary according to the size of the nuclear power program

Initially, the NRB should adopt an internationally accepted set of requirements for reactor licensing, such as those developed by the IAEA, the country of origin of the reactor supplier or those of countries that already have a fully developed nuclear regulatory regime. The NRB should address the following issues:

- Establishment of national nuclear safety regulations, design codes, and standards.
- Technical assessment, design review and licensing of reactor, fuel and waste management facilities in all phases of siting, design, fabrication, manufacturing, construction, commissioning, and operation.
- Issuance of permits and licenses at various stages of the execution of nuclear power projects.
- Licensing of operators.
- Inspection of design and manufacturing facilities to ensure compliance with the requirements of the approved codes, standards, criteria, and specifications.
- Capability to address emergency and abnormal events regarding the security and safety of nuclear power plants.

Cooperation with the International Atomic Energy Agency (IAEA)[1]

Under its Statute, the IAEA is authorized to assist any member state that is considering or has decided to introduce nuclear power to meet its energy needs; it has considerable experience of this through its assistance programs. For example, support can be provided for implementing the operational phase of a nuclear power plant (NPP) to the extent that the state has demonstrated that it has established the essential elements of a national framework. Advice and guidance on obligations and commitments can be provided during all phases of a program. The IAEA has recently prepared a number of guidance documents on infrastructure and other considerations for countries planning to launch a nuclear power program and stands ready to provide expert assistance in this area if requested.

With the exception of issues relating to commercial decisions, the IAEA can also assist by providing technical support to the owner/operator for the assessment of potential technology, the managerial approaches used in the implementation of a project, and issues related to ensuring the safe and economic operation of a NPP.

The IAEA also works to strengthen the capacity of member states to manage the development of their energy sectors, with the goal of promoting sustainable use of natural resources and increasing access to affordable energy services. A key aspect of this effort is the Agency's energy assessment services. Through these services, the IAEA trains local experts to develop and use energy planning models tailored to each country's special circumstances.

Assistance is also provided in developing comprehensive, national legal frameworks under the IAEA's legislative assistance program.

Specific IAEA support can also be sought in assisting the development of regulatory bodies to ensure that they are effective and fully competent to oversee the licensing of the facility, and by providing peer review services concerning all aspects of the nuclear power program.

Workforce Radiation Training

Radiation awareness training is required in the following areas:

- radioactivity and its sources;
- radiation health (biological) effects;

- radiation protection methods and regulations;
- measuring of radiation;
- exposure and contamination control;
- handling of radioactive wastes;
- establishment of radiation protection programs;
- radiation detection instruments;
- troubleshooting instruments;
- radiation instruments laboratory;
- radiation safety surveys; and
- releases and emergency response.

Economics and Finance

The first step is to demonstrate that the economics of nuclear power are within the acceptable range for inclusion in the national generation mix. Once the positive economics of nuclear power are demonstrated, the next step is to develop a financing model for the implementation of the technology, involving:

- Economic assessment of nuclear power.
- Funding of the nuclear power program:
  o financing of the pre-project development program; and
  o financing of project implementation.
- Methods of financing nuclear power plants.
- Final financial and economic assessment.

Education

Apart from nuclear safety and radiation protection, the majority of the technical expertise needed for the design, construction and operation of nuclear power plants is the same as for other large industrial or conventional power projects. Nuclear safety and radiation protection require the establishment of a management system[2] with a much higher level of quality and safety requirements as well as the use of more advanced software and hardware tools in the analysis, design and installation of plant components and structures. It is important that as part of the basic nuclear power infrastructure, human resources in the areas of design, construction, installation, commissioning and project management be developed in the country during the implementation of the first nuclear power projects. These services constitute a considerable portion of the project cost and their localization will reduce the cost of subsequent projects. The major specialties to be developed for a nuclear program are as follows:

Nuclear Plant Design Review Capability

This capability involves the following:

- nuclear and reactor physics;
- thermal hydraulics methodologies for nuclear and process systems design;
- piping analysis and design including thermal, seismic and water hammer;
- electrical, control and instrumentation;
- structural analysis and design;
- nuclear fuel performance;
- containment systems design, including reactor building structure;
- environmental qualification of equipment and systems;
- probabilistic safety analysis;
- radiation and shielding calculations and design; and
- human factors engineering principles.

It is reasonable to assume that the local involvement in the engineering of the first project will be limited, however the contract with the technology supplier should include classroom and on-the-job training for the transfer of design knowledge to local engineers and scientists.

In effective localization of design engineering, a team of up to 200 experts with knowledge of the design and licensing of nuclear power plant buildings, structures, systems and components should be developed either in government-owned agencies or within private sector engineering firms. With such a capability, and assuming significant standardization of reactor design, the role of the supplier of technology in the subsequent projects can be reduced to supervising and reviewing progress.

A smaller team of nuclear design and engineering experts should be developed during the construction and commissioning phases of the build to become part of the operation team of the in-service unit. The number of experts needed for this team depends on the operating policy of the plant. Some utilities maintain a very small in-house technical team and allow private sector companies to develop the necessary capabilities and provide the required services on a contract basis.

The educational institutions and programs required to develop the required engineering expertise are:

- Nuclear design and engineering courses in a specialized nuclear power department or engineering and science departments for the training of plant design engineers and scientists.

- Specialized programs offered by technical colleges in specific areas such as quality assurance, radiation monitoring and protection, instrument calibration, etc. for the training of technical experts.

Quality Assurance/Quality Management Capability

QA/QM activities are part of the organization's management system and require a formal process which includes preparation of procedures and training of personnel. The purpose of QA/QM is to support quality assurance in compliance with the applicable requirements in all phases of a nuclear power project such as design, procurement, construction, manufacturing, fabrication, commissioning, pre-operational and startup testing and – ultimately – operations. QA/QM activities cover the entire project; however, trained personnel will have different specializations and will be assigned to specific tasks for the duration of the project.

Training in QA/QM should lead to the development of three categories of QA/QM personnel for:

- planning and supervision of inspection and testing programs;
- performing and documenting inspections and tests in accordance with procedures; and
- evaluating the results and adequacy of procedures.

Project Management Capability

A successful nuclear project would have to put the plant into commercial operation in accordance with safety and licensing principles but also within the contract schedule and to cost. It is reasonable to assume that the overwhelming majority of the work force deployed at the site should be local and should have the basic skills to carry out their assigned tasks. In addition to general labor with limited or specialized skills, there is also the need for trained personnel to fill positions within the project management team. The expertise required includes planning and scheduling, QA/QM, equipment and materials supply, field engineering, construction and installation, and commissioning.

Operation and Maintenance Capabilities

The development and training of operation and maintenance staff should be through a planned program, which begins with classroom training at the start of the project and is followed by on-the-job training during the commissioning and testing activities of the project. Plat simulators are used to train the operators before the plant begins operation. Prior to reactor criticality, sufficient number of operators should have received their operating license from the NRB.

Local Institutions

Currently, there are no educational institutions that offer undergraduate qualifications in nuclear engineering in the UAE. However, The Masdar Institute of Science and Technology will offer a Master's degree in mechanical engineering (nuclear engineering track) starting in the Fall of 2009 and a Doctoral degree in the same field starting in the Fall of 2011.

The Masdar Institute of Science and Technology

The Masdar Institute of Science and Technology[3] is a private, not-for-profit, independent, graduate-level research institute developed in Abu Dhabi, UAE, with the support and cooperation of the Massachusetts Institute of Technology (MIT). The Masdar Institute is the centerpiece of the Masdar Initiative, a strategic undertaking of the Abu Dhabi government with the following objectives:

- economic diversification of Abu Dhabi from a fossil-fuel based economy to a knowledge-based economy;
- expansion of Abu Dhabi's position in evolving global energy markets;
- positioning of Abu Dhabi as a leading developer of advanced technologies; and
- positioning of Abu Dhabi as a major contributor towards sustainable human development.

Masdar Institute's Intellectual Platform

The Masdar Institute is critically important to the UAE because it establishes a world-class research institution that is unlike any other academic institution in the Gulf. The Masdar Institute is cultivating the educational environment, culture of R&D excellence and ties to government and industry that are essential to the development of indigenous innovation capacity and global competitiveness.

[127]

Knowledge capital is the know-how that results from the experience, information, knowledge, learning, and skills of individuals or groups, and provides a foundation for a long-lasting competitive advantage. Its development is essential to Abu Dhabi's strategic vision of leadership in the global energy market of the future. The Institute creates knowledge capital through world-class research and education as well as interaction with government and industry. Its ability to create knowledge capital is critically dependent on a culture that emphasizes technological innovation and research and development excellence. The development of this culture is enabled through movement toward cross-disciplinary collaboration with emphasis on both fundamental and applied research. The establishment of the Institute and its strong intellectual platform provides the ideal catalyst for this movement.[4]

The Masdar Institute's Vision and Mission

Figure 4.2
The Masdar Institute's Vision and Mission

Vision	Mission
•To be a world-class, graduate-level institution, seamlessly integrating research and education to produce future world leaders and critical thinkers in advanced energy and sustainability •To position Abu Dhabi as a knowledge hub and engine for socioeconomic growth	•Establish and continually evolve interdisciplinary, collaborative research and development capability in advanced energy and sustainability •Educate students to be innovators with the breadth and depth to grow technology and enterprise in the region and globally

In accordance with its vision and mission, the Masdar Institute will leverage its distinctive strategic advantages to provide its students with the technical knowledge and systems perspective critical to innovation and intellectual leadership.

Technical knowledge and systems perspective will be achieved through research and participating academics that are guided by the organizing principle of sustainability[5] and are aligned with three strategic areas: technology, policy and systems (see Figure 4.3). These interrelated foundational areas are defined as follows:

- TECHNOLOGY: the development of models, devices and materials that can be applied in achieving clean energy, energy efficiency and sustainability.
- POLICY: the development of plans and methodologies that guide national or industrial decisions and strategies related to sustainability.
- SYSTEMS: the development of integrated networks of sustainable technologies and policies.

Figure 4.3
Educational Program Strategic Design

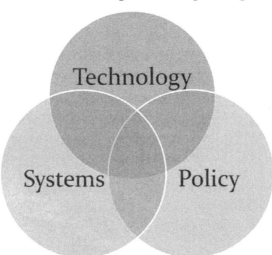

The Masdar Institute will focus on technology, policy and systems because these areas lie at the heart of renewable energy and sustainability trends of the future. For example, the extensive and diverse energy demands of residential, commercial and utility consumers made solar energy the most rapidly growing clean energy sector in 2006 and 2007.[6]

This growth has been enabled by policy, technology and systems innovations such as feed-in tariffs, new solar materials and devices and utility-scale concentrating solar systems. Similarly, the need for coordinated efforts in the domains of technology, policy and systems is demonstrated by multi-disciplinary clean energy trends such as electric vehicles, sustainable cities, and sustainable transportation.[7] The Masdar Institute will be a leader in the coordination of technology, policy and systems perspectives as it establishes leading-edge technologies that translate into larger networks of systems and policies. Technologies will be developed by engineers and scientists that advance the existing body of knowledge in advanced energy technologies and sustainability. Policies will be developed by engineers and scientists to create combined and mutually reinforcing regional and international perspectives at both the tactical and strategic levels. Systems will be developed by engineers and scientists that apply fundamental network design and analysis principles to assess the responses and feedback that arise from complex groups of discrete technologies and policies. Overlap of work in these areas is designed to establish the type of interdisciplinary collaboration that leads to groundbreaking ideas and discoveries[8] and builds the innovation culture and research and development expertise that the Institute seeks to establish. The Institute will maximize the overlaps in research and expertise through the pursuit of organizational structures and activities that adhere to a set of core principles:[9]

- Interdisciplinary research is essential to the creation of knowledge capital.
- The intellectual contributions of interdisciplinary programs are always greater than the sum of their parts.
- The valuable experience that interdisciplinary research gives students is an important factor in the Institute's desire to promote and support interdisciplinary programs.
- The development of new knowledge and the contributions that it makes to the Institute's educational commitments is an intellectually-rewarding experience.

- To realize the full potential and benefits of its research and educational contributions, the Institute will create an appropriate environment for participation, informal discussions, and networking, in addition to the development of rigorous intellectual and analytical capabilities.

Adherence to these principles will not only benefit research and education that integrate technology, policy and systems perspectives, but also research and education that are more focused on a single area. For instance, it is very difficult to directly relate some forms of technology research, such as the development of new materials, to either systems or policy. However, it is possible to ensure that the faculty and students researching and studying such an area (e.g. technology) have frequent interaction with those researching and studying in the other areas (e.g. systems and policy). Such interaction allows for the type of knowledge transfer that is considered best practice for industrial research and development organizations. When knowledge transfer occurs on a regular basis it is possible for breakthroughs achieved in one area to drive breakthroughs in the others. In an ideal scenario, this feedback mechanism functions like the interlocking gears in a high-performance machine.

Figure 4.4
Interlocking Nature of Technology, Policy and Systems

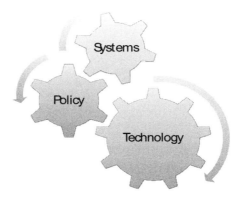

Based on the above, it is clear that the Masdar Institute must establish a culture of research excellence and innovation if Abu Dhabi is to make a successful transition to a knowledge-based economy with superior research and development skills capable of addressing renewable energy and sustainability trends both now and in the future. The graduate programs at the Institute will therefore integrate education, research and scholarly activities to prepare students to be innovators, creative scientists and researchers, and critical thinkers in multiple engineering and scientific domains. A combination of classroom learning and thesis-driven interdisciplinary research will equip students with the knowledge and skills necessary to address the world's most challenging problems in the realm of sustainability. To attain this goal, the Institute is committed to fostering freedom of education, learning and expression and striving to recruit, nurture and enable an outstanding community of students, faculty and staff.

Developed in cooperation with the Massachusetts Institute of Technology (MIT), the Masdar Institute emulates MIT's high standards and offers Master's and Doctoral-level degree programs. MIT is working with Masdar in establishing a sustainable, homegrown academic and scientific research institute with its own unique identity. The driving forces behind MIT's entrepreneurial and research excellence are attributed to four interrelated factors:[10]

- Science and engineering resource base (i.e. access to research funds).
- Industry funding of research.
- Quality of faculty.
- Organizational characteristics such as technology licensing office, entrepreneurship programs, interdisciplinary research.

Equally important is the context in which these factors operate at MIT:

- University mission that promotes entrepreneurship and economic development.

- Faculty culture.
- History and tradition.
- Location in an environment highly conducive to innovation.

The Masdar Institute will strive to emulate MIT's success through a commitment to addressing these critical factors via obtaining and further developing world-class academic programs, faculty, students, and research.

Figure 4.5
Masdar Institute Educational Program Mission

MIST Vision	MIST Mission	MIST Educational Program Mission
• To be a world-class, graduate-level institution, seamlessly integrating research and education to provide future world leaders and critical thinkers in advanced energy and sustainability • To position Abu Dhabi as a knowledge hub and engine for socioeconomic growth	• Establish and continually evolve interdisciplinary, collaborative research and development capability in advanced energy and sustainability • Educate students to be innovators with the breadth and depth to grow technology and enterprise in the region and globally	• Provide students with the knowledge, skills, and experience necessary for successful careers in industrial or academic roles in their chosen fields • Provide students with domain expertise and broad awareness in alternative energy and sustainable technologies and policies

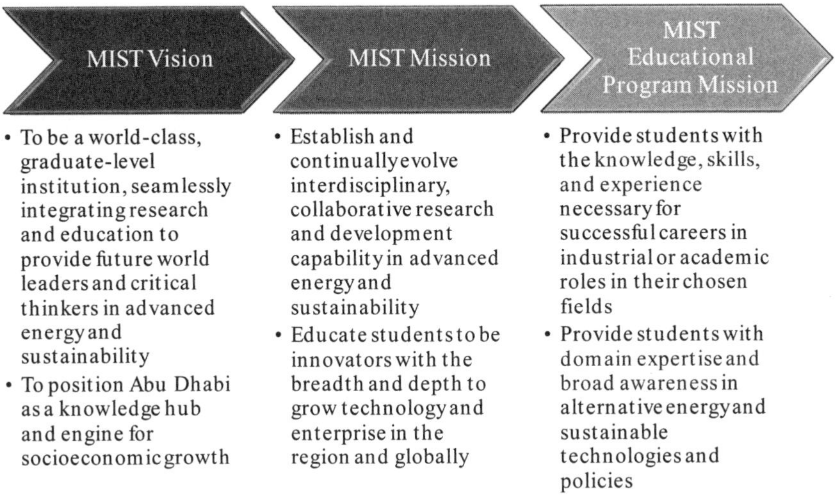

Academic Programs

Academic programs at the Masdar Institute are selected to be consistent with the university's focus on alternative energy and sustainability. The programs are not chosen to be vocational but rather strongly grounded in the science and engineering principles that are at the core foundations of not only energy and sustainability applications, but also other fields critical to enhancing UAE global competitiveness.

[133]

Each academic program is designed in consultation with a senior MIT faculty member who is an expert in the program domain. Each program consists of the following elements:

- *Required courses*, which are vehicles to convey the core material for an academic area.
- *Elective courses*, which serve the following purposes:
  - DEPTH: some electives provide greater depth into a given subspecialty within a general domain.
  - BRIDGES BETWEEN PROGRAMS: Some elective courses are meant to provide bridges between distinct programs, linking two areas.
  - CONTEXT: some electives are designed to place the technical content of the courses within a broader societal context. These might include discussions of the economic, political or cultural contexts of a particular technology. Electives focusing on contextual issues are of particular importance in a program that aspires to produce leaders in their respective areas.

By September 2011, the Masdar Institute will offer ten Master of Science Programs and two Ph.D programs. In line with the needs of the region, the programs offered will be those most conducive to developing highly skilled researchers and attracting substantial government and industry interest, particularly in advanced energy and sustainability research. The sequencing of programs is designed to immediately develop capabilities in domains of critical research importance, such as advanced materials, thermal fluids, building technologies, engineering systems, information technology, etc. As shown in Figure 4.5, the first five programs offered at the Institute will be Materials Science and Engineering, Mechanical Engineering, Engineering Systems and Management, Information Technology, and Water and Environment.

In order to earn a Master of Science degree from the Institute a student must have successfully completed a minimum of 192 course units, including 72 units for program core courses; 24 units for university core courses, and 96 units for the thesis. All students must be proficient in calculus and differential equations. Students may be required to take one or more preparatory course relevant to their academic program core courses.[11]

Figure 4.6

Masdar Institute Educational Program Implementation Timeline

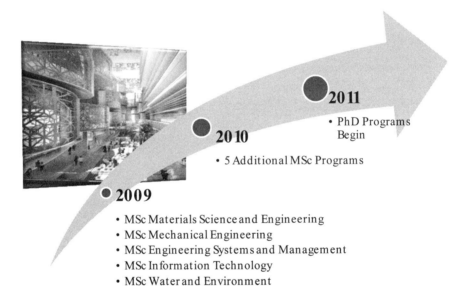

All students must take a minimum of two university courses, one of which is Sustainable Energy. As the university grows, educational tracks will be developed for each program and each track will consist of a defined set of elective program courses. The purpose of the tracks is to provide depth in particular program specializations relevant to student's research interests that are consistent with emerging trends in renewable energy and sustainability and several different tracks are discussed in the

Masdar Institute Intellectual Platform.[12] The elective courses will only be taken after students complete the required program core courses that are sufficient to prepare them for a qualifying examination for the program.

Students

The Masdar Institute is committed to admitting only the highest quality local and international graduate students with demonstrated intellectual and entrepreneurial talent. The Student Admission Committee is comprised of faculty from each of the Institute's academic programs as well as advisors from MIT.

Students that apply to the Masdar Institute must provide the following:

- Undergraduate transcripts.
- Statement of objectives.
- Two letters of reference.
- Standardized test scores:
  - TOEFL (Test of English as a Foreign Language);
  - GRE (Graduate Record Exam); and
  - CGPA (Cumulative Grade Point Average).

It is expected that all students admitted to the Institute will:

- Have a strong undergraduate Cumulative Grade Point Average (CGPA) with high marks in science, engineering and mathematics courses.
- Strong statement of objectives.
- Two letters of references that provide strong endorsement.
- TOEFL, GRE and CGPA scores on a par with those of the most selective graduate universities. The mean GRE scores for students applying to graduate engineering programs are as follows:[13]
  - Verbal – 470 (std. dev. 116);

o Quantitative – 718 (std. dev. 83); and

o Analytical Writing – 4.1 (std. dev. 0.9).

Based on these benchmarks, it is expected that students admitted to the Institute will obtain a minimum Quantitative score of 710. Verbal and Analytical writing scores will be evaluated on a case-by-case basis. Students must obtain a minimum TOEFL score of 577.

Most graduate universities do not publish a CGPA admissions target. However, the *Princeton Review* gives the following average undergrad CGPA for students accepted to the following upper, mid, and lower tier engineering programs (US benchmarks):

- Cal Tech (upper) – 3.80
- Georgia Tech (mid) – 3.60
- Northeastern (lower) – 3.40

Based on these benchmarks, the Masdar Institute will seek students with a *minimum* undergraduate CGPA of 3.60.

Enrollment Targets

Masdar Institute

The Masdar Institute will initially enroll only Master's students. Master's students will be expected to complete their degrees within two years. Starting in 2011, approximately six percent of each incoming class will consist of Ph.D students. The growth of the student body is shown in Figure 4.7.[14]

The Institute admitted a pre-class of 24 graduate students in October 2008. These students are part of the Class of 2009 and have been appointed as research assistants on projects designed by the Masdar faculty in Abu Dhabi.

Figure 4.7

Masdar Institute Enrollment Targets

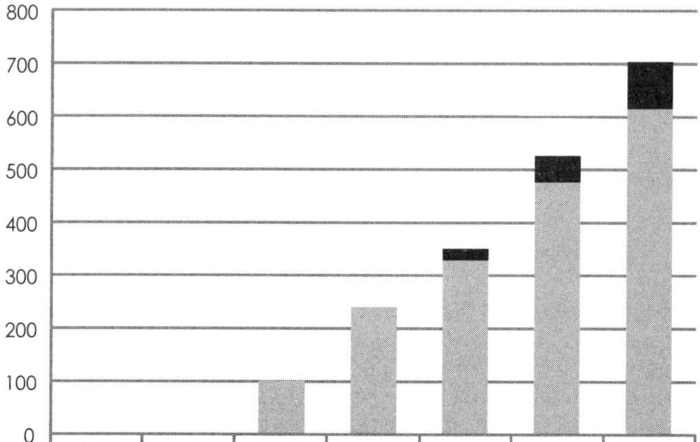

	2007	2008	2009	2010	2011	2012	2013
■ Total PhD Students	0	0	0	0	20	50	90
▨ Total Masters Students	0	0	100	239	330	476	616

MASDAR INSTITUTE FELLOWS PROGRAM

The Masdar fellows program will develop Emirati doctoral graduates in advance of the Masdar Institute Ph.D program initiation. Qualified Emirati students are being recruited for full support to pursue doctoral degrees at affiliated international schools.[15] Fellows are assigned to work on projects while studying for their degrees. They are required to spend a minimum of five years after completion of the Ph.D as Masdar Institute faculty members (if chosen through the regular application process) or Abu Dhabi Future Energy Company (ADFEC) employees. The Masdar Institute's goal is admission of fifteen Masdar fellows by September 2009.

Geographic Retention

It will be important to retain a large proportion of Masdar Institute graduates in the region to build regional knowledge capacity. As a benchmark, many students from the eight primary research universities[16] in

[138]

the Boston area remain in the region after graduation. In 2001, more than 310,000 alumni of these schools lived in the Boston area and this represented approximately 31 percent of their combined alumni population.[17] At some schools, however, the percentage of graduates who remain in the area is higher with nearly 75 percent of all University of Massachusetts Boston graduates, and half of all Boston College and Northeastern alumni living in the Boston metropolitan area.

The key driver of alumni retention for these schools, particularly for graduate-level students, is location in a thriving geographic region with ample high-technology employment opportunities. It is anticipated that Abu Dhabi will leverage the Masdar Institute and the broader Masdar Initiative to develop such an environment and therefore will attain the very high alumni retention rates noted for the Boston area universities. The UAE is already advantaged by the fact that the nation has very low relative emigration of individuals with technical skills and knowledge. The Masdar Initiative will be a key factor in reinforcing this trend.

Research

Masdar Institute research will be closely aligned with the university's academic programs. The overarching framework for both academic programs and research will be the technology, policy and systems model described above. The broader goal of Masdar Institute research is the transfer of technology and knowledge to industry and government. This will be accomplished via clearly defined university–industry linkage (UIL) mechanisms.

The following sections provide an overall Institute Research Framework and the additional details contained in the Institute Research Plan.[18]

Technology Research

Technology research is focused on the leading-edge renewable energy and sustainability themes shown in Figure 4.8.

[139]

Figure 4.8
Technology Focus Areas

Energy Efficiency and Energy Conversion	Advanced Materials	Water and Environment
•Photovotaic devices •Waste-to-energy •Nuclear energy •Marine energy •Geothermal energy •Bioconversion •Fuel cells •Batteries •Intelligent sensors •Building technologies •Solid-state lighting	•Photovoltaic materials •Thermoelectric materials •Functionally graded materials •Biomaterials •Nanostructured materials •Lightweight alloys •Materials processing and fabrication, including micro- and nano- fabrication	•Desalination devices •Water purification and filtration devices •Advanced membranes •Advanced metering for efficient water use •Air quality monitoring and control •Bio and Phytoremediation

Figure 4.9
Multidisciplinary Research in the Technology Domain

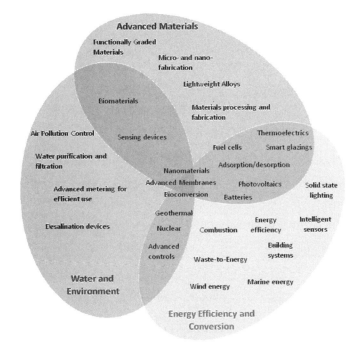

Masdar Institute technology research is multidisciplinary and distributed across the Energy Efficiency and Energy Conversion, Advanced Materials, and Water and Environment focus areas in the cross-disciplinary manner depicted in Figure 4.9.

Policy Research

Policy research will address sustainable development, sustainable energy and water policies in the Middle East and world wide. There are many challenges facing the future of the region: climate change, sustainable energy, water shortages, food shortages, etc. Therefore, policy faculty and researchers will reach out to all sectors, both local and global, and establish academic and research excellence in the areas outlined in Figure 4.10.

Figure 4.10
Masdar Institute Policy Areas

Systems Research

Systems research is inherently multi-disciplinary and will relate to the broader sustainability and energy objectives of the Masdar Institute and the Masdar Initiative. The systems perspective will be applied to technologies and/or policies within the focus areas shown in Figure 4.11 below:

Figure 4.11
Systems Focus Areas

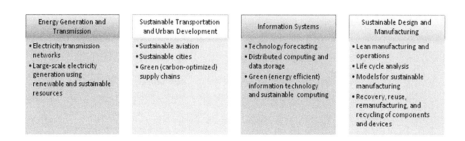

Energy Generation and Transmission	Sustainable Transportation and Urban Development	Information Systems	Sustainable Design and Manufacturing
• Electricity transmission networks • Large-scale electricity generation using renewable and sustainable resources	• Sustainable aviation • Sustainable cities • Green (carbon-optimized) supply chains	• Technology forecasting • Distributed computing and data storage • Green (energy efficient) information technology and sustainable computing	• Lean manufacturing and operations • Life cycle analysis • Models for sustainable manufacturing • Recovery, reuse, remanufacturing, and recycling of components and devices

Systems faculty will develop research collaborations with the public and private sectors, as well as international, regional and non-governmental organizations to address pressing real-world advanced energy and sustainability issues that have regional and global impact.

Agreement with MIT

MIT is assisting the Masdar Institute of Science and Technology in four integral areas:

- Joint collaborative research.
- Development of degree programs.
- Scholarly assessment of potential faculty candidates for the Institute.
- Support for capacity-building at Masdar Institute in terms of its organization and administrative structure.

The MIT/Masdar Institute Joint Executive Committee guides the development of the Institute during the life of the MIT contract.

Expanded Relationship with MIT

The Masdar Institute will continue to expand its relationship with MIT through mechanisms aimed at developing further research ties and knowledge-sharing between the Institute and MIT. The specific aims of the expansion will be to further enhance the following objectives:

- Create strong ties between the Masdar Institute and MIT in advanced energy and sustainability research.
- Translate MIT's culture of entrepreneurship and research excellence to the Masdar Institute.
- Create mechanisms for the Masdar Institute to engage in cutting edge, collaborative research that is of industrial importance.

MIT ENERGY INITIATIVE MEMBERSHIP

The Masdar Institute will create strong ties with MIT through the Technology and Development Program (TDP) and membership of the MIT Energy Initiative (MITEI). The Institute has joined MITEI as a founding member, and will be exploiting that relationship over the next three to five years to enhance its research programs. MITEI was established in 2006 as a call to action for MIT to tackle the global "energy crisis." The program provides an alignment of MIT campus sustainability objectives with the university's core academic mission.

Nuclear Engineering Research at Masdar Institute

Objective

This project will examine approaches for the evolution of the design of water-cooled nuclear power plants to address future needs of electricity, drinkable water and hydrogen through a highly efficient and environmentally friendly reactor (HEER). The HEER will have high power-conversion efficiency, with minimum production of spent fuel and waste. This advanced reactor will utilize innovative fuel designs to enable efficient fuel utilization and the ability to heat the coolant to a higher temperature in an integral vessel for the reactor and steam generation to maximize the power plant efficiency and safety. A reduction in the spent fuel production rate by a factor of two or more in comparison to today's reactors in Europe and Japan is anticipated.

Relevance to the Masdar Initiative

- Abu Dhabi's economic interests are clearly served by investigation of HEER for: (1) contribution to local power needs; (2) production of hydrogen for local refinery and transportation use; and (3) provision of heat for seawater desalination facilities.
- The Masdar Institute's capabilities are enriched via research that lays a foundation for advanced research in nuclear energy.

Scope of Work

Two of the major problems facing human development are the availability of affordable clean, environment-friendly energy and drinkable water sources. This is especially true for countries in the Arabian Gulf region where natural potable water sources are practically non-existent and the most widely developed energy source is oil.

Scientific research, like that envisioned by the Masdar Institute, can help to define a pathway for solutions to these two complicated problems. Recently, nuclear energy, with its almost non-existent greenhouse gas emissions, has been regaining ground in the competition for meeting the energy demand of most advanced countries (like the USA and France) and most populous countries (like China, India and Brazil). The vehicle for expanding nuclear energy enterprise is focused to a large extent on water-cooled reactors in order to benefit from the long-term experience with those reactors around the world—they constitute more than 80 percent of all existing reactors. While it is well-known that electricity can be produced from nuclear reactors, it is not as well-known that there have been a number of reactors that have produced desalinated water on an experimental scale, mostly in Japan.

The reactors in operation today were designed in the 1980s and do not take full advantage of advances in materials and information technology that could boost their power extraction and conversion performance, and therefore their economic performance. The main way that nuclear power has competed with alternatives is via the construction of very large units (over 1,200 MWe). Thus, the small- and medium-sized units have been considered less than large units for redesign. With the rise in the cost of other energy sources, the economic disadvantage of medium-sized nuclear units has disappeared. However, other obstacles against the deployment and use of nuclear energy as an alternative energy source in the Middle East remain. These are: the required technical capacity to operate the reactors reliably; finding a medium sized reactor with the right characteristics to co-produce power and water; storage and disposal of spent fuel; and the requirements to secure the fuel material to avoid diversion of its fissionable content for use in weapons applications.

The development of advanced nuclear power reactor designs that minimize these challenges is underway in different parts of the world, including at MIT. One approach is to design a power reactor that minimizes the need for frequent refueling throughout the reactor lifetime (40 or more years). This limits the need by the host country to amass a large inventory of fresh nuclear fuel (to be ready for refueling) or spent fuel (as the fuel would be discharged only infrequently). The fuel can then be supplied and later retrieved by the vendor country, making that reactor operation highly proliferation-resistant. This also minimizes any required shipments of spent fuel to international storage installations, or even national ones. While ordinary reactors are refueled every 12 to 24 months, it is possible to stretch the operation to a minimum of 48 months, with the fuel having a modest initial content of the U235 isotope, and even to ten or fifteen years with a higher initial enrichment in uranium or thorium. However, the maintenance and inspection needs of the reactor, which would have to be conducted without opening

[145]

during the fuel cycle, would need to be revisited and, using modern technology, converted to on-line operations.

The higher initial enrichment of the fuel needed to provide the 10-year operation is likely to make the fuel cost more expensive than the low-enriched fuel in today's reactor. However, if the fuel used in this reactor would enable a higher power density in the core, then the cost of the equipment, such as the vessel and containment, would be reduced. Hence, it would be important to investigate the use of advanced fuels to enable higher power density in the core. One such fuel is annular fuel, which has additional safety advantages, as has been documented in previous work at MIT. If the size of the reactor is of the order of 300 to 400 MWe, it might be possible to integrate the steam generators in the vessel that has the nuclear core, which will provide further safety and cost advantages.

Another approach for cost reduction would be boosting the thermal efficiency of the plant by raising the temperature of the coolant exiting the core. This would only be possible if a different material was used for cladding than the current material, with ceramic materials providing the best approach to withstanding the higher temperature without suffering accelerated corrosion. Attaining the higher temperature will enable the plant to meet demand for electricity with a reduced amount of fission energy, thus lowering the fuel cost while making it possible for the fuel to serve longer in the core.

The improved electricity generation efficiency and higher temperature of the coolant at the core exit both facilitate the use of advanced water electrolyzers to generate hydrogen and oxygen. Both of these products are useful in the operation of modern oil refineries, as a larger portion of the feedstock comes from heavier and more sour oils. Both heat and hydrogen are needed in refineries to enable distillation, thermal cracking and sweetening of oil feedstocks. Providing them from a nuclear plant coupled to a water electrolyzer would reduce the carbon emissions from the refinery, which would otherwise burn natural gas to provide the heat and hydrogen. Thus, it would be useful to explore the applicability of a high-efficiency nuclear reactor to provide the needed heat, electricity and hydrogen for refineries. The production of hydrogen – via High Temperature Electrolysis of Steam (HTES), for example – will be examined. But, the electrode materials degrade more quickly at higher temperatures. In the longer term, the hydrogen can also be used to power fuel cells that would meet power peaking as necessary during the day, whilst the hydrogen can be used to power Masdar's official hydrogen-vehicles.

To meet the water needs in the area, a high-efficiency, long operation reactor can be used as a power source for Masdar's Special Free Zone (SFZ). Some of the reactor power can be used to provide the SFZ with its electrical power needs, whilst the rest can be used to generate hydrogen. Waste heat from the HEER can be used to preheat gulf water streams for a hybrid Multistage Stage Flash (MSF)–Reverse Osmosis (RO) desalination plant that will provide the SFZ with its potable water needs. This will increase the water-to-power ratio, which will in turn increase the efficiency of the desalination plant which gets its electrical power from the HEER. With all three elements of electricity, water, and auto fuel in place, the HEER will truly help create a Self-Sustainable Zone SFZ (3SFZ) that is carbon free.

Conclusions and Recommendations

- Nuclear sector personnel must be a highly skilled component of the work force of countries that deploy nuclear power.
- A nuclear power program is a national undertaking.
- Nuclear power deployment needs go beyond nuclear expertise:
  - different types of engineers and scientists, managers of power plants, power sector lawyers, knowledgeable government officials (regulatory, atomic commission experts), procurement officials, energy traders, etc;

[147]

- o adequate specialized and integrated training is needed for all personnel.
- Universities are important to the education/training of experts.
- Significant investment is needed for infrastructure development:
  - o dedicated and well-maintained infrastructure crucial to success;
  - o international collaborations will be needed (IAEA and governmental).
- Research centers must be established for program sustainability.
- Most likely for the UAE, the first nuclear power plant(s) will be turn-key projects:
  - o provides time for needed manpower and infrastructure development;
  - o nuclear vendors can assist in necessary training initially.
- Local participation is important for a successful nuclear program:
  - o local industries and expertise must be integrated in plant construction;
  - o local experts must be retained and expatriates must be attracted.
- Transparency is important for both local and international support.

MANAGING THE NUCLEAR SECTOR: RISKS AND OPPORTUNITIES

5

Security Infrastructure and Non-Proliferation Standards

Charles D. Ferguson

Ensuring strong physical protection of nuclear power plants and adhering to rigorous non-proliferation standards complement and enhance each other. Both are necessary for continued peaceful use of nuclear energy. But these two activities have distinct differences, especially with regard to their consequences. The former protects against attacks on, or sabotage of, a nuclear facility. Importantly, attack or sabotage would not cause a nuclear explosion. Nonetheless, in a plausible worst case scenario, the damage could soar to hundreds of billions of dollars in the event of a massive release of radioactive material into urban environments. The loss of the nuclear power plant itself would result in up to several billion dollars' worth of damage. But far greater costs could result from radioactive contamination spurring a mass exodus from homes and even more widespread harm to the economy.

In comparison to an attack on a nuclear plant, the misuse of civilian nuclear facilities to make fissile material for bombs could threaten far greater destruction. A nuclear bomb is undoubtedly a weapon of mass destruction, and even a relatively "low yield" nuclear bomb could kill one hundred thousand or more people, depending on the population density of the targeted city. But nuclear power plants alone are not sufficient to lead to a nuclear weapons program. A nuclear power-producing state would need access to a reprocessing plant in order to separate plutonium resident in highly radioactive spent fuel that was discharged from the power plant.

A uranium enrichment plant can also supply weapons-usable fissile material. Thus, the state can make use of peaceful nuclear power without triggering fears of weapons programs as long as it refrains from acquiring nationally controlled uranium enrichment or reprocessing facilities and maintains rigorous safeguards to give its neighbors and the international community confidence that nuclear material has not been diverted into a weapons program and that clandestine enrichment or reprocessing facilities do not exist.

To understand what is needed to protect nuclear facilities, let's first examine physical protection infrastructure planning. Then, to understand how to deter use of peaceful nuclear programs for weapons purposes, we shall assess the status of non-proliferation standards.

Security Infrastructure

Design Basis Threat

Nuclear power plants face potential threats from attackers or saboteurs. These threats pose serious challenges to nuclear facility owners, operators, and security forces because there are various methods available to intelligent adversaries who wish to attack or sabotage a plant. While plant owners want to keep security costs at reasonable levels, they need to balance this financial constraint against the economic and social damage that could result in the event of a successful attack or sabotage. To determine how best to counter these threats, owners, operators, and security forces use the method of design basis threat (DBT).

The DBT first describes a state's evaluation of the current threat from actual or potential adversaries. The evaluation considers the adversary's attributes and characteristics, including their capacity to use various weapons, explosives, and employees of the targeted facility. Other important characteristics involve the ability – if any – to employ advanced infiltration tactics such as commando skills and other special operations.

[152]

The DBT also needs to evaluate the potential size of the attacking force. Next, security forces and plant operators need to evaluate what types of malicious acts could cause unacceptable consequences such as release of radioactive material to the environment or severe damage that would impair the ability of the plant to operate. This evaluation step then considers what protection measures and security forces are needed to significantly reduce the likelihood of an attack or sabotage causing those consequences. By following this procedure, plant operators and security forces can define performance requirements that both they and the plant must meet.

It is important to underscore the fact that the DBT is a dynamic tool. That is, the state, its intelligence agencies, and the plant's operations and security team should continually reevaluate the potential threats and the level of security performance. The threats could evolve so that today's attackers could acquire more sophisticated techniques in the near future. Also, the motivations of potential attackers could change from having a strong desire to target nuclear assets to having a low desire depending on the group's internal dynamics and external considerations, i.e., whether the group's leaders perceive that a target has been hardened against attack. The International Atomic Energy Agency (IAEA) continues to prepare guidance for states in their evaluation of the DBT.[1]

Defense-in-Depth

Attackers who want to maximize the damage from an attack on a nuclear facility would target those facilities containing large inventories of radioactive material. Such facilities include operating nuclear power plants, spent fuel storage pools (usually located in nuclear power plant sites), plutonium separation or reprocessing plants (along with associated facilities for spent fuel storage, liquid high-level radioactive waste storage, and liquid high-level radioactive waste processing), and research reactor sites (but these facilities tend to contain much less radioactivity than commercial power plant reactor sites). About 440 commercial nuclear power reactors (many power plants have more than one reactor) are currently operating in 31 countries; most power plants have one or more spent fuel storage pools; about 280 research reactors are running in 56 countries; civilian reprocessing plants are operating in France, Britain, India, Russia and Japan; and military reprocessing plants are found in Russia, India, Israel, Pakistan and North Korea (which has an agreement via the six party talks to disable and then dismantle its reprocessing plant). The totality of these facilities presents a relatively target-rich environment for terrorists.

However, terrorists tend not to attack well-defended targets. Most nuclear power plants and many other nuclear facilities, such as spent fuel storage sites and reprocessing plants, use defense-in-depth security systems. Consequently, terrorists would have to breach multiple layers of security if launching a ground-based attack. Commercial nuclear power plants have security perimeters surrounding the vital areas of the plant. To gain access to the innermost area, personnel undergo the appropriate background checks and are subject to continued monitoring of their behavior through personnel reliability programs. They are also required to follow the "two-man rule," which dictates that more than one authorized person must be present to access the most vital parts of the plant. Thus, unless the two persons have colluded, one can keep a watchful eye on the other. Even with additional layers of security, nuclear power plants and other nuclear facilities will remain potent symbols of a country's industrial might and thus will likely continue to interest some, but not most, terrorist groups in possible attacks.

Safety systems also contribute to a nuclear plant's defense-in-depth. Modern plants are designed to have backup safety systems to offer greater assurances of protection in the event of an accident. Such redundant systems can also help guard against plant damage in the event of sabotage or certain attacks. In particular, multiple layers of safety help prevent the release of highly radioactive fission products to the environment. The first layer is the design of the fuel pellet, usually made of uranium dioxide embedded inside fuel cladding. The cladding under normal and most accident conditions prevents the fission products from escaping from the fuel pellet. Emergency reactor core cooling systems help to protect against meltdown of the core. If these systems were not available and not working properly, a loss of coolant in an accident could lead to a release of fission products from the core. But even in the event of this severe accident, a final layer of protection is the containment structure, which is typically made of thick, reinforced concrete surrounding the reactor and many other pieces of major equipment such as main coolant pumps and steam generators. As long as the containment structure is not breached, radioactive material should remain trapped inside and thus away from the environment. Notably, Generation III+ reactors that have recently received regulatory approval have passive safety features that rely on gravity and natural circulation to keep the plant in a safe condition in the event of an accident. These newest reactors may provide added protection against sabotage.

Potential Adversaries

Dr. Jerrold Post, a renowned expert on the psychology of terrorism, has warned that "we are in the paradoxical position of having a clearer understanding of the interior of the atom than we do of the interior of the mind of the terrorist."[2] He has pointed out an additional motivational paradox: "On the one hand, to be motivated to carry out an act of mass destruction suggests profound psychological distortions usually found only in severely disturbed individuals, such as paranoid psychotics. On the other hand, to implement an act of nuclear terrorism requires not only organizational skills but also the ability to work cooperatively with a small team."[3] While he and other experts in terrorist psychology have not concluded that acts of nuclear terrorism can never occur, their analyses have uncovered few terrorist organizations that would covet such attacks. Soon after September 11, 2001, Dr. Post reiterated this point by

[155]

cautioning that those "who study terrorist motivation and decision making" are "underwhelmed by the probability of such an event [nuclear terrorism] *for most – but not all –* terrorist groups."[4]

Here, it is important to differentiate between nuclear and radiological terrorism. Nuclear terrorism refers to a terrorist group either acquiring and detonating an intact nuclear weapon from a state's arsenal or acquiring enough weapons-usable fissile material from states' stockpiles in order to make and detonate an improvised nuclear device, which is a crude, but potentially devastating, nuclear weapon. Radiological terrorism refers to two types of activities: dispersing ionizing radiation by using commercial radioactive sources or attacking a nuclear facility—often with the objective of causing a release of radioactive material to the environment. The former type of radiological terrorism is popularly called "a dirty bomb." But this term, while sure to spark the public's fears, gives a limited understanding of radiological weapons. A radiological weapon can disperse radioactive material without using a bomb or explosives as long as the material is in, or is placed in, a dispersal physical state. A full examination of radiological weapons is beyond the present scope of this paper. In this section on physical security of nuclear power plants, the focus is on the latter type of radiological terrorism.

There have been relatively few attacks or threats of attack against nuclear power plants or related facilities. As mentioned earlier, most, if not all, terrorists refrain from attacking hardened targets. Nuclear power plants typically project vigorous security. The terrorist group's psychological and cultural constraints can influence its decision to attack a hardened facility or to cause massive destruction or disruption. For example, religiously motivated terrorists do not want to disappoint their god or higher power by failing in an attack. Some of these terrorists may also feel constrained in using methods that could cause massive harm to civilians, for example, through release of radiation. In contrast, other politically motivated terrorists such as Al-Qaeda may feel that such attacks are justified against enemies who possess weapons of mass destruction.

[156]

Terrorist groups with strong links to national constituencies may feel constrained because they would not want to harm their supporters through release of radiation. However, individual members of political causes may decide to attack a nuclear power plant before it begins operations and thus before the attack has the potential to release radiation. For example, in December 1982, Rodney Wilkinson, a young white South African who had joined the African National Congress (ANC), detonated four bombs at South Africa's Koeberg nuclear power station while the plant was under construction.[5]

Radical environmentalists may despise nuclear power plants as symbols of humans' harming the environment through production of radioactive waste, for example. In January 1982, the Pacifist Ecological Committee fired five anti-tank rockets at the Creys-Malville Superphoenix breeder reactor in Lyon, France, while the reactor was still under construction.[6] In sum, the motivations of attackers or saboteurs can vary significantly. Despite the motivations, it is imperative to maintain the highest security standards to guard against all modes of attack or sabotage. In addition to non-state actors who may have motivation to attack nuclear facilities, a state may use military forces to attack an enemy's nuclear facilities. This possibility is considered along with other modes of attack.

Modes of Attack

An attacker can choose from a variety of attack modes: airplane crashes, truck bombs, commando attacks by land, waterborne attacks, cyber attacks, and military attacks. The ability of a nuclear plant to withstand any of these attacks depends on the plant's design, the robustness of security perimeters, the training of the operators and on-site guards, the responsiveness of emergency responders and external security forces, and, in the case of airplane crashes, the capabilities of airport and airline security personnel. Without revealing sensitive security information, let's briefly examine each attack mode.

Airplane Crashes

The terrorist attacks on September 11, 2001, on the World Trade Center in New York and the Pentagon in Washington, DC, demonstrated weaknesses in US airport and airline security and showed that terrorists can adapt and combine two well-proven techniques to cause massive damage. Although the major plot lines are well-known, it is worthwhile here to review the main themes to draw lessons learned. Four cells of attackers employed simple box cutters as weapons to overpower airline crews. Box cutters were chosen because they are small and thus not easy to detect, and at the time of the attacks they were not banned items on airplanes. Three of the cells had five men, and the other only four, but it was likely meant to include a fifth man. Each cell was small, but large enough to gain control over relatively small airline crews with then little or no defenses. Another essential element to the plan was the training of cell members at flight schools. Despite only expressing interest in flying a large, 757-type plane, and not in take-offs or landings, these cell members attracted relatively little attention although an FBI field agent was starting to sound the alarm a few months before the attacks. The relative openness of America at that time to allow this flight training and to allow many cell members to obtain identification cards such as driver licenses with little or no background checks was a major contributing factor to the success of the attacks. By this careful planning and exploitation of security loopholes, these terrorists used the well-proven technique of hijacking airplanes, which would have been reminiscent of numerous such events during the 1970s had not the attackers combined this technique with the deadly innovation of suicide bombing. As Al-Qaeda operatives, these attackers were willing to commit suicide to cause destruction on a scale never before seen from a set of coordinated terrorist attacks.

Could the 9/11 terrorists have caused even more destruction by targeting a nuclear power plant? According to the 9/11 Commission report, Mohamed Atta, one of the leaders of the 9/11 terrorists, considered crashing an airplane into a US nuclear power plant but thought preparing for such an attack would be difficult because of the restricted airspace around nuclear facilities.[7]

Many nuclear facilities are typically not hardened to withstand a crash of a large airplane. Although the nuclear industry is quick to point out that all US and other modern nuclear power plants use strong containment structures around nuclear reactors as a last line of defense to prevent a release of radiation in the event of an accident, containments were typically not designed to prevent breaching in the case of a premeditated attack, such as suicidal terrorists piloting a large airliner like a 757. Nonetheless, smashing a fast-moving large commercial airplane into a containment building is extremely challenging. Assuming that terrorists could in a post-9/11 world seize control of such a plane, they would need the ability to direct the plane onto a relatively small area to try to breach a containment structure. Independent experts have cautioned that airplane crashes might not damage containment structures but could harm auxiliary buildings, thereby causing considerable financial damage even if no radioactive material is released into the environment.

Could air defenses stop an airplane attack on a nuclear facility? Should such defenses be employed? Who would decide to fire the defensive missiles? While at first glance air defenses appear necessary for defense of a nuclear facility, as the preceding questions indicate, these defenses are fraught with difficult decisions. Very little time would be available to make the decision to fire defensive missiles. If a pilot had strayed innocently off course, a misuse of missile interceptors could doom hundreds of people to death. While French authorities did place missile air defenses outside of some of their nuclear facilities for a period of time after 9/11, US authorities decided against such action. Instead, the US response relied on strengthening security at airports through better screening of passengers, on hardening cockpit doors, and on stationing federal police officers onboard random flights as a deterrent.

Concerning security improvements in new plants, in February 2009, the US Nuclear Regulatory Commission (NRC) issued a rule "that requires applicants for new power reactors to assess the ability of their reactor designs to avoid or mitigate the effects of a large commercial aircraft impact."[8] In issuing this rule, the NRC clearly specified that operators of these reactors are not required to prevent aircraft crashes into their power plants; this responsibility resides with the US government. Moreover, the rule does not apply to existing power plants.

Truck Bombs

Truck bombs have been used to produce devastating effects in attacks on non-nuclear targets. For example, the two 1983 truck bombs that killed dozens of Americans in Lebanon: first the April 18 attack on the US Embassy in Beirut and then the October 25 attack on the Marine barracks. These truck bombs spurred the NRC to investigate the potential effects of such bombs against nuclear power plants. But despite research indicating devastating damage, the NRC decided against a regulation change. Not until the 1993 truck bomb detonation at the World Trade Center did the NRC include this mode of attack in its design basis threat. Ramzi Yousef, who had ties to Al-Qaeda, masterminded this attack. Nevertheless, US nuclear power plants did not fully install vehicular barriers until 1996, a year after the powerful truck bomb was detonated by domestic American terrorists in April 1995 at a federal building in Oklahoma City.

Commando Attacks by Land

A commando type attack has been included in the US NRC's design basis threat for many years. However, some outside the NRC who have knowledge of the DBT have expressed concern that the pre-9/11 DBT, and even the revised post-9/11 DBT, do not adequately take into account the potential consequences of an operation undertaken by a relatively large, sophisticated group of attackers. While the details of the DBTs remain classified, it is believed that the latest does not factor in an attacking force as large as the 19 well-trained men involved in the 9/11 attack. A group similar to the 9/11 attackers could form independent teams to launch near simultaneous attacks on a nuclear facility. Increased numbers of security guards may provide better security, but raising the number of guards increases security costs. More importantly, more guards may not necessarily mean better security. In a paper that challenges conventional wisdom, Scott Sagan expressed concern that more guards could lead to social shirking of responsibility in which a guard may not work as hard if he knows that there are more guards to cover for him.[9] Also, Sagan said that the appearance of greater security may lead the plant operators to take greater risks. In sum, he cautions against an enhanced security measure that would just involve hiring more guards. Layered access points and two-man rules are important means to protect against unauthorized access. Also, making sure that the nuclear plant has redundant safety systems can help mitigate the effects of an attack.

Waterborne Attacks

Nuclear power plants need external cooling water to ensure safe operation. Blocking cooling water to the plant could lead to reactor damage but this is unlikely owing to the existence of backup safety systems. But forcing the plant to shut down for an appreciable period of time could have serious economic and power-production consequences. Plants vary in their sources of external cooling water, which could come from oceans, rivers, or natural or manmade lakes. Some of these water systems are easier to access than others. For instance, a waterborne commando crew could try to sneak toward the cooling water intake by a surreptitious sea route. Although this mode of attack was reportedly not included in the NRC's pre-9/11 DBT, it has reportedly been added to the post-9/11 DBT.

Cyber Attacks

As nuclear power plants move toward more digital control systems, they may become more vulnerable to cyber attacks. Older plants usually employed analog control systems, which are less vulnerable to external hacking, although they could be susceptible to insider sabotage. In 1992, a technician at the Ignalina Nuclear Power Plant in Lithuania launched a computer virus into auxiliary control systems. He allegedly did this to draw attention to security weaknesses and expected to be rewarded for his efforts. Instead, he was arrested.

In January 2003, the slammer computer virus penetrated the network at the Davis Besse Nuclear Power Plant in Ohio. Fortunately, the plant was shut down at that time. The plant was susceptible because technicians had not installed a Microsoft security patch. In 2002, the NRC began a research program to identify ways to protect nuclear plants against cyber attacks.

Insiders

Personnel working at a plant have special knowledge about the facility's operations. If attackers could recruit insiders, they might gain the edge they need to launch a damaging attack. Insiders can help accelerate the attack and would likely not raise undue suspicion until possibly after it is too late to prevent major damage. Means to recruit insiders include finding those sympathetic to the attackers' cause, blackmail, or paying enough money to convince the insider to help with the attack. Insiders can identify vulnerable equipment or can take a more active role by actually disabling safety equipment. Disgruntled employees or those motivated by a political cause could also become lone actors. For example, as noted earlier, Rodney Wilkinson had access to the plant because of his job and thus could place limpet mines while minimizing his chances of being stopped. To reduce the risk of insiders assisting attackers, background security checks can try to spot those people who may be susceptible to recruitment. Prior to 9/11, the NRC required background checks of those personnel who would have access to sensitive parts of a nuclear plant.

Military Attacks

Nuclear facilities have been targets of military attack in the past. Military motivations to target such facilities can be fourfold: to damage the electrical power system of an enemy, destroy a major status symbol, degrade the capability for an opponent to make fissile material for nuclear weapons, or cause radioactive contamination of an enemy's territory.[10] There have been few incidents of such attacks. The first attack on a reactor occurred on September 30, 1980, when Iran's air force bombed a French-supplied, 40-Megawatt thermal nuclear research reactor under construction in Iraq. While this attack did not destroy the facility, on June 7, 1981, the Israeli air strike against the Osirak reactor did destroy it. Prior to this attack, Israel had employed diplomacy, coercion, and sabotage in its attempts to stop Iraq from obtaining the capability to make plutonium for nuclear weapons. During the Iran–Iraq War in the 1980s, Iraq bombed the commercial nuclear power plant under construction at Bushehr in Iran. More recently, in September 2007, Israel bombed a suspected nuclear reactor site in Syria. The US government reported that Syria was building a research reactor of the same or similar design to the North Korean reactor at Yongbyon, which has produced weapons-grade plutonium. North Korean technicians were also reportedly seen at the Syrian site.

Concerns about possible military attacks on civilian nuclear facilities prompted India and Pakistan to agree in 1988 to exchange annually lists of facilities that the military would refrain from attacking. While this confidence-building measure can help establish a norm against military attacks, states may decide to protect their nuclear facilities with air defenses and hardened burial, as Iran has done with its uranium enrichment plant at Natanz.

The International Nuclear Security Regime

While nuclear safety and security often support each other, sometimes they come into conflict. Safety thrives on openness and a willingness to air problems at a plant in order to quickly and effectively prevent a real problem from developing and to disseminate corrective actions to other plants. On the other hand, security relies on confidential information in order to guard plant vulnerabilities from adversaries. Although safety and security teams should learn from each other, at times these communities are intentionally or unintentionally walled off. Plant operators are gradually learning that they need to bring both teams together.

Another major difference between nuclear safety and security is that the former has well defined standards and decades-long vetting of how to implement these standards—although safety can always be improved. The international nuclear security regime, however, is still in relative infancy with as yet undeveloped rigorous and binding international standards. Nonetheless, a regime has been emerging. The mechanisms of this new regime include IAEA guidelines, the revised Convention on the Physical Protection of Nuclear Material (CPPNM), the International Convention on Suppression of Acts of Nuclear Terrorism, UN Security Council Resolution 1540, the Global Partnership Against the Spread of Weapons of Mass Destruction, the Global Threat Reduction Initiative (GTRI), and the Global Initiative to Combat Nuclear Terrorism. Here, the focus is on those mechanisms relevant to the protection of nuclear power plants and nuclear material at those and related facilities.

The IAEA underscores on its nuclear security website that "no single international instrument … addresses nuclear security in a comprehensive manner." Through its Nuclear Security Fund, which has limited resources, the IAEA has offered security assistance to many member states. Through its Nuclear Security Series of publications, the IAEA has served as an information source for member states. This series has the intent of helping states "establish a coherent nuclear security infrastructure." As of 2008, the series has only included four technical guides on "Technical and Functional Specifications for Border Monitoring Equipment," "Nuclear Forensics Support," "Guidelines for Monitoring Radioactive Material in International Mail Transported by Public Postal Operators," and "Engineering Safety Aspects of the Protection of Nuclear Power Plants against Sabotage." The IAEA is working with member states to develop fifteen other guides ranging from development of a design basis threat, to physical protection of nuclear material and facilities, nuclear security culture, security of radioactive waste, and combating illicit trafficking of nuclear and other radioactive materials. The IAEA is also working with member states to instill security culture in nuclear facility operations.

The CPPNM entered into force in February 1987 and has been the only international instrument on the protection of nuclear material. But for many years, it was limited to protecting material in international transport. To broaden its application, member states convened a conference in July 2005 to amend the convention to require parties to protect nuclear facilities and material in peaceful domestic use and storage. Moreover, the amendment requires enhanced cooperation among states in recovering lost or stolen nuclear and other radioactive material and mitigation of acts of radiological attack or sabotage. The amended CPPNM needs two-thirds of the 112 parties for it to enter into force; this process is expected to require many years.

One of the latest efforts to protect against nuclear threats is the Global Initiative to Combat Nuclear Terrorism. Launched at the 2006 US–Russia Summit, this initiative seeks to enhance international cooperation among partner nations to combat the global threat of nuclear terrorism. As of mid-2008, 75 countries had indicated interest in working on this initiative. But this is not a binding convention. Thus, countries are not required to participate in all or even any activities. Of relevance to physical protection of nuclear facilities, the initiative includes the principles: "enhance security of civilian nuclear facilities," "ensure adequate respective national legal and regulatory frameworks sufficient to provide for the implementation of appropriate criminal and, if applicable, civil liability for terrorists and those who facilitate acts of nuclear terrorism," and "promote information sharing pertaining to the suppression of acts of nuclear terrorism and their facilitation, taking appropriate measures consistent with their national law and international obligations to protect the confidentiality of any information which they exchange in confidence."[11]

Recommendations for Improved Security Infrastructure

- Integrate improved security features, including enhanced protection against large aircraft crashes, into new nuclear plant designs and fully implement these design features. To facilitate this enhancement, security assessment teams should work side-by-side with engineering teams who are designing plants. Plants should have greater protection against all modes of attack.

- Encourage safety and security teams to work closely together to develop and implement protections against the design basis threat. They also need to consult regularly with intelligence agencies to obtain the latest information about possible attackers in order to continually update the design basis threat. It is vitally important to instill a security culture among all workers at nuclear facilities.

- Conduct effective background checks of plant personnel to reduce the likelihood of an insider threat.

- Protect against cyber attacks, especially as the newer plants use more digital control systems.

- Develop and implement rigorous international security standards, but national authorities are still responsible for security at their facilities.

- Ratify the amended Convention on Physical Protection of Nuclear Material.

- Urge states to cooperate in information sharing on potential threats. An attack on a nuclear plant anywhere could affect the continued operation of nuclear plants everywhere.

Non-Proliferation Standards

For more than fifty years, the international community has recognized both the promises and perils of nuclear power and other peaceful nuclear technologies. The energy released from the fission of only one kilogram of uranium or plutonium could provide electricity for hundreds of thousands of homes or kill more than one hundred thousand people with one bomb.

The Non-Proliferation Treaty

The 1970 Non-Proliferation Treaty (NPT) has become one of the most universally adhered to treaties with only three states (India, Israel, and Pakistan) not having signed it. (North Korea left the treaty in 2003.) It is the fundamental pillar of the non-proliferation regime, which seeks to prevent the spread of nuclear weapons programs. The NPT has helped establish an international norm against nuclear proliferation. At its heart, the NPT has the objective of preventing nuclear war, promoting nuclear disarmament, and encouraging the use of peaceful nuclear energy. With those three goals in mind, the treaty contains two 'grand bargains': (1) non-nuclear weapon states (NNWS) have the right to acquire peaceful nuclear technologies as long as they maintain safeguards on their nuclear programs and do not seek to acquire nuclear explosive devices, and (2) nuclear weapon states (NWS) pledge to end the nuclear arms race, pursue nuclear disarmament, and work toward a treaty on general and complete disarmament. For ease of reference, here is the text of Articles I, II, III, IV, and VI:

Article I:

Each nuclear-weapons state (NWS) undertakes not to transfer, to any recipient, nuclear weapons, or other nuclear explosive devices, and not to assist any non-nuclear weapon state to manufacture or acquire such weapons or devices.

Article II:

Each non-NWS party undertakes not to receive, from any source, nuclear weapons, or other nuclear explosive devices; not to manufacture or acquire such weapons or devices; and not to receive any assistance in their manufacture.

Article III:

Each non-NWS party undertakes to conclude an agreement with the IAEA for the application of its safeguards to all nuclear material in all of the state's peaceful nuclear activities and to prevent diversion of such material to nuclear weapons or other nuclear explosive devices.

Article IV:

1. Nothing in this Treaty shall be interpreted as affecting the inalienable right of all the Parties to the Treaty to develop research, production and use of nuclear energy for peaceful purposes without discrimination and in conformity with Articles I and II of this Treaty.

2. All the Parties to the Treaty undertake to facilitate, and have the right to participate in, the fullest possible exchange of equipment, materials and scientific and technological information for the peaceful uses of nuclear energy. Parties to the Treaty in a position to do so shall also co-operate in contributing alone or together with other States or international organizations to the further development of the applications of nuclear energy for peaceful purposes, especially in the territories of non-nuclear-weapon States party to the Treaty, with due consideration for the needs of the developing areas of the world.

Article VI:

Each of the Parties to the Treaty undertakes to pursue negotiations in good faith on effective measures relating to cessation of the nuclear arms race at an early date and to nuclear disarmament, and on a Treaty on general and complete disarmament under strict and effective international control.

Recently, non-proliferation experts have renewed a debate on how to interpret Article IV's "inalienable right." Article IV does not specifically mention enrichment and reprocessing technologies. These technologies are inherently dual-use: either for producing fissile material for weapons or fuel for peaceful nuclear reactors. Enrichment increases the concentration of uranium-235, the fissile isotope or component of natural uranium. (Natural uranium contains 0.72 percent uranium-235, too low a concentration to use in nuclear weapons.) Low enriched uranium, which is defined as a uranium mixture with less than 20 percent uranium-235, can be used to fuel commercial nuclear reactors. The same enrichment technology can produce highly enriched uranium, which has greater than 20 percent U-235 and is weapons-usable. The greater the concentration of uranium-235, the more suitable the material is for weapons.

Reprocessing consists of a set of chemical techniques used to extract plutonium from irradiated or spent nuclear fuel. If the plutonium is completely separated from highly radioactive fission products, the resultant material can be used to power weapons or to make new fuel for reactors. The United States in recent years has been researching the potential for making reprocessing proliferation-resistant although independent scientists have raised concerns that the techniques under investigation can still raise the risk of diversion and use of weapons-usable plutonium and other fissile material such as americium and neptunium.

Some non-nuclear weapon states such as Argentina, Brazil, Canada, and Japan have interpreted Article IV to include these technologies. Iran has also interpreted the article in this way. It is important to realize the contingent nature of this "right." The article makes clear that states have the responsibility to refrain from obtaining nuclear explosive devices and to maintain effective safeguards on their civilian nuclear program.

Under the NPT, nuclear weapon states have the responsibility to pursue nuclear disarmament. These states need to recognize that by doing so they are upholding their end of the NPT bargain. But the balancing act is how to pursue nuclear disarmament while maintaining security assurances to allies. Alliance commitments, however, do not just depend on nuclear deterrence. The United States, in particular, could benefit from having honest discussions with its allies about the extent to which conventional capabilities can substitute for nuclear capabilities. A declaratory policy of no-first-use of nuclear weapons could relegate nuclear weapons to be only useful for deterring use of others' nuclear weapons. While the United States has yet to adopt this policy, China has done so. Russia has renounced its no-first-use policy since the Cold War because of an erosion of its conventional military capabilities. Although a full examination of this important issue is beyond this paper's scope, it is worth noting the importance of security assurances.

Security Assurances

Positive security assurances reassure allies that they will help ensure their common defense. A nuclear weapon state can offer an extended deterrent or "nuclear umbrella" relationship to states within such an alliance. This offer would comprise a promise to come to a state's defense if faced with threats from another nuclear-armed state. Negative security assurances from a nuclear weapon state reassure non-nuclear weapon states that they will not be attacked with nuclear weapons by the nuclear-armed state as long as they do not acquire nuclear weapons or align themselves with another nuclear-armed state.

Nuclear Safeguards

During the fifty-plus years since nuclear power was first commercialized in the 1950s, safeguards on peaceful nuclear programs have undergone significant evolution. According to the International Atomic Energy Agency (IAEA), "safeguards are measures through which the IAEA seeks to verify that nuclear material is not diverted from peaceful uses."[12] These measures comprise material accountancy as well as surveillance tools and techniques. When a state accepts safeguards on its nuclear program, it helps to reassure its neighbors of its peaceful intentions as long as the application is fully transparent and cooperative. Safeguards are not foolproof. A determined state could try to build

[170]

clandestine nuclear facilities. Thus, at best, safeguards serve as a deterrent, raising the bar to proliferation.

Throughout the evolution of safeguards, there has been a constant push and pull between the openness states have afforded to inspectors of their nuclear programs and the perceived need to protect states' sovereignty. This evolution has not been smooth, but instead has only happened because of crises. The first crisis arose from South Asia. India had acquired CIRUS, a modest-sized research reactor from Canada along with US-supplied heavy water. CIRUS had just the right thermal power rating to be able to make at least one nuclear bomb's worth of plutonium annually.

Perhaps the most transformative crisis was the revelation after the 1991 Gulf War that Iraq was perhaps within months of making a nuclear weapon. Despite being subject to safeguards on its nuclear program, Iraq exploited a loophole in its safeguards agreement. The loophole allowed Iraq to place undeclared nuclear facilities next to declared facilities. While the IAEA inspectors could have asked for access to the undeclared facilities, the IAEA Board of Governors has been very resistant to exercising the special inspections provision in the IAEA statute. The nuclear crisis in Iraq led the IAEA Board of Governors to approve the Additional Protocol to states' safeguards agreements.

The Additional Protocol stipulates that the IAEA must examine the entire fuel cycle, including uranium mining, enrichment, fuel production, reprocessing, and the handling and disposal of spent fuel. It also requires IAEA inspectors to determine if there are undeclared nuclear materials or activities in the states under examination. To facilitate this determination, the Additional Protocol allows short-notice inspections. Despite the popular press description of "snap inspections," the notification period is typically 24 hours, except when inspectors are at a site and can then ask for access within two hours to a facility on that site. While the Additional Protocol is not a miracle cure, it is a very valuable tool that would benefit the world if universally applied.

[171]

Although many states recognize the value of the Additional Protocol, even members of the IAEA Board of Governors have resisted bringing Additional Protocols into force. As of November 2008, 23 of the 35 Board members have Additional Protocols in force; an additional six members have signed it; and one more has had an Additional Protocol approved but not yet signed. That means that five Board members have neither signed nor approved or brought into force the Additional Protocol. Of the 188 members of the NPT, less than half – precisely 88 – have brought the Additional Protocol into force. The IAEA Director General has repeatedly called on states to adhere to the Additional Protocol.

Nuclear Suppliers Group Guidelines

As a result of India's "peaceful nuclear explosive" test in 1974, the United States and several other major nuclear supplier states formed the Nuclear Suppliers Group (NSG). The NSG has established and implemented guidelines for commerce in "trigger list technologies." According to the NSG, these technologies include: "(i) nuclear material; (ii) nuclear reactors and equipment therefor; (iii) non-nuclear material for reactors; (iv) plant and equipment for the reprocessing, enrichment and conversion of nuclear material and for fuel fabrication and heavy water production; and (v) technology associated with each of the above items." The updated 1992 guidelines required NSG members to sell these technologies only to states that have applied safeguards to all of their peaceful nuclear facilities. A second list of items includes dual-use technologies, "that is, items that can make a major contribution to an un-safeguarded nuclear fuel cycle or nuclear explosive activity, but which have non-nuclear uses as well, for example in industry." But the underlying tension is that the NSG seeks to encourage nuclear trade. As the US State Department's factsheet on the NSG points out, the guidelines "aim to ensure that nuclear trade for peaceful purposes does not contribute to the proliferation of nuclear weapons or explosive devices while not hindering such trade."[13]

The recently concluded US–India nuclear deal laid bare this inherent conflict. The United States sought to strengthen political and commercial ties with India, the world's largest democracy and a major developing nation. New Delhi made it clear that opening itself up to civilian nuclear trade was the litmus test of a stronger relationship with the United States. Because India had kept most of its civilian nuclear facilities outside of safeguards and had developed nuclear weapons, it was not eligible for trade from NSG members—although Russia had signed an agreement, prior to the tightening of the guidelines in 1992, with India to build two reactors. In exchange for placing more, but not all, of its nuclear reactors under safeguards, India convinced the NSG in September 2008 to make an exception. One of the main reasons New Delhi wanted nuclear trade was to alleviate its shortage of uranium. India has a very limited amount of indigenous uranium. Despite running low on available uranium, India decided to cut back on electric power production rather than reduce fueling weapons-grade plutonium production. Because of the relaxation of the NSG guidelines, India can now receive uranium and nuclear fuel from outside suppliers thus freeing India from the burden of having to sacrifice electricity for plutonium production. The political fallout from the US–India deal has yet to be fully felt. The new US administration may decide to place restrictions on the amount of fuel India can stockpile and may reaffirm that it will cut off civilian nuclear assistance if India conducts nuclear tests. Because India is one of the few states that never signed the NPT, the United States hopes it can manage the potential damage to the NPT regime. The US–India deal by itself, however, will not provide incentives for non-nuclear weapon states to break their NPT pledges.

Nuclear Fuel Service Proposals and Iran

Since the beginning of the nuclear age in the mid-1940s, there have been proposals to place sensitive parts of the nuclear fuel cycle under multinational controls. While no technical safeguards system is foolproof, if multinational corporations or multiple states owned and controlled all enrichment and reprocessing plants, the barrier to proliferation would be raised because nations in the fuel arrangement would be better able to monitor the actions of their partners. Multinational control of the sensitive parts of the fuel cycle could become a new norm that would help discourage, but not in itself prevent, a single state from developing a national enrichment or reprocessing program. This norm would circle back to the beginning of the nuclear age when the 1946 Acheson–Lilienthal Report and the Baruch Plan, the earliest such proposals, called for multilateral control of nuclear power.

Despite the approximately dozen proposals for fuel cycle assurances, these will not be sufficient to convince Iran to refrain from indigenous enrichment. Ultimately, the way to solve the Iranian nuclear problem will be through a political process involving negotiations among Iran, the United States, Russia, the EU, and the states in the Gulf region.

Nuclear Weapon-Free Zones

The NPT encourages the formation of nuclear weapon-free zones. According to the Arms Control Association, such a zone "is a specified region in which countries commit themselves not to manufacture, acquire, test, or possess nuclear weapons."[14] A large part of the globe is now covered in these zones, including Latin America, Southeast Asia, Central Asia, and Mongolia. Africa is also nearing ratification of a nuclear weapon-free zone. Each zone's treaty contains important protocols for the five nuclear weapon states to sign in order for those states to respect the provisions of the zone. The nuclear weapon states are called upon not to make nuclear threats to states within a zone as long as those regional states honor the requirements of the zone. But the nuclear weapon states have often issued declarations allowing them to reserve the right to use nuclear weapons in these zones. Moreover, these

[174]

states have also argued for the capability to transport nuclear weapons through airspace and the seas in these zones.

There have been calls for the creation of a Middle Eastern nuclear weapon-free zone or an even more comprehensive weapon of mass destruction-free zone.[15] The latter would ban chemical and biological weapons as well as nuclear weapons. But continuing concerns about Iran's nuclear intentions have stymied progress. Moreover, Israel's undeclared status has also blocked creation of such a zone. Furthermore, the investigation of Syria's alleged clandestine nuclear reactor construction both points to the need for such a zone and shows another impediment to its formation. The underlying political and security problems need to be addressed prior to verified and trusted implementation of a nuclear or WMD free zone in the region.

Recommendations for Improving the Nonproliferation Regime

- Properly interpret the NPT's "right" to peaceful nuclear technologies to make sure that it is contingent on the responsibility to refrain from obtaining nuclear explosive devices and to maintain rigorous safeguards on civilian nuclear programs.

- Make implementation of the Additional Protocol a requirement for each state's safeguards agreement. This should also be a condition of sale under the Nuclear Suppliers Group's guidelines.

- Ensure that the IAEA fully uses its available authorities, especially the authority under its statute to request special inspections of suspect activities such as those that have recently taken place in Iran and Syria.

- Make sure that any country found in non-compliance with its safeguards agreement by the IAEA's Board of Governors should suspend the suspect activity until the compliance problem is resolved.

- Encourage development of multinational fuel cycle facilities under rigorous multilateral control.

- Urge states to continue to form nuclear weapon-free zones and strengthen existing zones. Nuclear weapon states should honor the complete provisions of these zones.

- Encourage nuclear weapon states to adhere to their responsibility to pursue nuclear disarmament; all states should work seriously toward general and complete disarmament.

- Have nuclear weapon states commit to not threatening non-nuclear weapon states with use of nuclear weapons as long as the non-nuclear weapon state is not aligned with a nuclear weapon state; this is the so-called "negative security assurance."

- Urge nuclear weapon states to further strengthen positive security assurances to allied states, promising to come to their defense in the event of a nuclear threat.

6

International Cooperation and the Development of Nuclear Technology in the Middle East

Jungmin Kang

More than a dozen Middle Eastern countries have officially expressed interest in pursuing nuclear power programs since 2006, including Algeria, Egypt, Jordan, Kuwait, Libya, Morocco, Oman, Qatar, Saudi Arabia, Tunisia, Turkey, the UAE, and Yemen. Indeed, many of these countries have explored nuclear research in the past.[1] Although the development of peaceful nuclear energy programs is an inalienable right of all the parties to the Nuclear Non-Proliferation Treaty (NPT), there is global concern regarding issues of nuclear security and safety in those Middle Eastern countries that are seriously pursuing nuclear power as part of energy diversification strategies to cope with rising demand for electricity and water desalination. Therefore, the development of nuclear security and safety strategies will be a critical requirement for prospective nuclear Middle Eastern states not only in implementing nuclear power projects but also in easing world concerns. In this context, it is important for the states of the Middle East to cooperate closely with international organizations in the areas of nuclear fuel supply assurance, spent fuel management, nuclear non-proliferation, and nuclear safety. In terms of nuclear fuel supply assurance and spent fuel management, a number of the proposals for multilateral approaches to the fuel cycle that prevent the spread of sensitive technologies – or modified versions thereof – should be adopted by the states concerned.

Although the rationale for developing nuclear power in the Middle East seems largely to be to achieve energy diversification, security interests would be a motivation for some countries of the region, for example in response to Iran's nuclear program.[2]

Some aspects of nuclear energy programs – specifically within the nuclear fuel cycle – can provide countries with the know-how and nuclear material necessary for the operation of a nuclear weapons program. Uranium enrichment provides the means to produce highly enriched uranium (HEU), in itself a path to nuclear weapons development, while reprocessing provides the means to separate plutonium from the spent fuel discharged from nuclear reactors—another route to nuclear weapons.

Status of Nuclear Energy Development in the Middle East

Algeria

Algeria has conducted significant research in the realm of nuclear physics since the 1970s. Algeria has two research reactors and a nuclear waste storage facility, but has no fuel fabrication, uranium enrichment, or reprocessing facilities. In November 2006, Algeria announced its plan to pursue a nuclear energy program in order to exploit the country's substantial uranium resources. As of May 2007, Algeria officially reports total "Reasonably Assured Resources" of 26,000 tons of uranium of an initial purity of 0.1 to 0.5 percent, while the Stockholm International Peace Research Institute (SIPRI) estimated Algeria's total uranium resources at about 56,000 tons uranium in 2005.

Egypt

As of April 2007, Egypt had more than 850 highly qualified scientists in various fields of nuclear science and engineering, supported by about 650 technical staff who work on nuclear research and development (R&D) projects, according to the country's Atomic Energy Authority (AEA).

Egypt has not reported its uranium resources, but has a light water research reactor – ETRR-1 with a maximum thermal power of 2 MW – and a Multi-Purpose Reactor (MPR) that produces 22 MW of steady thermal power. In September 2006, Egypt announced that the country intended to resume its nuclear power program and that the first nuclear power station would be constructed within a decade.

Jordan

Although some Jordanian universities offer physics programs with courses in nuclear physics, the domestic educational program in nuclear physics seems to lack the means to train a workforce in nuclear energy development and management. In May 2007, it was estimated that Jordan had 80,000 tons of uranium, and that its phosphate reserves contain an additional 100,000 tons of uranium. Jordan has no nuclear reactors or nuclear fuel cycle facilities. In February 2007, Jordan announced plans to build its first nuclear power plant by 2015 for electricity generation and desalination. If Jordan proceeds with its plan to deploy a nuclear power plant, it will require adequate nuclear expertise from abroad.

Kuwait

Kuwait has no central nuclear research body or regulatory agency; neither does it have any identified uranium deposits, nuclear reactors, or nuclear fuel cycle facilities. In 2006, as a member of the Gulf Cooperation Council (GCC), Kuwait expressed an interest in developing a joint plan for nuclear energy generation in the near future.

Libya

Established in 1983, the Tajoura Nuclear Research Centre (TNRC) in Tripoli is the main nuclear research and training facility in Libya, tasked with conducting fundamental and applied research, advanced studies and training in peaceful applications of nuclear energy. There are no uranium resources in Libya, which has one Soviet-designed, TNRC-operated research reactor—IRT-1 with 10 MW of thermal power capability. In 1968, Libya signed the NPT, which was ratified under Muammar Gaddafi in 1975. Libya acceded to a safeguards agreement with the IAEA in 1979. Throughout 2003, however, Libya had conducted clandestine nuclear activities, having first expressed interest in a nuclear weapons program in the late 1960s. Since Libya rescinded its WMD programs in 2003, it has agreed to transfer to the United States sensitive design information, nuclear weapons-related documents, and most of its previously undeclared enrichment equipment. At present, Libya's activities seems to be limited to research for the future application of nuclear techniques in the areas of desalination, agriculture, and nuclear medicine in the short-term, with the longer-term purpose of building a nuclear power plant for desalination and electricity generation.

Morocco

Although Morocco has established research centers and nuclear engineering associations, it is still in the early stages with regard to the development of its nuclear energy program. Morocco has no uranium resources, but is one of the leading exporters of phosphates—from which uranium can be extracted. As of May 2006, Morocco was in the process of commissioning its first nuclear research reactor – a TRIGA-Mark II with 2 MW of thermal power – designed and manufactured by the US company General Atomics. In March 2007, the Moroccan Minister of Energy and Mines, Mohammed Boutaleb, stated that Morocco encourages the use of nuclear technology for peaceful purposes.

Oman

Research on nuclear medicine has been conducted by Omani scientists, but no other nuclear research has been conducted. Oman has no uranium resources, nuclear reactors or nuclear fuel cycle facilities. In 2006, as a member of the GCC, Oman expressed an interest in developing a joint plan for nuclear energy generation in the near future. With the rationale of meeting increasing demand for electricity, water desalination and agriculture, Oman participated in commissioning a feasibility study in December 2006 on employing nuclear technology for peaceful purposes in accordance with international regulations and standards.

Qatar

Aside from some work on environmental radiation, nuclear research has not been overtly conducted in Qatar. Qatar has no uranium resources, nuclear reactors or nuclear fuel cycle facilities. Like Oman, as a member of the GCC, Qatar supported the joint plan for nuclear energy generation and participated in the commissioning of the related feasibility study.

Saudi Arabia

Saudi Arabia has collaborated with China, Germany, Iraq, Jordan, Pakistan, Switzerland and the United States in the areas of nuclear research, reactor operation, isotope analysis, and radiation protection. Some nuclear energy-related research has been conducted by the Department of Nuclear Engineering of King Abdulaziz University in Jeddah since late 1970s. Saudi Arabia's Atomic Energy Research Institute (AERI) was established in 1988 within the King Abdulaziz City for Science and Technology (KASCT) with specific responsibilities, including: drafting and supervising the implementation of a national atomic energy plan; conducting research in the field of nuclear sciences; and training and developing the national specialists in the field of nuclear research. Saudi Arabia has no uranium resources, nuclear reactors or nuclear fuel cycle facilities. Owing to extremely limited fresh water resources, Saudi scientists have studied the viability of introducing nuclear power for water desalination and electricity production since 1978. Saudi Arabia is currently collaborating with other GCC countries in pursuit of an Arab joint nuclear program for peaceful purposes, as mentioned above.

Tunisia

The National Center for Nuclear Science and Technology (CNSTN), established at the technology center at Sidi Thabet in 1993, is the main nuclear research and training institution in Tunisia. The CNSTN conducts nuclear research, but not specifically energy applications. Tunisia has no uranium resources, but is one of the leading exporters of phosphates. As of 2000, Tunisia had one nuclear fuel cycle facility under study – intended for use in recovering uranium from phosphates – with a design production capacity of 120 tons of uranium per year. As of 1999, Tunisia had planned the construction of a light water TRIGA research reactor with a thermal power of 2 MW. Tunisia intends to develop nuclear research and technology for desalination and electricity production purposes.

Turkey

Established in 1961, the Çekmece Nuclear Research and Training Center (CNRTC) is the first of its kind in Turkey. Several Turkish university programs also offer courses and degrees in nuclear physics and nuclear energy-related fields. Turkey's nuclear educational and research programs are both detailed and comprehensive, and contribute to a knowledge base in the country for creating the technical workforce needed to develop and manage a nuclear energy program. The Turkish Atomic Energy Authority (TAEK) is the country's main regulatory organization governing nuclear energy programs and safety. Turkey is said to have total "reasonably assured resources" of 9,129 tons of uranium of an initial purity of 0.05 to 0.1 percent and about 380,000 tons of thorium reserves. Turkey has two fuel facilities in operation, the Conversion Nuclear Fuel Pilot Plant, with a production capacity of 0.1 tons of uranium per year, and a Pellet Production Nuclear Fuel Pilot Plant. Turkey has one research reactor in operation—a TRIGA mark II with 250 kWh of thermal power. In the past, Turkey has operated two pool-type research reactors: a TR-1 that generated 1 MW of thermal power and was shut down in 1977; and a TR-2 that generated 5

[182]

MW of thermal power and was shut down in 1995. In 2007, Turkey announced plans to begin construction of a nuclear power plant.

UAE

At present, the UAE has no central institution for nuclear research or radiation management, although some courses are offered by the university system. The UAE has no uranium resources, nuclear reactors or nuclear fuel cycle facilities. Although the UAE's interest in nuclear energy is fairly recent, it has shown serious interest in deploying nuclear power plants. In addition to the joint GCC plan, the UAE is moving forward with its own plans for nuclear power and is working with the IAEA on a feasibility study covering nuclear power and desalination. In August 2008, the Emirates Nuclear Energy Corporation (ENEC) invited global firms to bid for a contract to manage its nuclear power program. In January 2008, the UAE and France signed a nuclear cooperation agreement, and in December 2008, the UAE signed an agreement formalizing cooperation with the United States on the peaceful use of nuclear energy.

Yemen

Very little nuclear-related activity has been conducted by Yemeni institutions. In 1999, the National Atomic Energy Commission (NATEC) was established for the purpose of building radiation protection infrastructure for peaceful applications of nuclear energy. NATEC has developed a "National Nuclear Security Action Plan" in collaboration with the IAEA with aims to be a leader in the Middle East in the field of nuclear security. Yemen has no uranium resources, nuclear reactors or nuclear fuel cycle facilities. In December 2006, the President's Science and Technology Advisor, Dr. Moustafa Bahran, announced that the country intended to begin working on producing nuclear energy in 2007. In June 2007, Yemen's President Ali Abdullah Saleh promoted the development of a nuclear energy in his election campaign.

[183]

Developing Nuclear Power Takes Time

It would take a long time for most Middle Eastern countries to develop adequate national infrastructure and to educate and train workforces to support nuclear power programs.

Besides, the deployment of a commercial nuclear power plant (NPP) itself is a long-term national project and takes time: approximately 6–7 years in construction of a NPP; 40–60 years in operation of a NPP; and a few decades in decommissioning a NPP after reactor shutdown.

As Middle Eastern countries examine nuclear power, effective guidance and oversight by the international community, including the IAEA and the developed countries, will be important.

Nuclear Fuel Supply Assurance

As countries consider the future of their nuclear power programs, the issues of nuclear fuel supply as well as spent fuel disposal will play very important roles in the choices they make. The provision of reliable, uninterrupted supplies of fresh fuel for nuclear power plants – without undermining the nuclear non-proliferation regime – is a key factor for the success of nuclear power projects. For nuclear fuel supply assurance, a multilateral approach to the nuclear fuel cycle is an important concept in dissuading countries from developing national enrichment capabilities.

Twelve Proposals on Nuclear Fuel Supply Assurance Summarized by the IAEA

Nuclear fuel supply assurance is a concept that has existed since the dawn of the nuclear age. Tariq Rauf and Zoryana Vovchok of the IAEA recently summarized twelve proposals for a multilateral approach to the nuclear fuel cycle which help strengthen the nuclear non-proliferation regime.[3] The twelve proposals are briefly outlined below:[4]

Proposal # 1

In September 2005, the US government committed up to seventeen metric tons of highly enriched uranium (HEU) from excess stocks to be down-blended to low enriched uranium (LEU) to support assurances of reliable nuclear fuel supplies for states that forgo enrichment and reprocessing. The US envisions a central role for the IAEA to act as an intermediary for the purpose of securing the supply of nuclear services and materials. The aim of the US proposal is that supplier states and the IAEA establish a reliable mechanism to resolve problems of disruption in supply, should they occur.[5]

Proposal # 2

In February 2006, Vladimir Putin, then President of the Russian Federation, outlined potential international centers for the provision of nuclear fuel cycle services – including enrichment – on a non-discriminatory basis and under IAEA safeguards. The purpose of the Russian proposal is to create the prototype of a global infrastructure that will give all interested countries equal access to nuclear energy, while stressing reliable compliance with the requirements of the non-proliferation regime. Russia is prepared to establish an international center on its territory. Any country could pay for a membership stake in the international centers which would guarantee it a supply of enriched uranium at world prices.[6]

Proposal # 3

In February 2006, the US Department of Energy (DOE) announced the formation of the Global Nuclear Energy Partnership (GNEP) initiative, aimed at fostering international collaboration to promote the safe use of nuclear power, reduce the threat of nuclear proliferation and decrease the volume and radiotoxicity of nuclear waste. Nuclear fuel supply assurance is one of the elements that the GNEP aims to bolster. The GNEP proposes countries that agree to forgo enrichment and reprocessing should have reliable access to fuel for civil nuclear power reactors at reasonable cost. Under the GNEP, a consortium of countries with advanced nuclear technologies would provide nuclear fuel on a lease basis to countries that refrain from operating their own enrichment and reprocessing facilities, and would take back the spent fuel discharged. Based on the GNEP, the United States seeks to establish and sustain a "cradle-to-grave" fuel service.[7]

Proposal # 4

In May 2006, the World Nuclear Association (WNA) published a report titled: "Ensuring Security of Supply in the International Nuclear Fuel Cycle" that summarized a three-level mechanism to assure enrichment services:

- Level I: Basic supply security provided by the existing world market.

- Level II: Collective guarantees by enrichers supported by governmental and IAEA commitments.

- Level III: Government stocks of enriched uranium product.

The WNA proposes a collective guarantee of supply from primary enrichers with the expectation of equal fulfillment of disrupted supplies by all other participants. The WNA proposal formed the basis of a Reliable Access to Nuclear Fuel (RANF) proposal.[8]

Proposal # 5

In June 2006, the six enrichment service supplier states – France, Germany, the Netherlands, Russia, the UK and the US – proposed the concept of a multilateral mechanism for the Reliable Access to Nuclear Fuel (RANF) to guarantee that if one supplier failed to meet its contract obligations, other suppliers would ensure supply. The RANF proposal is based on the WNA proposal, and is composed of two levels: a basic assurances level and a reserves level. At the basic assurances level, suppliers would agree to substitute for each other to cover possible supply interruptions. The customer countries are encouraged to obtain suppliers on the international market rather than pursuing sensitive fuel cycle activities. Acceptance of customers would depend on IAEA assurances that all NPT obligations had been met. At the reserve level, supplier countries could provide stockpiles of LEU that would be made available if the basic assurances were to fail.

Proposal # 6

In September 2006, the Japanese government proposed an "IAEA Standby Arrangements System for the Assurance of Nuclear Fuel Supply" under the auspices of the IAEA, which incorporates both an information system – based on data provided voluntarily by suppliers – to contribute to the prevention of interruptions in nuclear fuel supplies, and the backup feature for supply assurance proposed in the RANF proposal. The Japanese proposal envisages that supplier countries would notify the IAEA by registering their nuclear fuel supply capacity in terms of current stock and capacity in the following areas: uranium ore supply capacity, uranium reserve supply capacity (including recovered uranium), uranium conversion capacity, uranium enrichment capacity, and fuel fabrication capacity. The proposal sees the IAEA concluding bilateral "standby arrangements" with participating countries, handling administration, and acting as an intermediary in cases of supply disruption.[9]

Proposal # 7

In September 2006, former US Senator Sam Nunn, speaking on behalf of the Nuclear Threat Initiative (NTI), proposed the establishment of a Nuclear Fuel Bank to provide fuel supply guarantees to consumer countries upon compliance with NPT obligations. The NTI committed US$ 50 million to the IAEA to help create the Nuclear Fuel Bank with a condition that one or more Member States contribute an additional US$ 100 million in funding or its equivalent value in LEU. The proposed concept of the NTI is to purchase a supply of the LEU stockpile of the

Nuclear Fuel Bank, which could be handed out to qualified consumer countries if supply arrangements were disrupted. The US$ 150 million covered by the NTI proposal could be utilized to purchase 60–100 metric tons (MT) of LEU, depending on future prices of uranium oxide and enrichment services.[10]

In August 2008, the US government committed US$ 50 million to the IAEA Nuclear Fuel Bank; Norway committed US$ 5 million in February 2008; the United Arab Emirates (UAE) committed US$ 10 million in August 2008; and the European Union committed up to 25 million Euros (approximately US$ 32 million) in December 2008. As of the end of 2008, just US$ 3 million must be raised in order to meet the NTI condition.

Proposal # 8

In September 2006, the UK government proposed a concept of so-called "enrichment bonds"—a voluntary scheme for reliable access to nuclear fuel. The aim of the UK proposal is to develop the RANF concept in practice. An enrichment bond would involve an agreement between supplier countries' governments, the recipient country, and the IAEA. According to the enrichment bond agreement, the supplier countries would guarantee that their enrichment corporation will not withhold supplies to consumer countries for political reasons. The IAEA would certify the customer countries with regard to meeting NPT and safeguards obligations. Germany and the Netherlands are cooperating with the UK on this scheme.[11]

Proposal # 9

In January 2007, Russia adopted the necessary legislation to establish an International Uranium Enrichment Center (IUEC) on the site of the Angarsk Electrolysis Chemical Complex (AECC) to provide guaranteed access to uranium enrichment capabilities to the Center's participating countries on the condition that they do not develop uranium enrichment capabilities on their territories. According to the proposal, the IUEC would be established as an open, joint-stock company, whose participants under Russian legislation would be entitled to receive dividends from the results of its activity. Russia committed not to transfer to IUEC participants uranium enrichment technology or information that could be considered of importance to national secrecy. On May 10, 2007, Kazakhstan signed-up to support the Russian initiative to establish the IUEC on Russian territory. Ukraine also signed the agreement on June 6, 2007.[12]

Proposal # 10

In April 2007, the German foreign minister H.W. Steinmeier proposed a concept of the multilateralization of the nuclear fuel cycle, including the creation of a multilateral uranium enrichment center by a group of interested countries under the control of the IAEA. This center would provide a guaranteed source of nuclear fuel to consumer countries whose nuclear fuel purchasing contracts were disrupted. The German proposal requires the establishment of a new international enrichment venture that operates an enrichment plant on a commercial basis as a new supplier in the market, located in an area ceded by a country to the enrichment consortium. The core elements of the German proposal are as follows:[13]

- A host country would have to cede administration and sovereign rights over a certain area of its territory to be defined by the IAEA.

- Participating countries would be able to establish one or more commercial enrichment plants, on the basis of agreements between the IAEA and those countries.

- The IAEA must draw up a list of criteria which would guarantee the release of deliveries of LEU from the area.

Proposal # 11

In April 2007, Austria proposed the concept of a multilateralization of the nuclear fuel cycle via a Nuclear Fuel Bank, based on a twin-track multilateral mechanism. The first track would optimize international transparency, going beyond current IAEA safeguards obligations. According to the Austrian proposal, all participating countries should declare to both the IAEA and each other all their existing nuclear programs and future development plans, including activities involving nuclear material, equipment and related technologies. The increased transparency resulting from their nuclear declarations would provide greater clarity and enhance overall confidence. The second track would ensure participating countries' equal access to and control of sensitive technologies, enrichment and reprocessing. This could be achieved by placing all nuclear fuel transactions – including for enrichment and reprocessing – under the auspices of the Nuclear Fuel Bank. Once in place, the international Nuclear Fuel Bank would ensure that the nuclear fuel cycle is operated and controlled by all participating countries so that they would no longer need to pursue national enrichment and reprocessing programs.[14]

Proposal # 12

In June 2007, the EU submitted a non-paper to the IAEA Secretariat and the 2007 NPT Preparatory Committee meeting, proposing criteria for the assessment of a multilateral mechanism for a reliable of fuel supply system providing access guarantees and relevant multilateral provisions without disturbing market conditions. The EU proposal offered a list of criteria rather than proposing a multilateral mechanism for fuel supply assurance. The criteria included: proliferation resistance, assurance of supply, consistence with equal rights and obligations, and market neutrality. It also called for technical issues of safety and security to be considered.[15]

Common Aim and Requirements of all the Proposals

The summarized proposals above all aim to achieve the objectives of both nuclear weapons non-proliferation efforts and secured access to nuclear fuel for all interested countries. At the same time, most of the proposals require potential consumer countries to give up indigenous enrichment capabilities and be placed under strict international monitoring regimes, including IAEA safeguards, in return for being granted guaranteed access to nuclear fuel.

It is worth noting that most of the twelve proposals come from potential supplier countries, and none from the countries that presumably seek assurances. Supplier countries might have their own interests in driving for the establishment of multilateral nuclear fuel supply assurance mechanisms. It would be only natural for consumer countries to demand new nuclear fuel supply assurance schemes that meet their own interests.

Middle East International Enrichment Consortium

Assuming Middle Eastern states refrain from indigenous enrichment, a regional fuel supply assurance scheme with the participation of one or more suppliers might be a practical alternative for consumer countries seeking supply assurance

In that context, a proposal by the Saudi Foreign Minister Prince Saud al-Faisal on October 28, 2007 regarding a "ME Enrichment International Consortium" deserves attention. The proposal, or so-called "GCC Initiative," envisages the establishment of a joint enrichment center, based in a neutral country, to supply nuclear fuel to nuclear power plants in the region. The GCC Initiative is designed to secure the supply of nuclear fuel for member states of the Middle East Consortium without access to enrichment technology, thus placing enrichment activities under a strict international monitoring regime.

The GCC Initiative would provide fuel for nuclear power plants deployed in the region – including in Iran – therefore centralizing enrichment activities and placing them under IAEA safeguards systems to prevent the militarization of civilian programs and the consolidation of a nuclear arms race in the region.

On November 3, 2007, Iran, however, officially rejected the proposal of the GCC Initiative, arguing that it had no intention of stopping its

[191]

enrichment activities on Iranian soil or abandoning its rights. Previously, in August 2005, the UK, France and Germany (the so-called "EU-3") offered Iran guaranteed nuclear fuel supply assurances in return for giving up indigenous enrichment activities, but the offer was rejected.

In February 2007, Russia proposed uranium enrichment for Iranian nuclear power plants on Russian territory, aiming to curb the spread of enrichment technology by preventing new countries from accessing the technology. Iran rejected the Russian proposal on the grounds of financial cost and loss of control over its national enrichment program.[16]

Stockpile of Fresh Fuel

Besides multilateral approaches to the nuclear fuel cycle, national fuel stockpiles could be an alternative for the purposes of fuel supply assurance. According to Frank von Hippel, a professor of public and international affairs at Princeton University, the cost of the stockpile strategy is comparable to the extra cost of building a small national enrichment plant instead of buying enrichment services on the international market. Moreover, a national stockpile of fresh fuel is far superior to a national enrichment plant, from a non-proliferation perspective.[17]

Spent Fuel Management

Current Spent Fuel Strategies

The management of spent fuel discharged from nuclear reactors is one of the greatest challenges facing the nuclear power industry. Two spent fuel strategies have been adopted so far. The first involves the reprocessing of spent fuel, whilst the second relies on interim storage of spent fuel with a view to either direct disposal or reprocessing. The reprocessing strategy separates plutonium, which is either stored for possible future use in fast reactors or recycled as mixed oxide fuel (MOX) in light water reactors (LWRs).

Reprocessing

Countries that have adopted the reprocessing strategy include France, Japan, India, Russia and the UK. Countries that have sent their spent fuel to France or the UK for reprocessing but which have now decided to adopt the strategy of interim storage are Belgium, Germany, the Netherlands, and Switzerland.

Reprocessing is uneconomic. As a 2003 Massachusetts Institute of Technology (MIT) study revealed, the option of reprocessing spent fuel and recycling the separated plutonium in MOX fuel cost roughly four times that of the non-MOX option (based on estimated costs in the United states).[18]

Frank von Hippel of Princeton University similarly concluded that reprocessing spent fuel and recycling the separated plutonium in MOX fuel just once would double the cost of the back-end of the fuel cycle, based on a report to the Prime Minister of France in 2000.[19] Furthermore, he estimated that the cost of reprocessing would be ten times more than that of the on-site storage of the spent fuel (considering a fast reactor recycle).[20]

Besides the lack of economic merit, reprocessing carries with it a proliferation risk. Separated plutonium can be transported easily and is much more vulnerable than spent LWR fuel that is self-protecting for more than a century.[21]

Environmental benefits derived from reprocessing are not significant, since spent MOX fuel is not reprocessed and high-level waste (HLW) that contains all the fission products and transuranics other than plutonium must be placed at geologic disposal sites.

Furthermore, reprocessing does not guarantee nuclear fuel security. Recycled plutonium (once in MOX fuel) replaces less than twenty percent of the uranium-oxide fresh fuel.

On-Site Dry-Cask Storage

Compared with reprocessing, on-site dry-cask storage of spent fuel is more economical. A 2007 Idaho National Laboratory report estimated a reference cost for dry storage of spent fuel at US$ 120 per kilogram of heavy metal (kgHM), with a range of US$ 100–300 kgHM.[22] Harold Feiveson, a senior research scientist at Princeton University, similarly estimated approximately US$ 100–200 kgHM for the costs of the dry-cask storage.[23]

On-site dry-cask storage of spent fuel is proliferation resistant. For example, a pressurized water reactor (PWR) spent fuel assembly requires a twenty-ton container to transport it, followed by remote reprocessing behind thick walls to recover the plutonium.[24]

In terms of safety, on-site dry-cask storage is safer than the pool storage option as consequences of accidents and attacks on dry-cask-stored fuel would be less than from attacks on reactors or storage pools by several orders of magnitude.[25]

On-site dry-cask storage of spent fuel is durable and requires little space. The dry-cask can sustain more than sixty years of operation and only requires 0.18 km^2 for 40,000 metric tons (MT) of heavy metal (HM) of spent fuel.[26]

Proposals on Regional and International Spent Fuel Storage/Disposal

In principal, countries have to dispose of their spent fuel domestically. However, owing to difficulties in siting geological repositories, interest in regional/international spent fuel storage/disposal increased significantly in the 1970s and early 1980s. In 1977, the IAEA reported that regional fuel cycle centers were feasible and would offer considerable economic and non-proliferation advantages. In 1982, the IAEA concluded its International Fuel Cycle Evaluation (INFCE) project in which expert groups suggested the establishment of international plutonium storage and international spent fuel management centers.[27]

In the mid-1990s, the concept of the International Monitored Retrievable Storage System (IMRSS) was proposed by Wolf Hafele in Germany. The IMRSS envisioned international sites where spent fuel – and possibly excess separated plutonium – could be stored and monitored for an extended period but could be retrieved at any time for peaceful use or ultimate disposal.[28]

In the mid- to late-1990s, several proposals for nuclear energy cooperation in the Asia-Pacific region were developed, including those noting that spent fuel management was an important issue to be addressed regionally. Tatsujiro Suzuki, a senior research scientist in the Central Research Institute of the Electric Power Industry in Japan, summarized a comparison of various proposals for regional nuclear cooperation offered during the period, and concluded that there were potential areas of cooperation where common needs and interests exist, such as nuclear safety and management of the back-end of the fuel cycle.[29]

In the 1990s, a commercial group called Pangea was seeking an international geologic repository for both spent fuel and radioactive wastes. Envisioning a facility for disposing of 75,000 MTHM of spent fuel/HLW, Pangea initially selected Australia for its proposed repository, but is now seeking other sites around the world after confronting political opposition there.[30]

During the late 1990s and early 2000s, two proposals involving depository sites in Russia were presented. One is based on a call from the Non-Proliferation Trust (NPT) for the establishment of a dry cask storage facility in Russia that would accept 10,000 MTHM of spent fuel from abroad. The other is a concept offered by the Russian Ministry for Atomic Energy (MINATOM), which suggested a plan for an international spent fuel service involving offering temporary storage with later return of the spent fuel, or reprocessing of spent fuel without return of plutonium or radioactive wastes to customer countries.[31]

In 2003, Mohamed ElBaradei, Director General of the IAEA, suggested multinational approaches to the management and disposal of spent fuel and radioactive waste. In 2005 – commissioned at Dr. ElBaradei's suggestion in 2003 – the IAEA published a report on "Multilateral Approaches to the Nuclear Fuel Cycle" in which the IAEA concluded that such approaches are required and are worth pursuing, on both economic and security grounds.[32]

In January 2006, Russian President Vladimir Putin announced a Global Nuclear Power Infrastructure (GNPI) initiative to provide the benefits of nuclear energy to all interested countries in strict compliance with non-proliferation requirements, through a network of international nuclear fuel cycle centers (INFCCs) which reprocess and dispose of waste under IAEA safeguards.[33]

In February 2006, the Bush government proposed the Global Nuclear Energy Partnership (GNEP) to provide fuel supply guarantees and execute take-back arrangements. The goal of the GNEP is to establish and sustain "cradle to grave" fuel services or leasing arrangements over time and at a scale commensurate with the anticipated expansion of nuclear energy by helping to solve the nuclear waste challenge.[34]

Benefits

Regional or international spent fuel storage/disposal could provide the following benefits to a host country and to participating countries:[35]

- An alternative for countries that seem have no positive prospects for domestic siting of interim storage or disposal facilities, mainly owing to political resistance from local communities.

- Achieving economies of scale for geologic repositories. A large-capacity repository could offer economic benefits for the host and participating countries, allowing partners to achieve substantial economies of scale by sharing fixed capital costs, operating costs, and financial liabilities.

- Creating a new revenue source for a host country. The host country that accommodates a regional/international repository could receive significant income through payments by participating countries.

- Reducing proliferation risk by avoiding unnecessary accumulation of separated plutonium. Some countries want to develop reprocessing capacity largely because they lack the means to dispose of their spent fuel. A regional/international repository would offer an alternative place for spent fuel and thus avoid unnecessary reprocessing.

Challenges

To implement regional or international spent fuel storage/disposal strategies, the following obstacles must be overcome:[36]

- ETHICAL ISSUES IN THE REGION. There is a public perception that the countries which benefit from nuclear power generation should bear the burden of storing and disposing of their radioactive wastes. This argument raises ethical issues that oppose the concept of a regional or international repository. To obtain public and political support, an arrangement for such a repository should be based on a fair and equitable sharing of benefits between a repository host and other participating countries.

- COMPLICATING NATIONAL SPENT FUEL/HLW MANAGEMENT POLICIES. A regional or international repository could detract from national spent fuel and radioactive waste management programs. If such a facility is not forthcoming, or is delayed in implementation, national needs for spent fuel management could become critical without national solutions being in place.

- INCREASING TRANSPORTATION REQUIREMENTS IN THE REGION. A regional or international repository will involve frequent transportation of spent fuel/radioactive waste from participating countries to a host country, and increasing concern over nuclear accidents during the transportation that may lead to radioactive release into the environment. Proliferation risks owing to the diversion of materials during transport are also a concern.

Nuclear Non-Proliferation

The world's concerns regarding Middle East countries moving toward nuclear power stem from a suspicion that some states may try to use their nuclear facilities for the purposes of weaponization in the future, considering the geopolitical instability of the region.

To defuse such concerns, Middle Eastern countries must strictly adhere to the global nuclear non-proliferation regime; specifically, they must accept both the full range of IAEA safeguards and the Additional Protocol to ensure their peaceful use of nuclear energy. As shown below in Table 6.1, only half of the states of the Middle East have signed the Additional Protocol, despite having accepted IAEA safeguards.

Table 6.1
Safeguards Agreements and the Additional Protocol
in the Middle East (January 6, 2009)[37]

Country	Full Safeguards	Additional Protocols
Algeria	In force on Jan. 7, 1997	Approved on Sept. 14, 2004
Egypt	In force on Jun. 30, 1982	Not signed
Jordan	In force on Feb. 21, 1978	In force on Jul. 28, 1998
Kuwait	In force on Mar. 7, 2002	In force on Jun. 2, 2003
Libya	In force on Oct. 4, 1979	In force on Aug. 11, 2006
Morocco	In force on Feb. 18, 1975	Signed on Sept. 22, 2004
Oman	In force on Sept. 5, 2006	Not signed
Qatar	Approved on Sept. 24, 2008	Not signed
Saudi Arabia	Signed on Jun. 16, 2005	Not signed
Tunisia	In force on Mar. 13, 1990	Signed on May 24, 2005
Turkey	In force on Sept. 1, 1981	In force on Jul. 17, 2001
UAE	In force on Oct. 9, 2003	Not signed
Yemen	In force on Aug. 14, 2002	Not signed

The full IAEA safeguards (INFCIRC/153) are effective in allowing inspections of declared nuclear material and facilities, but are not adequate in identifying countries conducting secret nuclear programs, while the Additional Protocol (INFCIRC/540) allows effective inspection of nuclear fuel cycle-related research as well as development activities not involving nuclear material, even if undeclared.[38]

As a result of the revelation of secret nuclear programs in Iraq and North Korea, and the experience the IAEA gained in South Africa, the Additional Protocol was created in 1997 in order to strengthen the effectiveness and efficiency of the IAEA safeguards system by complementing routine environmental sampling with broad access, short-notice inspections and remote monitoring, thus ensuring the absence of undeclared nuclear materials or activities.

As David Albright and Andrea Scheel of the Institute of Science and International Security argue, the Additional Protocol needs to be in force in all Middle Eastern countries since Iraq, Iran, Syria, Algeria, and Libya have all evaded detection of their clandestine nuclear programs despite permitting conventional inspections by the IAEA.[39]

Besides submitting to IAEA safeguards, the Middle East countries also need to strengthen their export control systems to control nuclear-related dual-use items during import and export, and to enter into the international export control regimes, including UN Security Council Resolution 1540 on the non-proliferation of weapons of mass destruction.

Adopted by the United Nations Security Council on April 28, 2004, UNSCR 1540 was aimed largely at preventing the spread of weapons of mass destruction (WMD) to non-state actors, and specifically to terrorist groups. UNSCR 1540 also aims to ensure that states have in place measures to prevent non-state actors engaging in proliferation-related activities.[40]

Nuclear Safety

Environmental impacts resulting from leakage of radioactive materials after serious accidents in reactors or in a spent fuel storage pools are not confined to the country or region involved. The April 1986 Chernobyl accident caused more than 100,000 residents from 187 settlements to be permanently evacuated owing to contamination by caesium-137 (Cs-137) of 2 mega curies (MCi) that was released from the core which contained about 80 MT of spent fuel.[41]

Another serious nuclear accident occurred at the Three Mile Island Unit 2 in 1979. Although radiation releases from the accident had no perceptible effect on cancer incidence in residents near the plant, the Three Mile Island accident involved a partial core meltdown. It was the most significant accident in the history of the American commercial nuclear power industry, resulting in the release of an estimated 43,000 curies of radioactive krypton (but less than 20 curies of hazardous iodine-131).[42]

Since Three Mile Island, reactor safety technology has improved greatly, so likely core damage frequency is 1:100,000 (reactor-years), based on claims of advanced light water reactor (LWR) designers. Considering the fact that terrorists might find nuclear facilities attractive targets, however, further nuclear plant safety measures will be needed to prevent potentially catastrophic core damage.[43]

Despite efforts to bolster plant safety, the world's nuclear power community devoted much less attention to the safety of spent-fuel pools, which are vulnerable to sabotage or terrorist attack. Although no such events have occurred thus far, we cannot exclude a possibility of a loss of coolant or a fire in the spent-fuel pools, the consequences of which would affect a large area.

For example, a typical US reactor spent-fuel pool today contains about 400 MT of spent fuel comprising around 35 MCi of Cs-137. If 10–100 percent of the Cs-137 in a spent-fuel pool, (i.e. 3.5–35 MCi) were released by a fire into the atmosphere, 37,000–150,000 km^2 would be contaminated above 15 Ci/km^2, 6,000–50,000 km^2 would be contaminated to greater than 100 Ci/km^2 and 180–6,000 km^2 to a level greater than 1,000 Ci/km^2. This means that even for the 3.5 MCi releases, the areas calculated as contaminated above 100 Ci/km^2 are 3–8 times larger than the area contaminated to this level by the 2 MCi release from the Chernobyl accident. To reduce both the likelihood and consequences of a spent-fuel pool fire, it is proposed that, after a decade-long transition period, all spent fuel be transferred from wet to dry storage within five years of discharge.[44]

Conclusions

Although developing peaceful nuclear energy programs is the inalienable right of all parties to the NPT, global concerns over nuclear safety and security issues surround the issue of Middle Eastern countries pursuing nuclear power for energy diversification and water desalination.

For their nuclear power projects to succeed, and to ease international concerns, Middle Eastern countries must strongly commit to abide by the global nuclear non-proliferation regime; specifically, they should accept both the full IAEA safeguards and the Additional Protocol to ensure their peaceful use of nuclear energy and high standards of nuclear safety.

In this context, it is important for these states to cooperate closely with international societies in the areas of nuclear fuel supply assurance, spent fuel management, nuclear non-proliferation, and nuclear safety.

CASE STUDIES OF CIVILIAN NUCLEAR POWER PROGRAMS

7

Peaceful Use and Acceptance of Nuclear Energy in Japan: A Personal Perspective

Junko Ogawa

Whenever I speak overseas, I am always asked "Why does Japan promote nuclear power when it was a victim of the atomic bomb?" Considering the terrible experiences that nuclear weapons forced upon the Japanese, it would seem natural that Japan should attempt to survive without relying on nuclear power. However, Japan is a country with extremely few natural resources. Energy continues to occupy a position of great strategic importance for countries today, just as it has done in the past, and is often the cause of international disputes. In other words, the stable supply of energy plays a significant role in maintaining peace. This is why Japan decided to put its disastrous experiences behind her and start to develop nuclear technology for peaceful uses. Most Japanese people would not hesitate to show their disapproval of nuclear technology for military use. It is precisely because of the experiences that Japan has suffered at the hands of nuclear weapons that the country finds itself in a unique position to be able to champion the peaceful use of nuclear power. Ultimately, Japan's part is that of a role model, offering an ideal towards which the rest of the world should aim.

In this paper, as well as looking at the progress Japan has made in its peaceful uses of nuclear power and explaining the current situation of its development and use, I would also like to describe my ideas and experiences pertaining to my specialist field, namely, activities to promote the understanding of nuclear power amongst the general public.

[205]

Japan's Model of Peaceful Use of Nuclear Energy

The Atomic Energy Basic Law and Early Nuclear Development

From the end of World War II until the signing of the San Francisco Peace Treaty in 1952 that brought it back into the international community, Japan was prohibited from developing nuclear technology. In 1953, when research into atomic energy was finally starting to be discussed among Japanese scientists, US President Eisenhower delivered his "Atoms for Peace" speech at the UN General Assembly, in which he made proposals for the peaceful application of nuclear technology. By this time, preparations for atomic research were already underway in Japan. 1955 saw the establishment of the Atomic Energy Basic Law, which stipulated that nuclear technology be limited to peaceful use and be guided by the principles of "democracy," "independence," and "disclosure." In the same year, the Atomic Energy Commission and the General Administrative Agency of the Cabinet's Atomic Energy Bureau were established to implement atomic energy policies. Today, atomic energy is governed by: the Cabinet Office's Atomic Energy Commission and Nuclear Safety Commission; the Ministry of Economy, Trade and Industry's Agency for Natural Resources and Energy; and the Research and Development Bureau of the Ministry of Education, Culture, Sports, Science and Technology.

In 1956, two nuclear research institutes were established: the Japan Atomic Energy Research Institute and the Atomic Fuel Corporation. These two institutes were later integrated to become the Japan Atomic Energy Agency (JAEA), which is the largest atomic energy research institute in the country today, with a workforce of 4,200.

In 1957, the Japan Atomic Energy Research Institute set up JRR-1, the first small research reactor in the town of Tokaimura, Ibaraki Prefecture, the birthplace of atomic energy in Japan, where initial education and research was carried out. Since then, a total of about twenty research reactors have been constructed (most of them were in operation for more than thirty years before they were either decommissioned or overhauled). Later, the Atomic Fuel Corporation established the Power Reactor and Nuclear Fuel

[206]

Development Corporation, developed fast reactor cycle technology and techniques to treat highly-radioactive waste, and constructed the Monju fast-breeder reactor, which in 1994 reached criticality. Monju is currently in the process of being converted to a safer system as a result of a sodium leak in 1995.

In the private sector, the creation of the Japan Atomic Power Company (JAPC) led to the commercialization of nuclear power generation. When JAPC constructed a British-designed 166 MW Caulder Hall-type reactor at its plant in Tokaimura, it marked the first time that commercial energy had been generated by the private sector. Following this, nuclear power stations were constructed throughout the country by Japanese electrical power companies, which preferred BWR and PWR light water reactors. BWRs are used extensively throughout Japan including at the JAPC Tsuruga Power Station Unit 1, Tokai No.2 Power Station, and by the Tokyo Electric Power Company (TEPCO), Tohoku Electric Power Company, Chubu Electric Power Company, Chugoku Electric Power Company, and Hokuriku Electric Power Company. Electricity produced by BWRs currently accounts for approximately half of the nuclear energy generated in Japan. In 1970, Kansai Electric Power Company became the first company to use a PWR in its Mihama Unit 1. It was followed by the Shikoku Electric Power Company, Kyushu Electric Power Company, Hokkaido Electric Power Company, and JAPC.

Figure 7.1

JAPC Tokaimura No.1

Participation in the International Atomic Energy Agency (IAEA): Promoting the Peaceful Use of Atomic Energy

When Japan first embarked on its journey toward atomic energy development, the experience of World War II divided scholars into two camps with conflicting opinions. The first maintained that they wanted nothing to do with nuclear energy, while the second were drawn by the prospect of using the enormous energy potential for peaceful means. The first to propose the peaceful use of nuclear energy was parliament and the government, which in 1954 passed a budget for atomic energy. In response to this, the Science Council of Japan put forward three conditions for the peaceful use of atomic energy to be realized: (1) the rejection of research into nuclear weapons; (2) independent operation by the citizens of Japan; and (3) the complete disclosure of all information. These aforementioned three principles formed the basis of an independent atomic research body unparalleled by anything abroad. Parliament accepted the scientists' demands, incorporating the three principles in the Atomic Energy Basic Law, creating a spirit that was embraced by the scientists involved in the research and development of atomic energy.

Internationally, too, from the perspective that never again should the tragedy wrought by nuclear weapons be allowed to recur, President Eisenhower, in his "Atoms for Peace" address, proposed the establishment of "an international body whose aim is the peaceful use of atomic energy." In 1957, the International Atomic Energy Agency (IAEA) was established as a specialist UN agency dealing with the research and development, technological cooperation, and safeguarding of atomic energy for peaceful uses. The two main missions of the IAEA are: (1) promotion of the peaceful use of atomic energy; and (2) verification that atomic energy is not used for anything other than peaceful means—in other words, nuclear safeguards.

Japan has been an integral part of the IAEA ever since attending the preparatory congress for its establishment, and still remains an important member country within its structure today. Japan was particularly influential in the area of safeguarding nuclear technology when, in 1958, it became the first country to comply with an inspection of its research reactors, providing impetus for the implementation of such inspections among the hitherto-reluctant member states of the IAEA council. All of Japan's domestic research reactors subsequently underwent inspections, followed by all commercial reactors, starting with the JAPC Tokai Power Station. Complying with a safeguard inspection, in practical terms, means the following:

- Maintenance of records of changes in the amount of nuclear material in the facility by type and quantity, as well as inventory amounts and transferred amounts.

- Collection of material accounting reports.

- Authorization for inspectors to enter the facility to conduct inspection of facility records, taking of stock and non-destructive inspection and measurement of nuclear material, taking of samples for analysis, installation of surveillance cameras, the restriction of transportation of nuclear material by sealing, and actual confirmation of inventory amounts on an annual basis.

- Comprehensive evaluation of the existence of any violations based on results of the above.

The nuclear fuel plant where I was previously employed used to process uranium inside the factory, making it important for the facility to comply with inspections. I can recall the team of IAEA inspectors who would come into the factory and shut it down for a week during the summer and winter vacations while they meticulously investigated the amount of uranium in each part of the process, and confirmed the records kept by the facility. Japanese atomic energy facilities felt naturally duty-bound to undergo IAEA inspections, and I believe that their quiet dedication coupled with the cooperation which they showed the IAEA helped to boost the cause of atomic energy for peaceful purposes, as well as increasing Japanese integrity among the international community.

There was, however, a large loophole in the inspections that safeguarded nuclear power. The inspections were recognized as being powerless against undeclared nuclear facilities and materials. In response to this, the additional protocol was adopted to make the inspections more effective. The characteristics of the additional protocol were designed to detect undisclosed activities, and made it compulsory for countries to disclose information regarding all of their nuclear energy research and development facilities, including those not using nuclear materials. They also bound states to allow access to all buildings on a nuclear facility site and gave the inspectors the right to take samples from a broad range of locations within the environment and carry out unannounced inspections. It was hoped that the signing of these additional protocols would achieve greater efficacy in suppressing the development of nuclear weapons.

The additional protocols had been adopted by 131 states as of March 2009, including Japan and Canada, as well as one body (EURATOM). It is hoped that more countries will sign up to the additional protocols.

The Non-Proliferation Treaty and the IAEA

Sixty-four years have passed since the atomic bombs were dropped on Hiroshima and Nagasaki, and in that time the use of nuclear weapons as a means of mass destruction has not been repeated. In this regard, the introduction of the Non-Proliferation Treaty (NPT) in 1970 is regarded as playing a significant role. The NPT confirmed the IAEA as the security body charged with overseeing the rules governing nuclear technology, which include the prevention of the transfer of nuclear weapons technology to other countries, the prevention of nuclear technology development in fields other than peaceful uses; and the prevention of non-nuclear powers from producing nuclear weapons. The original signatories of the NPT were the five countries which had established themselves as nuclear powers prior to 1967 (America, Russia, Britain, France and China) and 189 countries without nuclear weapons capabilities. The IAEA and NPT thus formed the foundations of the entity which would uphold the peaceful use of atomic energy. In 2005, in recognition of the accomplishments its activities – conducted over more than half a century – had achieved in greatly reducing the threat of nuclear war, the IAEA was awarded the Nobel Peace Prize. Internationally, this led to the profile of the IAEA being raised, and its endeavors being more highly regarded.

At the IAEA 2008 General Assembly, Director General Dr. Mohammed ElBaradei concluded his address by stating:

> The IAEA operates on the premise that all counties will accept our safeguarding activities. In order for our activities to be accepted by all countries, movement toward nuclear disarmament is crucial. Member countries must consider how they will respond to this, and it is clear that implementation will require funding to be enhanced, but when compared to the security of mankind and the benefits that its development will bring about, the amount of money is trivial. I hope that member countries can see things with a broad perspective, and consider what is best in the long-run.

Japan's Atomic Energy Policy: From the Basic Act on Energy Policy to the Nuclear Power National Plan

At 96.1 percent, Japan's dependence on imported energy resources is incomparably higher than that of other developed countries, meaning that securing a stable supply of energy is of critical importance. Nevertheless, there was never a law covering a long-term energy policy. Thus, in light of the development of other Asian countries – which is predicted to make securing future energy supplies more difficult – and following the recognition of the harmful environmental effects that energy usage causes, in 2002 the Diet passed the Basic Act on Energy Policy, representing the first time the country has attempted to formulate an energy strategy. The Basic Act on Energy Policy emphasized three core elements of energy procurement; stable supply, compatibility with the environment, and economic feasibility; in other words, energy which can be supplied regularly, energy which produces low amounts of carbon dioxide and other greenhouse gases, and energy which is appropriately priced.

[211]

It was in this spirit that the Basic Energy Plan – the guidelines for implementation – was formulated in 2003. The plan contained the following three points: firstly, that atomic energy should be made the principle source of electricity; secondly, that advances should be made in facilitating the reprocessing of spent fuel from nuclear power stations, with the plutonium obtained from this spent fuel being used in current reactors (referred to as the "pluthermal plan" in Japan); and third, that an adequate system be constructed for the recycling, treatment and disposal of radioactive material produced in nuclear power stations.

From its inauguration in 1955 onwards, the Atomic Energy Commission has put in place the policies governing the medium- and long-term development and use of atomic energy. However, in 2005 the Framework for Nuclear Energy Policy was drawn up reflecting the decisions of the cabinet on the aforementioned Basic Act on Energy Policy and Basic Energy Plan. Whilst implementing policies for the projects of government ministries and agencies, and policies to advance atomic energy, the commission was also charged with the governance of nuclear utilities and local public bodies which were connected to nuclear power, as well as being expected to maintain a mutual understanding with the general public. The Framework for Nuclear Energy Policy was more specific than the Basic Energy Plan in its handling of atomic energy. Its basic objectives were: (1) for atomic energy to supply at least 30–40 percent of the nation's entire electricity demand from 2030 onwards; (2) to promote the nuclear fuel cycle; and (3) to realize the commercial use of fast feeder reactors.

The country's new energy strategy, the "Nuclear Energy National Plan," was the real driving force that brought about the practical realization of the basic objectives contained in the Framework for Nuclear Energy Policy. It is a common guide for all the people who are engaged in nuclear energy in the country today. The Nuclear Energy National Plan is even more detailed in its action plans, demonstrating the country's resolve to be at the forefront of the advancement of policies. The following ten items constitute the framework of the Nuclear Energy National Plan:

[212]

1. The realization of new construction and expansion of nuclear power stations in the age of electric power liberalization. Targets include the new construction and expansion of thirteen units by 2015, growth in capacity of 90 percent, and electricity generated by nuclear power accounting for 41 percent of domestic supply.

2. Ensuring the safety of existing reactors so they can continue to operate. This means the introduction of an inspection system to respond to the individual circumstances of each reactor, and the creation of policies dealing with aging reactors which have been in operation for over 30 years.

3. The deployment of strategies to secure energy resources. In addition to the development of uranium mines in Kazakhstan through a Japan-Kazakhstan nuclear agreement, these strategies also include cooperation on other nuclear energy-related technological issues.

4. Advancement of the nuclear fuel cycle and strategic strengthening of related industries. This includes the operation of the Rokkasho reprocessing plant and the use of fuel, including plutonium, in existing reactors.

5. The early realization of the fast-breeder reactor cycle. Demonstration reactors will be realized by 2025, and commercial reactors developed before 2050.

6. Securing sufficient human resources and technology for the next generation. This includes projects for the next generation of light water reactors, training for technicians in the field, support for the transfer of skills, and the creation of nuclear energy education programs in universities and other institutes.

7. The deployment of international support for the nuclear energy industry. This includes support for countries such as Kazakhstan, Vietnam and Indonesia, which plan to introduce nuclear power plants.

8. Active participation in the construction of international frameworks whose objectives are the expansion of nuclear power plants and nuclear nonproliferation. By fully utilizing its experience and technologies,

[213]

Japan should look to contribute to and cooperate in the establishment of international frameworks (e.g. US-proposed Global Nuclear Energy Partnership, and Japanese-proposed Fuel Supply Assurance).

9. The fostering of trust between the government and local communities, and through public hearing and PR activities. Direct dialogue with local residents should be undertaken in order to boost trust, promote understanding of nuclear energy among women and the next generation, and to support the prosperity of local areas.

10. The strengthening of measures for radioactive waste. This includes reinforcing the commitment to secure locations for the final disposal of highly radioactive waste.

The Current Situation of Nuclear Power Generation

As of the end of 2007, Japan had a total of 55 operating reactors in seventeen locations, with a combined capacity of 49,580 MWe. Japan ranks third in the world behind the US with 104 reactors, and France with 59. There are currently two reactors under construction with a capacity of 2,285 MWe, and eleven reactors – with a capacity of 14,945 MWe – in the preparation stage of construction. These include JAPC Tsuruga Units 3 and 4, with a total capacity of 3,076 MWe, which are due to begin operation in 2014 and 2015, respectively.

In FY 2007 (April 2007 to March 2008), nuclear energy accounted for approximately thirty percent of Japan's total electricity consumption, showing that it has come to play an important role. The frequency of automatic shutdowns in Japanese nuclear power stations (unplanned reactor shutdowns which result from unanticipated trouble) was low compared to other countries. However, the capacity factor of 61 percent for Japanese plants is low compared to the 80–90 percent achieved by other major nations. One of the main reasons for this is that following the Niigataken Chuetsu-oki Earthquake of July 2007, the world's largest nuclear power plant – the Tokyo Electric Power Company's Kashiwazaki-

Kariwa Nuclear Power Station (with a capacity of 8,210 MWe) – was completely shut down.

It can thus be seen from the above that the issues confronting Japan are: (1) an improvement in capacity factor; (2) beginning full reprocessing; (3) progression of the pluthermal plan; and (4) selection of locations for the disposal of highly radioactive waste which is produced during the reprocessing phase.

The Nuclear Fuel Cycle

As a country heavily dependent on imported energy resources, Japan has a basic policy of operating a "nuclear fuel cycle," which means reprocessing nuclear power plants' spent fuel and efficiently recycling the recovered uranium and plutonium for use as fuel.

Based on this policy, on the premise of ensuring safety, various initiatives are taken by the government and utilities as they acquire the understanding and cooperation of the general public including local residents. However, globally, there are countries which do not operate policies to reprocess spent fuel. In such countries, highly radioactive waste is disposed of directly. The specific circumstances in each country dictate which policy is adopted. Japan, France, Russia and Britain have adopted a reprocessing policy, while Germany, Canada, Sweden and Korea follow a policy of direct disposal of spent fuel.

Natural uranium is put through a process of enrichment which increases the concentration of Uranium-235 to produce uranium for use as a nuclear fuel. Because the reprocessing of spent fuel and enrichment could produce highly enriched uranium and plutonium, which could potentially be misused to create the materials for use in nuclear weapons, Japan is the only country in the world, other than those countries with a nuclear weapons capability, permitted to operate the enrichment process. This international recognition is testament to the fact that Japan has for over fifty years developed and used nuclear energy for peaceful means alone. It is hoped that Japan will continue to strive to provide a model for the rest of the world in the administration of its enrichment and recycling programs.

Use of Radiation

Together with the development of energy, the use of radiation has grown into an industry in its own right. Radiation exists in various forms, including X-rays, Gamma rays, electron rays and quantum rays. The application of radiation in fields such as medicine, industry, agriculture and academia, is referred to as "use of radiation." While exposure to large doses of radiation is harmful to the living tissue of humans and animals, there are many positive characteristics of radiation which can be exploited, including the ability to penetrate material, the capability to concentrate energy, the ability to annihilate cancer cells, and the ability to change the characteristics of chemical material after it is exposed to radiation. However, for these applications to be realized, rules and regulations regarding the safe handling of, and protection from, radiation need to be put in place. In Japan the aim is – under a safety administration based on the related law, to oversee the research and development of radiation use – to promote broader use, and to contribute to developments in science and technology to bolster various industries.

Fostering Organizational Culture Related to Peaceful Use of Nuclear Energy (From the Activities of the Atomic Energy Society of Japan's Ethics Committee)

In the White Paper on Nuclear Energy in 2007, the Atomic Energy Society of Japan uses the term "organizational culture" in connection with the peaceful uses of nuclear energy. Culture refers to human actions which are repeated through our lives on a daily basis, and for the Japanese people involved in nuclear energy, it is fair to say that the spirit of nuclear energy for peaceful use is the culture incorporated into their everyday actions. I would like to introduce part of a regulation drawn up by the Atomic Energy Society of Japan's Ethics Committee for the sake of the researchers and technicians involved in nuclear energy. This code of ethics is continuously reviewed, added to or abandoned as changing circumstances may demand, reflecting the principles of nuclear energy researchers and technicians in a way unparalleled by anything else in the world. The Atomic Energy Society of Japan is an academic society which counts many of the people who work in the nuclear industry among its members. Many of Japan's nuclear policies, as well as R&D and technical development, come to fruition through participation in the Atomic Energy Society of Japan. There are also a significant number of members who are active internationally.

[Preamble of the Atomic Energy Society of Japan's Code of Ethics]

We the members of the Atomic Energy Society of Japan (AESJ) amply recognize that nuclear technology brings tremendous benefits to humans but also raises the possibility of catastrophe. Based on that premise of recognition, with pride and a sense of mission of being directly engaged in the peaceful use of atomic energy, we energetically pursue human welfare and sustainable development while conserving global and local environments through the use of atomic energy.

Whenever we conduct atomic energy research, development, utilization, and education, under the principle of information disclosure, we the members of the AESJ make constant efforts to enhance our knowledge and skills, to keep pride and responsibility in our work, to keep a spirit of self restraint, to maintain a harmonious relationship with society, to comply with laws and regulations, and to secure nuclear safety. In order to implement these ideals, we the members of the AESJ have established herein fundamental canons of attitude and conduct.

Fundamental Canons [extract]:

1. We shall restrict the use of atomic energy to peaceful purposes while endeavoring to solve the problems confronting humans.

2. We shall hold the safety of the public paramount in the performance of our professional duties and through our conduct strive to obtain the public trust.

3. We shall strive to improve our own professional competence and simultaneously to improve the professional competence of persons involved.

Conduct Guide [extract]:

1 -1   Fundamental Policy for Nuclear Energy Use:

The Society's specialist fields for the peaceful use of nuclear energy do not limit themselves to nuclear energy for the generation of electricity, but

[217]

extend to cover a myriad of areas including medicine, agriculture, industry, and fields connected with the technical applications of radiation and isotopes. Members act in a way that will contribute to the welfare of mankind through use of their specialist technology, while at the same time acknowledging that in addition to the benefits their work can offer mankind, the potential for bringing about a disaster also exists.

1- 2   Use Limited to Peaceful Means:

The use of nuclear energy is limited to peaceful means. As a point of honor and dignity, members shall take no part whatsoever in the research, development, construction, acquisition or use of nuclear weapons.

1 -3   Caution Regarding Nuclear Proliferation:

In recognizing that nuclear technology has the potential to be used for the research, development, and construction of nuclear weapons, members shall take the utmost care to ensure that the results of their work do not contribute to nuclear proliferation.

In light of the above Code of Ethics, the Ethics Committee of the Atomic Energy Society of Japan examines ethics-related cases that are observed in nuclear energy research, development and business operations, and publish the results of the examination. The Committee has the responsibility to sound the alarm regarding members whose conduct is deemed inappropriate.

Public Acceptance (PA) of Nuclear Energy in Japan

What is Public Acceptance and Why is it Needed?

Public Acceptance (PA) refers to acceptance by a region of the various commercial activities related to nuclear power, after distribution of information to the general public and building mutual understanding through communication. It can also denote broader acceptance of nuclear energy policies by the general public. In the Framework for Nuclear Energy Policy explained in the previous section, it was stated that realizing coexistence between nuclear energy institutes and the general public or local community is a prerequisite for the promotion of nuclear energy policy. In addition to being a 'megatechnology' which is very difficult to understand, nuclear energy must also overcome fears stemming from the general public's confusion of nuclear energy with nuclear weapons, and accidents such as Chernobyl, where radioactive material was leaked. To attain the goal of public acceptance of nuclear facilities in their region, the general public must not only be provided with information, but the government and utilities must forge personal connections with the people to create a relationship of trust. Since Japan is an extremely heavily populated country with a population of over 120 million people, receiving the approval for national nuclear energy policies is no easy task. Furthermore, Japan is a mature democratic country, where bonds between households in local communities are not as strong as in the past. There is a diverse range of lifestyles and values, and everyone freely voices their opinion. Gone are the days when communities followed the opinion of the local leader, meaning that PA activities which do not respect individuals will not result in the creation of a relationship of trust. For nuclear energy businesses, gaining the trust of local residents is essential for the accomplishment of their future plans.

In 1986, at the time of the Chernobyl accident in the former Soviet Union, the directors of the company I was working with tasked me with creating a new division to respond to local people's concerns and inquiries about nuclear energy. This was the start of my journey in nuclear energy PA activities. I have since undertaken many types of work related to PA. In this section, I would like to discuss my experiences, focusing on the following:

- my participation in public conferences, which forms part of the PR/public hearing activities of the Japan Atomic Energy Committee;

- the activities of the global organization Women in Nuclear (WiN), which targets women and members of the younger generation; and

- an outline of the energy awareness survey recently conducted by the Japan Atomic Industrial Forum.

The Atomic Energy Commission of Japan: Public Participation Conferences

The Atomic Energy Commission of Japan, which is responsible for drawing up nuclear energy policies, holds public participation conferences (PPCs) as a means of gathering opinions directly from the public and reflecting them in future policies. The PPCs, which take place several times per year in various locations, are organized by the AEC as part of their "forum to hear the opinions of the public." They are planned by a nine-member commission, and have included a journalist, a leader of a women's group, a critic, and a university professor. I myself have been involved with the PPCs since they began, as a commission member.

Figure 7.2
AEC Public Participation Conference

The commission members discuss many topics, beginning with how to incorporate the opinions heard at PPCs, along with a general summary of the PPCs held across the country, and deciding the location and contents of the next PPC. The meetings are attended by the head of the Atomic Energy Commission, as well as other Commission members, as observers, making the PPCs a place for direct dialogue between the

Atomic Energy Commission and the general public. They are run based on the promotion of common understanding.

The event that led to the creation of the PPCs was the public vote held in the spring of 2001 regarding the establishment of a pluthermal reactor (using mixed oxide [MOX] fuel) in Kariwa Village, Niigata Prefecture. Faced with the public rejection of the pluthermal reactor, the Atomic Energy Commission began to think that "even if our perspectives differ, we will need a forum to allow calm, realistic discussion on the kind of lifestyle that residents hope for, and how to meet energy supply needs on top of that. It will also provide an opportunity for discussion on how to provide information on nuclear energy, which we will listen to as we proceed." The idea of the PPC was born from a desire for the general public to understand and accept the path nuclear energy should take.

Since the first PPC, held in Kariwa Village in January 2002, there have been a total of seventeen PPCs, up to and including the session in Kyoto in January 2008. Past conference locations have included major metropolitan cities and cities in regions where nuclear power facilities are located. Some examples include Tokyo, Fukushima, Saitama, Tsuruga, and Fukuoka. Topics on the agenda are based on the theme of "Are you receiving the information you need?" They are set to correspond to the timing and location of the conferences, to promote interest and participation among the local community in the siting area. Past topics include: "What is the electric power crisis?"; "Opinions on the long-term nuclear energy plan"; and "Radiation use."

The program is conducted in various ways: sometimes with a two-stage format, where the panelist provides the topic and opinions are exchanged with participants; and sometimes as a panel discussion. The format is decided based on the circumstances of each conference. Participants are given the chance to voice their opinions at every conference, with time allotted for responses to these opinions at some conferences.

There are usually around 200 participants at each conference, and while it cannot be denied that many of the participants are from nuclear power-related companies, about 30 percent of those who attend do so out of an interest in energy and environmental problems; these include women and young people.

According to participant surveys, almost 90 percent of participants consistently stated that they felt "very satisfied" or "mostly satisfied" by the proceedings, while the most common reason given for dissatisfaction was: "There were many technical terms that made it difficult to understand." Indicators such as this should prove that information provision should be facilitated by easier-to-understand language.

Women in Nuclear (WiN)

Women in Nuclear (WiN) Global is an international NGO for women working in nuclear energy created in Europe in 1993. It currently boasts approximately 2,300 members from 68 countries, regions and international organizations. Its main objectives are: to contribute to activities that promote understanding of nuclear energy; improve the qualifications of its members; and fosters international exchange. WiN Japan (WiN-J) is the Japanese branch of this organization, and was established in 2000. It has 130 full members and 80 patron members.

I am often told that it seems strange to differentiate between men and women in this day and age, but it is interesting to look at the number of women who are active in scientific technology fields, and especially in the field of nuclear energy. In Europe, where the WiN was first established, that number is 20 percent; in Japan, it is a mere two percent (according to the ratio of female to male members of the Atomic Energy Society). If nothing is done to rectify these lopsided figures, this one-sidedness will become even more pronounced. Another thing to consider is society's reluctance to accept the nuclear power industry, which is especially high amongst women. It seems fair to say that if women accept the technology, then society as a whole will follow suit. However, as long as those involved in nuclear energy are all men, the gap between the

[222]

industry and society will persist. Any industry dominated by men gives women the impression of being instinctively rigid, bureaucratic, and cold. This feeling acts as a kind of mental barrier, making any hope of establishing a relationship based on trust a distant prospect. The occupations of WiN members are not limited to PR, but form a vast array, including technicians and researchers. All members are willing to meet female members of the public, listen to their fears and questions about nuclear energy, and speak to them in their own words. Through such methods, they hope to promote the idea that there are many approachable women working in nuclear energy, and to change the image of nuclear energy from the domain of male technicians to an industry which women feel is less cold. Thus, the intended recipients of WiN's promotional work on the understanding of nuclear energy are women and the young.

As an example of WiN Global's activities, I would now like to present a report from the 16th annual WiN Global conference, held in May of this year, before going on to introduce an activity unique to WIN-J: holding face-to-face dialogues between women who work at nuclear power plants and female members of the general public, on topics relating to energy and nuclear power.

The 16th Annual WiN Global Conference held in Marseilles, France, 2008

Every year, WiN Global holds its annual conference in a different country. This year, the conference was held in Marseilles, France (May 26–31) with "Arising Key Competencies for Nuclear Energy: A Challenge and an Opportunity for Diversity Development" as its main theme. It was attended by 250 people from 30 countries, including eleven members of WiN-J.

France is an advanced nation in terms of nuclear energy, with 59 reactors producing 80 percent of the entire country's electricity. It is a country which draws a great deal of interest from those in nuclear power, as it possesses almost all processes in the nuclear fuel cycle. It also leads the world in terms of women active in the nuclear power industry, considering the following: the leader of the industry's largest business is a woman; it was the first country to have a woman become director of a nuclear power plant; and there are numerous women occupying high-ranking positions across the industry.

The meeting began with a keynote speech by the director of the French Atomic Energy Commission entitled, "The Current State of Nuclear Energy and Innovative Solutions for the Future: The Prospects for Nuclear Energy." It went on to cover a multitude of current global topics in the field of nuclear power, giving participants the opportunity to deepen their understanding of their own field of expertise while broadening their knowledge of areas outside them as well. The topics covered included: the issue of nuclear power skill acquisition training; the younger generation and energy; the environment; the nuclear energy training curriculum; lessons learned from international transfer of technology and contractors during restructuring of nuclear power companies; and the site selection process for highly radioactive waste disposal. I also took part in the conference, giving a presentation on the first time occasion that a large-scale nuclear power station had experienced a large earthquake, entitled, "Lessons from Earthquake Damage sustained by Kashiwazaki-Kariwa Nuclear Power Station."

Figure 7.3
Participants at the 16th Annual WiN Global Meeting
Marseilles, May 2008

Later, representative WiN associations from 24 countries described government nuclear power policies, development and applications, the nuclear power PA, and the state of WiN activities in their own countries. I believe that WiN Global is the only international meeting where this many female representatives of the nuclear power industry can meet at once to be heard.

As well as being the host venue for the conference, southern France is also known for having many nuclear power facilities. As part of the technical tours which accompany the conference itself, I visited the French Atomic Energy Commission's Marcoule research facility and the Cadarache complex, where an international nuclear fusion test plant is due to be built. Additionally, the WiN-J participants took a tour of the MELOX plant which manufactures the MOX fuel that, in the near future, will be used in Japan.

During the conference, participants formed a tight-knit community, eating and sleeping under the same roof, talking for hours on end, absorbing the local culture, introducing one another to the culture of their own countries, and creating strong bonds of trust. Far from a simple exchange of information, this conference nurtures a global network in the spirit of mutual cooperation. To countries which adopt nuclear power in the future, I recommend they make use of networks of women when networking globally.

The Speech at Omaezaki and the Conference for Exchange between Women: "Let's talk About Energy and Nuclear Power"

In Omaezaki City in July 2007, WiN-J held a speech and conference for mutual exchange between women. Omaezaki is home to the Hamaoka Nuclear Power Station. Since its establishment in 2000, WiN-J has held speeches and conferences for exchanges between women in six regions across Japan. It has provided a forum for close exchange between female members of the general public and members of WiN-J, who are experts in the field of nuclear power. The features of the meetings are its small groups and the fact that it allows any topic to be raised and discussed; the active participation in these open dialogues leads to heightened awareness and deeper understanding of problems concerning nuclear power and energy.

Figure 7.4
Omaezaki WiN-J Conference, 2007

First to speak was a male celebrity guest who appeared in the television commercials for the electric power company which supplies the region. The title of his speech was: "Our Lives and Energy, and Actions Requiring Courage." His experiences as a celebrity made it a very interesting speech which succeeded in drawing in participants. Next, participants were split up into groups of ten (each including 2–3 WiN-J members) and held discussions at their tables. During this activity, the members encouraged others to give their frank impressions of the speeches they had heard and to ask any questions they had regarding nuclear power. WIN-J members then gave explanations on the spot to assuage the women's concerns and fears, in a two-way dialogue in the hope of boosting mutual understanding. Some of the positive comments received included: "The questions I had wanted answering were resolved today. Today's meeting was extremely useful. I thought it was wonderful to see so many women working hard in the nuclear power industry." While the table discussions were useful in promoting understanding through dialogue between WIN-J members and the general public, they were also beneficial in helping participants recognize that many different attitudes exist amongst the public. The experience of talking candidly with people who have different perspectives proved extremely valuable.

PA activities are not undertaken for the sake of one single firm, but for the continuance of nuclear power technology as a whole, and the common legacy of mankind. Regardless of whether there are objections to nuclear power, for countries which have opted to use it, PA activities are a common courtesy to their people, as well as a responsibility. Transparency is important for disclosure of information and the progress of nuclear power policies, in that the people can see the kinds of exchanges that have taken place. For nuclear reactors, which will continue to operate for up to 60 or even 80 years into the future, transparency is necessary in order to establish public trust in the government and utilities.

Energy Awareness Fact-Finding Survey (The Japan Atomic Industrial Forum)

The government, local governments, and institutes connected to nuclear power often conduct surveys of the awareness of the general public and local residents. They do so to establish a point of reference for nuclear power policies and corporate governance. I would now like to introduce the results of questions concerning nuclear power, taken from the "Energy Awareness Survey" conducted in September 2008 by the Japan Atomic Industrial Forum. This survey was conducted nationwide, as well as in regions near nuclear power station siting areas. Nationwide respondents numbered approximately 2,200 people, while there were around 250 from nuclear power station siting areas.

The first question, which concerned the comfort level of their current lifestyles and energy use, was: "Do you want to continue to have a comfortable convenient lifestyle?" Approximately 70 percent of the responses were "Yes" and "For the most part, yes." The following results were also obtained: 66 percent replied that they were trying to save energy in their everyday lives; 38 percent stated that they would rather enjoy their present lifestyles than sacrifice for the sake of the future; 22 percent responded that it was not necessary to lower their standard of living to conserve energy. From the responses, it seems that most Japanese people are aware of issues relating to energy conservation.

While around 53 percent of respondents considered nuclear power necessary, the same percentage of respondents were concerned by it. This reveals that over half of the population holds the ambivalent view that, "nuclear power is necessary, but it causes concern." This is a typical human response, demonstrating that PA activities involving people do not always go as predicted. Nevertheless, PA activities cover the necessity of nuclear power as well as its safety aspects, but explanations of safety tend to become complicated and technical, meaning that in order to reassure public concerns, more innovation is still is needed.

Responses indicating areas of concern regarding the safety of nuclear power were, in order of frequency: (1) the occurrence of an accident (63 percent); (2) the fact that radiation cannot be seen or felt by humans (52 percent); (3) that in the case of an accident, the effects on humans and the environment is unknown (50 percent); (4) that the government and electrical power companies may cover up or falsify facts (43 percent).

The previous section stated that women are significantly more negative toward nuclear power than men. This can be seen clearly from the following results: 52 percent of men said they were scared of nuclear power, compared to 68 percent of women; 47 percent of men were concerned by nuclear power, compared to 58 percent of women; 50 percent of men supported nuclear power generation, compared to 25 percent of women; 55 percent of men thought nuclear power was an excellent method of generating electricity, compared to 37 percent of women. Maternal feelings are deep-rooted in women, since they are the gender which bears and raises children. This means they consider safety to be the top priority above all else, for the sake of their children's health and future. I personally consider "health and the future" to be precisely the reason nuclear power is the most favorable option.

This survey also compared the results obtained from areas with nuclear power stations and those without. This showed that 40 percent of people living near a nuclear power station considered the issue of nuclear power an immediate problem, compared to 4 percent in other areas, highlighting the increased interest from people living close to a facility. It could also be said that people living near nuclear power facilities had a more favorable opinion of nuclear power; specifically, 56 percent of people living near nuclear power stations considered nuclear power an excellent method of generating electricity, compared to 45 percent of those who did not. Other results which reflect the favorable opinion of residents in areas with nuclear power stations are: 29 percent think that more nuclear power should be used (compared to 19 percent who do not). Electric power companies use various approaches to provide information in areas with nuclear power stations, so it appears there is a connection between the extent of resident understanding and favorable image.

Human Resource Training in the Field of Nuclear Power

In the field of nuclear power, from an energy security viewpoint, it is extremely vital to maintain and develop the industry and nuclear power generation technology. In the 30 years that have passed since the 1970s, reactor after reactor has been built in Europe and the Americas; meanwhile, in Japan our progress was lesser, if not steadier, in building new nuclear reactors. It is for this reason that Japan's nuclear power technologies are being handed over more steadily in comparison to those of Europe and the Americas; it is why they show an impressive strength in a world nuclear market brought to life by the "nuclear renaissance."

However, in the next 20–30 years, the construction of domestic nuclear power stations is expected to plummet; indeed, the amount of research funds and number of engineers allocated for nuclear power-related areas has fallen as well. Estimates predict that demand for replacement construction of nuclear power stations currently in operation will arise sometime around the year 2030. However, we currently face a very serious problem: how to maintain technology/industry/human resources until that time comes?

At universities in these last two or three years, it is said there are many students who have begun choosing to study nuclear power again. However, there is no denying that the nuclear power industry lacks the allure to capture the attention of many students; students who are especially sensitive to contemporary social standards. There is a need to revitalize the workplace, create a sense of purpose, and reform our workplace into one that lives up to society's expectations.

In the universities, high expectations are held regarding the implementation of specialized education, beginning with basic nuclear engineering, that will create human resources which: carry technical innovation; exhibit creativity; and gain knowledge and education on many subjects. Also vital are active, unified efforts by related parties toward the effective use of internships, the graduate school partnership system, and nuclear research facilities.

As for utilities, contractors, and national and local institutions, it is hoped that they will drive efforts for human resource training, such as provision of a lateral system of technical qualifications for repair and maintenance of nuclear power facilities, and creating networks for training facilities and curricula for earning qualifications. Also, vertical coordination between utilities and contractors will not be enough; it is also necessary to for the nuclear power industry to consider moving forward as a whole, including the possibility of lateral coordination between businesses, as well as contractors.

Nuclear Power Use in Asia and Japan's Contributions

I have participated in the efforts of the Forum for Nuclear Cooperation in Asia (FNCA) since 2001. Japan is a central part of this organization, which operates in neighboring Asian countries (China, Korea, the Philippines, Vietnam, Thailand, Malaysia, Indonesia, Bangladesh, and Australia). It gathers information on planned and current use of nuclear power technology in these countries, performs information exchange, and contributes to and assists the promotion and establishment of the peaceful use of nuclear power. It does so through activities designed to proliferate the nuclear power technology which Japan has cultivated.

Of the member nations of the FNCA, only Japan, Korea, and China have introduced systems for nuclear power generation; nuclear development in the other nations is focused on research reactors, mostly for radiation use. The FNCA offers broad assistance in eight fields: nuclear PR; agricultural use (nuclear breeding, bio-fertilizers); medical use (medical cyclotrons/PETs, radiotherapy); industrial use (electron accelerator use); nuclear safety culture; radioactive waste management; human resource education and training; and research reactor use (Tc-99 generators, neuron activation analysis, research reactor basic technologies). Recently, these countries have successively shown signs of introducing nuclear power generation; thus, it is thought that information exchange on the nuclear power generation acceptance activities of each country will become more relevant in the FNCA's future.

Conclusion: How I began Working in Nuclear Energy

In 1995, I traveled to Gothenburg, Sweden, to attend the first-ever WiN Global conference. Mrs. Agneta Rising, who was then the President of WiN Sweden (and later became the 2nd WiN Global President), delivered these words in her speech: "With the advent of energy, came freedom to women everywhere." I was deeply moved by those words, and I believe this is the point where I truly began working in nuclear power.

For women in Japan, it was only 50 or 60 years ago that life would mean getting married and having children; in other words, a Japanese woman's kitchen was her life, and she had very few options available to

[231]

her. Even now, there are many countries that suffer from internal conflict, poverty, lack of education, overpopulation, and a lack of medical facilities. In each and every one of these regions there is, without fail, a lack of energy. The women there are hounded every moment by childbirth, childcare, housework, and manual labor. Fortunately for those in developed countries, energy is widely available; housework, transportation, and communications can be left to energy-driven devices, and people may have a wide variety of choices for their lives, which they may make themselves. The days pass peacefully within such plenty. Energy is unmistakably one of the prerequisites for sustained peace, and a high level of freedom in a woman's life can be used to measure such peace. Nuclear power is irreplaceable in this valuable provision of energy. This is the idea which has driven me in this line of work.

Whether for better or worse, mankind has developed nuclear power technology, and it is impossible to ignore. Therefore it is imperative that we use these technologies as best we can, to ensure the most people possible live in happiness, whilst wildlife and nature may continue to thrive. In closing, I am certain that the people of the future will look back with gratitude on the fact that, in the 21st century, mankind chose nuclear technology.

At the Crossroads: Germany's Peaceful Nuclear Program

Kirsten Westphal

G ermany is the fourth largest producer of nuclear energy in the world.[1] This fact is often overlooked owing to Germany's decision to opt for a nuclear phase-out. On April 26, 2002, the "Act on the structured phase-out of the utilization of nuclear energy for the commercial generation of electricity" (*Gesetz zur geordneten Beendigung der Kerenergienutzung zur gewerblichen Erzeugung von Elektrizität*) became effective. It made fundamental amendments to the 1959 Atomic Energy Act; instead of aiming to promote nuclear energy, the purpose of the Act is now to phase out its use in a structured manner. According to this Act the last reactor will be shut down by 2021/2022. In the intervening period, Germany must undertake significant efforts to transform its energy system.

This paper will provide an overview of Germany's civil nuclear program. First, it will explore the history of the program, detailing the various nuclear installations in Germany and the current energy mix in general. The main part of the paper will discuss the 2002 amendment of the Atomic Energy Act that changed the objective of the law of 1959. It will provide an analysis of German energy policy-making and major factors that affect the decision-making process. It will also shed light on relevant structures in the German electricity sector. After giving an overview of the general hierarchies and principles of German energy policy, it will identify the pros and cons in the debate on the nuclear phase-out. The analysis of the domestic background will be complemented by a focus on external factors. Finally, the paper will provide an outlook for Germany's energy system and the structured phase-out of nuclear power.

History of Germany's Civil Nuclear Energy Program

World War II was instigated by Germany and ended with atomic bombs being dropped on Hiroshima and Nagasaki. Germany's peaceful and civil nuclear program therefore must be understood against this background. A number of German nuclear scientists had moved to the United States during the time of Adolf Hitler's Nazi regime and had been involved in developing technologies there as well as in the Manhattan project. After the war, West Germany was not allowed to develop and build a reactor or to engage in uranium processing.[2]

US President Eisenhower's "Atoms for Peace" speech to the General Assembly of the United Nations on December 8, 1953 resulted in a global push for nuclear technologies. These ideas helped to shape the International Atomic Energy Agency (IAEA) statute, but also laid the path for a policy change toward Germany.

On October 23, 1954 the Paris Agreements were signed that came into force in May 5, 1955. West Germany – the Federal Republic of Germany (FRG) – became a member of the North Atlantic Treaty Organization (NATO) and the Western European Union (WEU). Most importantly, along with these developments the FRG waived the right to produce nuclear, biological and chemical weapons. In return, the FRG regained full sovereignty and was allowed to launch a peaceful nuclear energy program. The government established a Federal Ministry for Nuclear Issues to promote the technology. On October 31, 1957 the research reactor of the Technical University of Munich, the so-called "atomic egg," was put into operation as the first reactor in Germany.

The peaceful German nuclear program must be viewed as being part of, and governed by, the framework for European integration. The

European Atomic Energy Community (EURATOM) became part of the Treaties of Rome on March 25, 1957, that came into effect on January 1, 1958. The aim of EURATOM was not only to create a market for nuclear energy and technology but also to act as a centerpiece for economic welfare, stability and trust-building among the then six members of the European Community.

On December 23, 1959, the Atomic Energy Act became the basis for construction and operation of nuclear power plants in the FRG. A German Atomic Forum (*Deutsches Atomforum e.V.*) was established by associations, authorities and companies to advance the peaceful use of nuclear energy. This was necessary because where manufacturers and the government wanted to push the program, the energy industry was slightly reluctant because of high capital costs and related risks. Thus, the government approved the first of four federal nuclear programs to provide the necessary funds. In 1966 the research center *Kernforschungszentrum Jülich* generated the first chain reaction in Germany within a high-temperature reactor.

The real implementation of nuclear power, however, was only achieved between 1967 and 1975, when the majority of Germany's operating nuclear power plants were commissioned. In order to balance investment risks, the energy industry promoted extensive use of electricity in private households for cooking and heating.[3] In 1967, the in-situ testing program to deposit radioactive waste material at the Asse salt mine was launched. In 1972, Germany's first commercial nuclear power plants Stade and Würgassen began operation. In 1974, the world's first 1,200 MW unit was activated in Biblis. Its commission in 1969 was a breakthrough for the industry, owing both to its capacity and the fact that a second electricity company, RWE, became engaged in the commercial use of nuclear alongside Preussen Elektra (the first electricity company to build a nuclear power plant).

What is important, however, is the fact that public resistance and opposition increased in parallel to the massive expansion of nuclear power in Germany at the beginning of the 1970s. The development of an opposition movement against the nuclear industry – and in particular against the nuclear power plant of Brokdorf – was closely related to the development of the German Green Party, or the "Greens." The Greens have used the slogan "Atomic power—no thanks" ever since. This also forced the other parties of the German Bundestag to clarify their positions on the issue. In the German Social Democratic Party a fundamental discussion took place, resulting in a 1977 resolution that foresaw more restrictive licensing practices. This increased planning time and the costs of building new nuclear power plants. As a consequence, the nuclear power program stagnated. In 1982 the foundations for the first large-scale uranium enrichment plant in the Federal Republic of Germany were laid in Gronau. In 1984 the spent fuel interim storage facility at Gorleben became operational, where nuclear waste with negligible heat generation is stored.

The accident in Chernobyl on April 26, 1986 was the starting point of the demise of nuclear power in West Germany. The Social Democratic Party opted to abandon nuclear energy, as did the unions. Public opinion was also against nuclear energy. This skepticism and resistance prevented a new generation of nuclear power plants being commissioned. At this time the industry was about to begin using plutonium, a plan which consequently failed in West Germany.

In the East, the German Democratic Republic (GDR) launched its nuclear program in 1955—this was only possible after the Soviet Union had granted permission for the GDR to do so. In 1956, the Council of Ministers decided to build the first nuclear power plant in Rheinsberg. The reasons behind the GDR's nuclear program were manifold and included: the international euphoria inspired by nuclear power; a flow of trained personnel returning from the USSR; and the availability of its own uranium reserves. The resultant expansion of electricity generation was less driven by demand than by political planning based on the idea of lignite-to-electricity and nuclear power. At the beginning of the 1960s, the ambitious nuclear energy program was controversial. The Council of

[236]

Ministers finally decided in 1965 to import the major components of nuclear power plants from the Soviet Union. There is consensus among historians that this led to the stagnation of the GDR's nuclear program. Moreover, the nuclear program in East Germany was affected by quality defects and bottlenecks in capacity.

From a macroeconomic point of view, the withdrawal of East German industry from the nuclear program was logical. As a result, only six reactors were operating in the GDR, with an installed capacity of 2,142 MW, in comparison to 22,375 MW in West Germany in 1990.[4] Moreover, the 1986 accident in Chernobyl acted as a catalyst for the environmental movement in the GDR, which also gained political weight at the round table meetings that took place during the transition period in the GDR in 1989–1990.

In the new unified Germany, the interests of the anti-atomic movement and the energy industry aligned with regard to the GDR's nuclear power plants: the German utilities were not willing to take over the Soviet-equipped nuclear plants and they were shut down in 1991. The GDR had also maintained a nuclear waste repository in Morsleben and radioactive waste material was stored in the facility from 1971 to 1998. A plan-approval procedure for the decommissioning of this facility is currently under way.

After the unification of the Federal Republic of Germany and the German Democratic Republic in 1990, the political landscape did not change drastically. As political parties merged, the split between the two big German parties over the civil use of nuclear power became even more pronounced: The Social Democratic Party (SPD) began to favor a shut-down of nuclear power plants in several federal states. The ruling Christian Democratic Party/Christian Socialist Party (CDU/CSU) coalition with the Liberal German Party (FDP), however, remained in favor of the peaceful use of nuclear energy. In the first half of the 1990s a series of round table talks took place to settle the issue and to reconcile the views of the parties. However, these talks ended in failure.

Meanwhile, the anti-atomic movement had drawn massive popular support as a result of increasing animosity towards the waste deposit facilities in Gorleben and the transport of the cask for storage and transport of radioactive material (CASTOR) to the facility.

With the elections in 1998, the Social Democratic Party and the Greens came to power and the coalition treaty aimed to abandon nuclear power in its respective legislative period. The resulting structured nuclear phase-out was part of an agreement between the major energy utilities and the federal government achieved on June 14, 2000. This agreement was then formally signed on June 11, 2001, and appeared in the amended Atomic Energy Act that came into force April 26, 2002, sixteen years after the nuclear catastrophe in Chernobyl.

In summary, debates concerning the role of nuclear power in Germany have always been heated and laced with ideology. Together with discussions on the future of German coal mines and open pits, the issue of the future of nuclear power has always been at the heart of German energy policies and politics. These two topics have not only had a direct influence on energy policies but also on domestic politics.

Before examining the details of the nuclear phase-out, we shall first discuss the German energy mix.

The German Energy Mix

The German primary energy mix is comparable to other highly industrialized states and dominated by fossil fuels. Oil accounts for 36 percent, natural gas 23 percent and coal 23 percent. In terms of overall primary energy consumption, nuclear accounted for twelve percent and renewables for six percent in 2006.

A major feature of the German energy system is the high proportion of import dependency (see Figure 8.1). Germany imports more than 60 percent of its coal consumption, almost 80 percent of its gas consumption and almost 99 percent of its oil consumption.

Figure 8.1

German Import Dependency (%)

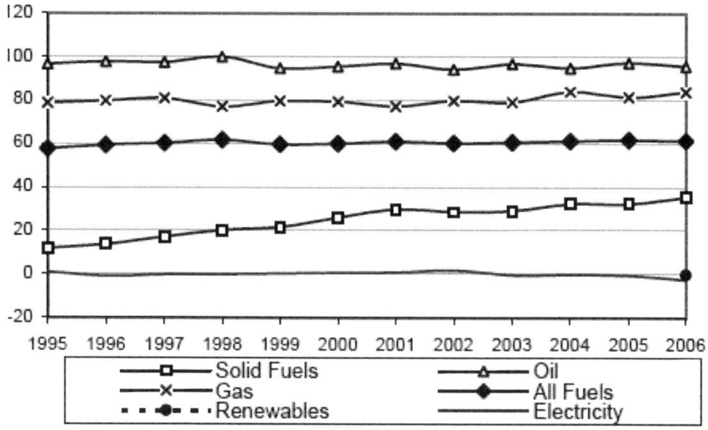

Source: European Commission 2008

The major energy supplier for oil and gas to Germany is Russia, with 33 percent and 45 percent of the overall imports respectively. All of Germany's uranium requirements are also imported. Lignite is the only domestic energy source in which Germany is self-sufficient. The overall import dependence was 74.2 percent in 2005.[5] This dependency decreases with the rising share of renewables in the energy mix. Uranium is imported from France, Canada, the UK and the United States.

Germany's import dependency has become an issue after Germany and other Western European states were hit by gas supply cuts as a consequence of the Russian-Ukrainian gas disputes in 2006 and 2009.

Coal and lignite are the major sources for electricity generation in Germany (see Figure 8.2). Gas has a share of 12 percent, which is likely to increase. The share of renewables has increased significantly over recent years; their share should further increase to twenty percent in 2020 according to the EU's 20+20+20 initiative. This initiative foresees a 20 percent reduction in greenhouse gas emission by 2020, and an increase in the share of renewables by 20 percent over the same period. Moreover, the EU aims to increase energy efficiency by 20 percent. The strategic

[239]

20+20+20 goals were first launched in the EU Commission Energy Action Plan in January 2007, and were subsequently approved by the EU Council in March 2007.

Figure 8.2
German Electricity Mix (% TWh, 2006)

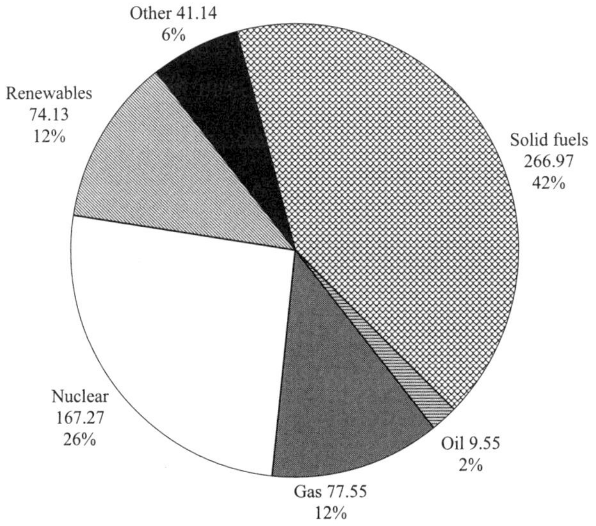

Other 41.14
6%

Renewables
74.13
12%

Solid fuels
266.97
42%

Nuclear
167.27
26%

Oil 9.55
2%

Gas 77.55
12%

Source: European Commission 2008.

German energy policy with regard to energy demand, climate change and environmental issues is highly structured and shaped by EU policies. While these objectives are formulated at the EU level and translated into EU legislative acts, the EU releases no directives regarding the mix of energy sources. The national energy mix is, and will remain, a national concern. The national energy mix is mostly a consequence of different corporate strategies: i.e., it is the private energy industry that decides whether to build gas- or coal-fired power plants or windmills, etc.

The gross domestic consumption of electricity in Germany was 617.5 terawatt hours in 2007.[6] 44.3 terawatt hours of electricity were imported and 63.3 terawatt hours were exported. These facts reflect that Germany's grid is part of the Union for Coordination of Transmission of Electricity (UCTE) grid and the European Union's internal electricity market. 1,240 power plants exist in Germany, of which 890 generate electricity for the overall supply and 350 supply electricity for industry.

Net electricity consumption slightly increased by 0.3 percent in 2007. Consumption is structured as follows: 47 percent of electricity is used in industry; private households consume 26 percent; commerce, trade, services and public institutions consume 24 percent; and transport three percent.

The share of nuclear energy has decreased since the start of the nuclear phase-out in 2002—from 27.6 percent in 2003 to 22.1 percent in 2007 (see Figure 8.3).

Figure 8.3
Germany's Electricity Mix (2006–2007)

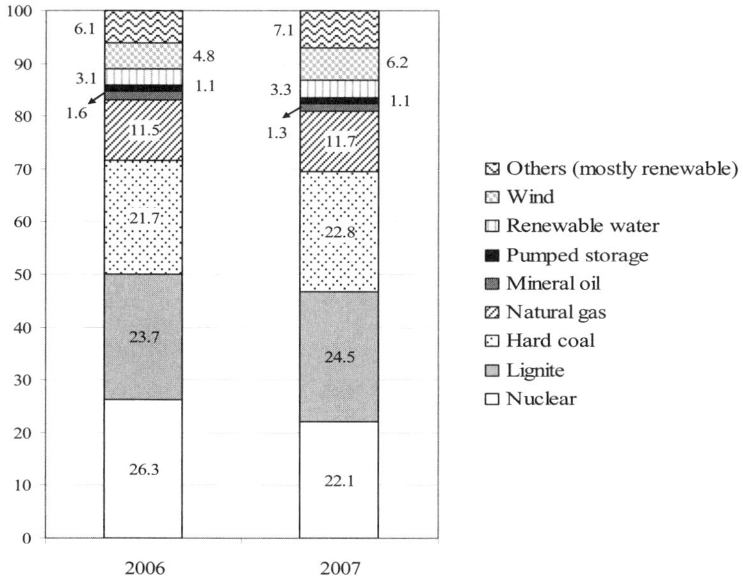

Source: Author's calculations, based on Schiffer, 2008.

Currently, seventeen nuclear power plants are operating in Germany with an installed capacity of 21.497 megawatts (see Table 8.1).

Table 8.1
Nuclear Power Plants in Germany (2007)

Nuclear Power Plant	Type	Rated Capacity MW, gross	Electricity Generation GWh, gross, 2007
KKB Brunsbüttel	BWR	806	2.601
KBR Brokdorf	PWR	1.480	12.013
KKU Unterweser	PWR	1.410	9.530
KKK Krümmel	BWR	1.402	5.689
KWG Grohnde	PWR	1.430	11.460
KKE Emsland	PWR	1.400	11.594
KWB A Biblis	PWR	1.225	0
KWB B Biblis	PWR	1.300	935
KKG Grafenrheinfeld	PWR	1.345	10.901
KKP-1 Philippsburg	PWR	926	7.277
KKP-2 Philippsburg	PWR	1.458	11.777
GKN-1 Neckar	PWR	840	5.187
GKN-2 Neckar	PWR	1.400	11.114
KKI-1 Isar	BWR	912	7.041
KKI-2 Isar	PWR	1.475	12.009
KRB B Gundremmingen	BWR	1.344	11.053
KRB C Gundremmingen	BWR	1.344	10.353

Notes: PWR: pressurized water reactor (Druckwasserreaktor); BWR: boiling water reactor (Siedewasserreaktor).
Source: (http://www.kernenergie.de/r2/en/Gut_zu_wissen/KKW/?navanchor=2210015).

[242]

Figure 8.4
Nuclear Power Plants in Germany

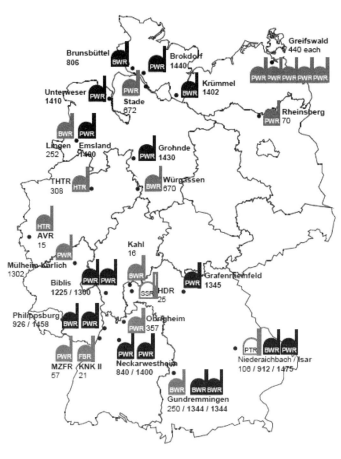

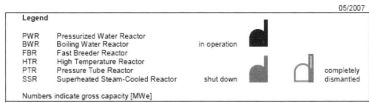

Source: Federal Ministry for the Environment, Nature Conservation and Nuclear Safety, 2007.

The overwhelming majority of nuclear installations is based in the western parts of Germany (see Figure 8.4), whilst the Eastern nuclear plants have been shut down. In the former GDR the only facility in operation is an interim storage and a planned final waste storage in Morsleben.

Table 8.2

Decommissioned Nuclear Power Plants (2007)

Name, Location	Rated Capacity (gross, MW)	Operating Period
HDR Großwelzheim	25	1969–1971
KKN Niederaichbach	100	1972–1974
KWL Lingen	268	1968–1976
KRB-A Gundremmingen	250	1966–1977
MZFR Karlsruhe	57	1965–1984
VAK Kahl	16	1961–1985
AVR Jülich	15	1967–1988
THTR Hamm-Uentrop	308	1983–1988
KKW Mülheim-Kärlich	1.302	1986–1988
KKW Rheinsberg	70	1966–1990
KGR 1–5, Greifswald	5 x 440	1973–1990
KNK II Karlsruhe	21	1977–1991
KWW Würgassen	670	1971–1994
KKS Stade	672	1972–2003
KWO Obrigheim	357	1969–2005

Source: (http://www.kernenergie.de/r2/en/Gut_zu_wissen/KKW/?navanchor=2210015).

The power plants decommissioned thus far have often been experimental, prototype and demonstration facilities built in the 1960s and 1970s, but also include the five units of the Nuclear Power Plant Greifswald, which were closed for general safety reasons (see Table 8.2).

Agreement on a Nuclear Phase-Out

The "Act on the structured phase-out of the utilization of nuclear energy for the commercial generation of electricity" of April 26, 2002 created new basic conditions in Germany for the use of nuclear power. It was based on the above-mentioned agreement between the federal government and the power utilities of June 14, 2000 (signed on June 11, 2001). The amended Atomic Energy Act is now geared toward the structured phase-out of nuclear power, rather than its promotion. The act includes six main points:

- a ban on the construction of new nuclear power plants;
- restriction of the residual operating life to 32 years from the commissioning of the plant;
- each nuclear power plant has a corresponding residual electricity volume;
- electricity volumes of older nuclear power plants can be transferred to newer plants;
- legal stipulations for regular safety reviews; and
- the financial security required for each nuclear power plant to cover possible damages is increased ten-fold, to €2.5 billion.

The major point of the agreement is that no new nuclear power plants will be built in Germany. The agreed starting point for a step-wise phase-out of the operation of nuclear power plants is an average total operating lifetime of 32 years. The restriction to 32 years has no technical justification but is based on a political decision in conjunction with a balancing of the benefits and risks of nuclear power by the legislator. In light of the fact that there are a number of exceptions, the real lifetime for some of the reactors will most likely be 35 years, as some critics emphasize. Moreover, it is possible to transfer the remaining electricity volumes that have been calculated for each power plant from older to newer ones. Therefore, the exact date for the nuclear phase-out cannot be given, but it should be around 2021/2022.

The decision on a structured phase-out has also had an impact on other policy areas. Up to 2000, Germany continued to grant *Hermes-Bürgschaften* (surety bonds) for the export of nuclear technology. Since April 26, 2001, when the new environmental guidelines came into force, these surety bonds have been dedicated to secure exports of renewable energy technologies.[7]

Moreover, the transport of nuclear waste for reprocessing abroad was stopped in 2005. For an (unlimited) transition period the waste is now stored in interim facilities that have been opened at each nuclear power plant site. So far, no clear decisions have been taken on a final repository for all kinds of radioactive and nuclear waste, and in particular heat-generating waste. As long as there is no operational final repository, radioactive waste must be put into interim storage. Spent fuel, for example, is stored in interim storage facilities in close proximity to the nuclear power stations.

There is also a moratorium on final waste disposal. As agreed between the federal government and the utilities with respect to the phase-out of nuclear energy on June 14, 2000, the underground exploration of the Gorleben salt dome was interrupted on October 1, 2000, for three to ten years. The idea is to use this period to efficiently clarify conceptual and safety-related disposal issues. Among these are the control of gas generation due to the corrosion of the casks and the decomposition of the waste. Moreover, the suitability of salt as host rock compared with other rocks such as clay and granite is to be analyzed.

This brings us to the political background and decision-making process in Germany.

German energy policy is informed and guided by a strategic triangle of energy policy objectives (see Figure 8.5). Any prudent energy policy must be directed to secure the steady, adequate and uninterrupted supply of energy at reasonable prices and in environmentally and climate-friendly conditions. Therefore the three objectives of supply security, economic efficiency and environmental and climate compatibility form the angles of the triangle. However, the objectives may be conflicting, and this requires a balanced approach. The policy outcome depends on the hierarchy of the concerns and on agenda-setting.

Decision-Making and Energy Policy in Germany

Figure 8.5
Triangle of Energy Objectives

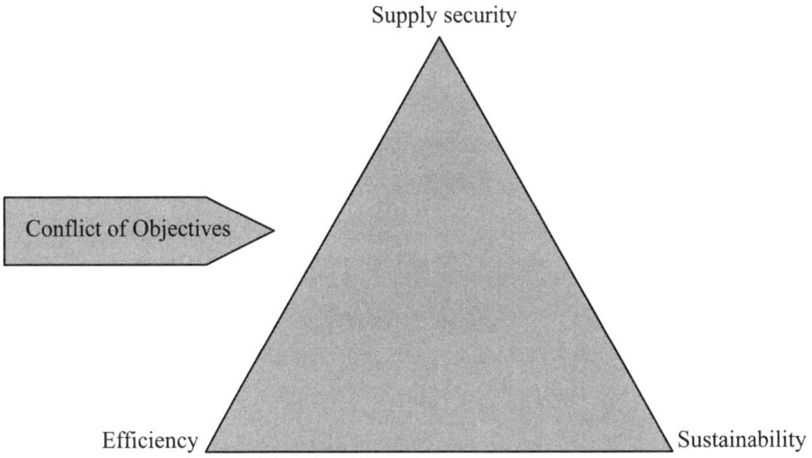

Source: Author.

One may argue that the German institutional architecture reflects, to a certain degree, the strategic triangle (see Figure 8.5). Germany has no ministry of energy, and some complain about the allocation of responsibilities in the energy policy between the Federal Ministry of Economics and Technology and the Federal Ministry for the Environment, Nature Conservation and Nuclear Safety, and perceive it as a hindrance to an effective energy policy. In fact, however, besides the two federal ministries mentioned above, a number of ministries and authorities deal with energy policy, including the Foreign Ministry, Federal Ministry of Education and Research, the Federal Ministry of Transport, Building and Urban Affairs and the Federal Ministry for Economic Cooperation and Development. This reflects the fact that energy is an issue that cuts across traditional bureaucratic boundaries.

Figure 8.6
Major Players in German Energy Policy

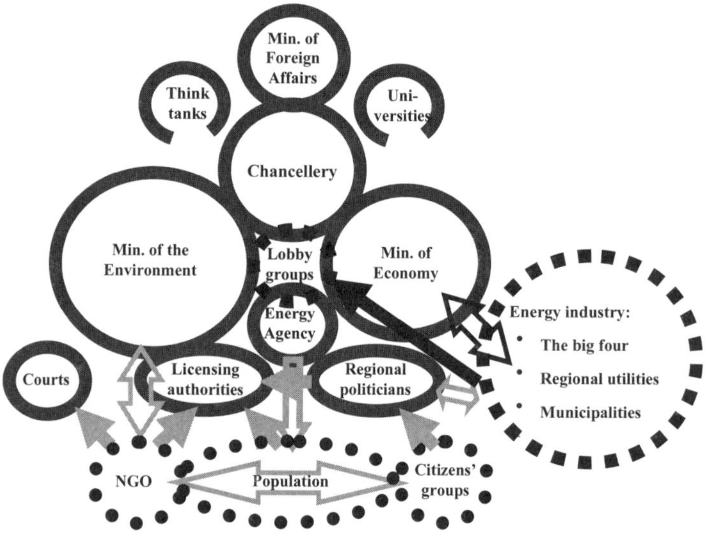

Source: Hobohm, 2008.

Moreover, the very nature of Germany as a federation gives the federal states (*Bundesländer*) a certain say in energy policy, too. As regards nuclear power, for instance, for construction and operation of a nuclear power plants two licenses, eventually granted together, were needed. The German states, the *Länder*, evaluated the new plant's expected safety. After this procedure, it was possible to split the respective partial licenses for siting and construction of essential civil structures; construction itself; handling and storage of fuel elements (in particular initial fuel loading); nuclear commissioning; and finally continuous operation.[8] A regional planning procedure was performed before the beginning of the nuclear procedures. Competent authorities performed the Environmental Impact Assessment (EIA) and reviewed water utilization, emissions protection and local environmental conservation. One public hearing was held and all application documents were disclosed at nearby locations for two months.

In Germany, there are roughly fifteen main acts, ordinances and criteria dealing with nuclear energy.[9] The German Energy Agency, founded in 2000, acts as the center for renewable energies and energy efficiency in the country.

Without doubt, energy policy is a field where large and well-organized lobby groups play a role. Despite the fact that more than 1,000 companies are active in the German electricity market, the market is dominated by an oligopoly. The majority are small municipal companies (*Stadtwerke*), but generation and transmission are dominated by four big companies that produce and sell almost half of Germany's 630 TWh/year of electricity. The leading company is E.On, with more than a seventeen percent share. The others are RWE, EnBW and Vattenfall.[10] All four have close ties and (cross-ownerships) with the gas industry and/or the coal or lignite industry. Electricité de France, with its huge share of nuclear power, owns more than a third of EnBW.

As has been described, energy policy is a policy area where non-governmental organizations (NGOs) and citizen pressure groups are most active in Germany on different levels (federal, state and local levels). A number of referenda have been linked to projects in the energy industry and to power generation, etc. There exist various conflicts between the views of societal groups, political parties and the energy industry on the diversification of energy sources (see Table 8.3).

The table illustrates the fact that energy policy is discussed from different angles and in differently informed fora. In general, energy supply and climate change have moved up on the political and societal agenda in Germany since 2006. The reasons are manifold, but three major factors have raised the importance of the issue for the public: price jumps in energy have affected the views of the population and led to more public sensitivity on the issue; the Russian-Ukrainian gas dispute in 2006 raised the public's awareness of Germany's high import dependency when Europe was hit by gas shortages; and, last but not least, the UN Inter-governmental Panel on Climate Change (IPCC) reports released in 2007 drew public attention to the ongoing debate over climate change. These three issues reflect the points of the strategic triangle and highlight the necessity of formulating a balanced approach toward energy and climate security.

Table 8.3

Energy Sources and Political Standpoints

	GOVERNMENT	ENERGY INDUSTRY	POPULATION
Nuclear Energy	grand coalition stand-off	extension, delay tactics	50–50 split over extension, climate change: pressing issue
Coal/ Lignite	yes, but CCS needed	first choice	"not in my backyard"
Gas	security, storage, dependence on Russia	gas to power is expensive, heat market shrinking	dependency, price hikes; if fossil used, gas is best option
Renewables	ambitious goals	rethinking	silver bullet

Source: Author (based on Hobohm, 2008).

In the political arena, the major topics discussed have been climate change and resource constraints as well as the reliability of Russia as the main energy supplier to Germany. Moreover, coal and the move toward carbon dioxide capture and storage (CCS) is being discussed intensively.

The extension of nuclear power has long been a "non-policy issue" because of the stand-off in the grand coalition.

The population has been hit by oil and gas price developments in 2002–2008. In 2008, the Germans spent around €50 billion on electricity and heating.[11] This is around nine percent of GDP. Electricity prices have increased significantly: major customers paid €19/MWh in 2000 compared to €41.58/MWh in 2007. End-consumers paid €0.13/KWh in 2000 compared to more than €0.19/KWh in 2007. This has resulted in increased awareness of energy-saving measures and technologies. Other issues raised by the media have been the high dependency on Russia and personal connections between energy industry and politics. Concerns surrounding these issues have been fuelled by high energy prices and criticism from Brussels that competition has not been properly established in Germany. In turn, the energy industry has been concerned with the changing legal situation governing nets and grids and vertically-integrated companies. Since the first EU directives in 1997 on the electricity market, the legislative and regulative environment for companies has been changing constantly.

As regards the strategic triangle, the primary objective was environmental compatibility and sustainability while the Social Democratic–Green Party coalition was ruling between 1998 and 2005. The grand coalition since 2005 has perpetuated this. As mentioned above, generally the electricity mix is an outcome of the choices of the electricity companies. The political instruments used in energy policy range from regulatory measures (such as restrictions, thresholds and sanctions) to fiscal instruments (such as tax exemptions, subsidies and taxes) and land-use planning.[12]

On August 23, 2007 the government approved an "Integrated Energy and Climate Program" at a meeting in Meseberg. This was a decisive step because it took up the EU Commission's "Climate and Energy Package" of January 10, 2007 that formulated the 20+20+20 targets. The "Meseberg Program" was also planned in preparation for the Word Climate

Conference in Bali and was approved during Germany's presidency of the G8 in 2007 and backed by the Heiligendamm process. The Meseberg package[13] foresees a reduction of greenhouse gas emissions by 30–40 percent by 2020 against the level of 1990. Thus far, Germany has reduced these emissions by 18 percent.

The Integrated Energy and Climate Program encompasses two packages: the first package of December 5, 2007 is composed of fourteen legislative proposals; the second smaller package of June 18, 2008 consists of six regulative and legislative proposals, they include: amendment to the combined heat and power act, building standards (Energy Saving Ordinance [EnEV]), the Renewable Energy Sources Act (EEG) / Renewable Energies Heat Act (EEWärmeG), the grid extension acceleration law and other measures to improve energy efficiency (e.g. new metering systems, etc.).

Regarding nuclear energy, however, the grand coalition is paralyzed by a stand-off. So far, on the pending issues of final storage, no final decision has been reached.

In January 2009, the Federal Ministry for the Environment released a Road Map ("New Thinking, New Energy: Ten Principles for Sustainable Energy Supply")[14] that approves the Nuclear phase-out. However, the financial and economic crises, and the recent gas crisis between Russia and Ukraine has strengthened arguments for a step back from the nuclear phase-out. The elections to the federal parliament in September 2009 may see a conservative–liberal coalition that is in favor of the extended use of nuclear power in Germany.

Pros and Cons of the Debate in Germany

Security and safety issues have always been at the heart of the debate over nuclear power. Germany has a good reputation in applying international safety standards; the member states of the EU are subjected to the safeguards system of the European Atomic Energy

Community (EURATOM). The EURATOM system is verified by the International Atomic Energy Agency (IAEA). This encompasses the fuel cycle and measurements of radiation. Moreover, containment and surveillance provisions for safe entombment are covered. Also, as member of the Non-Proliferation Treaty (NPT), Germany must maintain verification and safeguard standards. Germany has also used the civil reprocessing installations in France (Le Hague) and Great Britain (Sellafield).

Germany is a signatory of the Joint Convention on the Safety of Spent Fuel Management and on the Safety of Radioactive Waste Management.[15] The Joint Convention aims to achieve and maintain a high level of safety world-wide in spent fuel and radioactive waste management through the enhancement of national measures and international co-operation, including – where appropriate – safety-related technical co-operation. Compliance with the provisions of the Joint Convention is dealt with within the scope of review meetings taking place at least every three years. For these meetings reports are discussed which are made available to all signatory countries.[16] This task is assigned by the Federal Ministry for the Environment.

One of the major pending issues is radioactive waste storage. The Federal Office for Radiation Protection is fulfilling executive tasks for radioactive waste management, federal custody of nuclear fuels and transportation of radioactive material pursuant to the Atomic Energy Act, the Radiation Protection Ordinance and the Law on the Transportation of Dangerous Goods. Consequently, it operates federal facilities for the disposal of radioactive waste and is responsible for the execution of federal custody. Moreover, it is the licensing authority for the transportation and storage of nuclear fuels; it decides, for example, issues relating to the transport of spent fuel elements from nuclear power plants and approval of decentralized storage facilities at power plant sites.[17]

[253]

The provisions of the amendment to the Atomic Energy Act of April 2002 commit the operators of nuclear power plants to maintain in-situ interim storage facilities nearby nuclear power plants. According to the ban on reprocessing in France and Great Britain from June 30, 2005, and with the central interim storage facilities currently already in operation, it is calculated that further transports in the Federal Republic of Germany, such as from nuclear power plants to central interim storage facilities, will not be required.[18] The high-level radioactive waste (HAW, in vitrified waste containers) produced in the process of reprocessing German fuel elements in La Hague and in Sellafield must be transported back to Germany from France and Great Britain. It can only be stored in the Gorleben Transport Cask Interim Storage Facility (Gorleben TBL), since only this interim storage facility holds the appropriate license for intermediate storage. The CASTOR (special storage casks for spent fuel elements) transports from France and Great Britain to the interim storage facility in Gorleben are always accompanied by public demonstrations. In sum, the following waste streams need to be taken back by the utilities operating nuclear power plants (Table 8.4).

Table 8.4
Nuclear Waste Streams to Germany

From France	Vitrified fission product concentrate (CSD-V, compacted waste), CSD-C (originally cemented waste) medium-level radioactive vitrified products; and CSD-B (originally bituminised waste).
From Great Britain	Vitrified fission product concentrate (originally additionally medium-level radioactive heat-generating cemented waste and low-level and medium-level cemented waste). On account of substitution only one single waste stream will be returned to Germany: HAW vitrified waste canisters.

[254]

Table 8.5

Planned Transports from Reprocessing (July 2008)

	Radioactive waste	Number of casks[*]	Anticipated period of transport
AREVA NC	HAW vitrified waste canisters (CSD-V)	33	2008–2011
Sellafield, Ltd.	HAW vitrified waste canisters	21	2012–2015
AREVA NC	High-pressure compacted waste (CSD-C) radioactive waste	150 (max.)	2012–2025
AREVA NC	MAW vitrified product (CSD-B)	20 (approx.)	2015–2017

Notes: *cask: such as CASTOR, TGC36.

Source: Bundesamt für Strahlenschutz, 2008.

The major ongoing discussions concern final storage.[19] Germany's disposal policy is based on final storage within the country's borders; radioactive waste from Germany cannot be exported to other countries. Ultimately, a permanently safe disposal option in deep geological layers is required for waste with negligible heat generation originating primarily from the decommissioning of nuclear power plants, but also from research establishments, industry and, in small amounts, from the medical sector. Over 88,000 cubic meters of radioactive waste with negligible heat generation is currently being stored in temporary facilities and state collection depots. Its disposal is a national responsibility that has been given to Germany's Federal Office for Radiation Protection.

Since 2002, with the approval of the Konrad repository, such a system has been operating in Germany. It is the first repository for radioactive waste with negligible heat generation to be approved under Germany's Atomic Energy Act. The Konrad repository was approved by the Lower Saxony Ministry of the Environment in 2002, after a planning approval procedure lasting twenty years. The Konrad approval was also certified at supreme-court level on March 26, 2007. On that day, Germany's Federal Administrative Court in Leipzig rejected, without appeal, objections to the Konrad repository. The decision about the final storage facility for waste with high heat generation is still due to be taken. The moratorium on Gorleben is still in place.

The major concerns about nuclear power in Germany are related to safety and security issues that range from radiation and contamination to nuclear catastrophe, and are linked to all stages of the fuel cycle.[20] Studies in Germany have been published that show the rate of cancer among children is higher in the vicinity of nuclear power plants.[21]

Security concerns are also linked to the proliferation of nuclear weapons technology and materials and the danger of nuclear power plants and installations becoming targets for terrorists.

From an ecological point of view the major argument in favor of nuclear power plants has been the relatively low level of carbon dioxide emissions over the whole lifecycle of nuclear power plants. However, the issue of the final storage of highly radioactive waste with significant thermal dissipation has not been solved and creates a problem for future generations. In a strict sense, this is not compatible with ethical and democratic principles.

From an economic point of view, nuclear power plants have high investment costs, but the operation is cheap—in particular, the costs for uranium account for only a small share of the running costs. This is true because many costs are externalized to the public or benefit from tax exemptions. Companies are obliged to establish reserves for decommissioning, deconstruction of sites and waste storage. These reserves are exempt from taxes. Moreover, the fact that the energy sector has been liberalized, privatized and deregulated creates a new economic and financial situation for the nuclear power industry. It would be extremely difficult to generate the necessary funds to build or replace a nuclear power plant if a future German government opted for a continuation of the nuclear path. The major factor is that of public opinion, which could translate into new government coalitions. German public opinion with regard to nuclear energy has traditionally represented that of Europe as a whole: with only one quarter in favor of greater use of nuclear power.[22] Eighty-one percent of the population is in favor of the extended use of renewables for power generation. By February 2007, only a slight majority (51 percent) of the German population was still backing the nuclear phase-out. This trend was reversed, however, in July 2008 when a slight majority (54 percent) was revealed to be in favor of nuclear power beyond 2021.[23]

There are several external factors that must be taken into consideration when it comes to the question of the electricity mix in Germany and the possible extension of the life of nuclear power plants. The developing internal EU electricity market is a significant factor for German supply. There are many signs that the energy mix of the 27 EU member states will vary owing to geographic and geological differences but also because of diverging public attitudes toward different technologies and energy sources. Geographic and geological differences have an influence on the renewables available such as wind, solar, hydro or geothermal energy. These sources are unequally spread and member states have varied opportunities to use such energy sources. However, this can be a comparative advantage; to a certain extent, the same applies to public acceptance of nuclear energy or coal-fired plants, for instance. In any case, the future development of electricity markets becomes more vague and uncertain with liberalization and the evolution of the EU internal market. This places a question mark over the "renaissance" of nuclear energy in Europe. The high capital costs mean an amortization time of about twenty years. Private companies tend to opt for technologies with a shorter span of amortization. Investments in new nuclear power plants presuppose a strong involvement of private and public actors that provide and guarantee the necessary credit.

Figure 8.7

German Voltage Grid and UCTE Neighbors

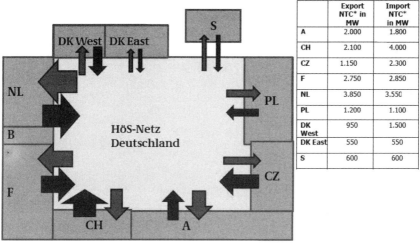

	Export NTC* in MW	Import NTC* in MW
A	2.000	1.800
CH	2.100	4.000
CZ	1.150	2.300
F	2.750	2.850
NL	3.850	3.550
PL	1.200	1.100
DK West	950	1.500
DK East	550	550
S	600	600

Notes: "HÖS-Netz Deutschland" is the German voltage grid; NTC: net transport capacity.
Source: DENA, 2008.

This wider spectrum of the energy mix reflects an integrated electricity market that functions under the principles of solidarity and competition that dominate EU policy. This will raise discussions in Germany over whether it makes sense to abandon certain technologies internally but to then import nuclear power from other states. In a number of neighboring countries, use of nuclear power has been extended. France and Finland are building new nuclear power plants. The German maximum voltage grid is part of the Union for the Coordination and Transmission of Electricity (UCTE) grid. Therefore, nuclear-generated electricity is crossing Germany's borders. Several EU countries are sticking to nuclear energy and are even thinking about increasing and substituting capacities: i.e., the Netherlands, the UK, Poland, Sweden, the Czech Republic, Slovakia, Lithuania (in cooperation with Estonia and Latvia), Bulgaria and Romania. Not surprisingly, Eastern European

[258]

countries in particular favor nuclear energy as a means to reduce dependency on Russia. Moreover, they are forced to face increasing demand of four percent per annum, double that of Western Europe. Furthermore, their infrastructure is old and out-dated. This has repercussions for Germany. Six nuclear reactors (3.5 GW) operate in the Czech Republic and two more reactors (1.5 GW) are currently planned.

France is the second largest nuclear power in the world with 59 nuclear power plants and an overall capacity of around 64 GW. Seven nuclear power plants are operating in Belgium and providing almost 60 percent of national electricity consumption (a discussion about their extension is under way). One nuclear power plant is operating in the Netherlands (its life-time has been extended to 2033). Ten nuclear power plants are running in Sweden and are providing 48 percent of electricity (there are ongoing discussions about modernization programs for these reactors). Switzerland has two nuclear power plants with a capacity of 1.1 GW, and plans to replace them with a capacity of 1.6 GW.

With regard to the external dimension, Germany seeks to introduce standards and measures for better control of the whole nuclear power cycle. It aims to multilateralize access to nuclear fuel and enrichment cycles.[24] Yet, the coalition stand-off complicates Germany's "nuclear diplomacy" for domestic reasons, but also owing to foreign perceptions. This makes it more difficult to effectively promote its broader aims.

Electricity or Efficiency Deficit?

In 2008 the German Energy Agency warned of an electricity deficit. The nuclear phase-out implies that more than 25 percent of electricity-generating capacity must be replaced step-by-step up to 2020. Energy intensity has significantly decreased in Germany over the past fifteen years. The same is true for electricity intensity—that is, the relation between gross domestic production and gross domestic electricity consumption. The demand for electricity has increased though, with the growing economy.

[259]

The political decision on the nuclear phase-out has consequences for electricity supply. Based on decreasing intensity, economic growth and the share of renewables, experts predict an electricity supply gap in the range of 3.5–20 percent by 2020.[25] It is very difficult to say which scenario is most likely because electricity demand is dependent on a variety of factors. In particular, it is very difficult to predict the increase in efficiency in consumption and generation. Nuclear power plants traditionally provide the base load because of high inflexibility in the processes. In the EU, there is ongoing debate about the level of base load that is required. Future developments in the energy mix, as well as increased efficiency in supply and demand, place a question mark over this technology, as do uncertain levels of consumption and demand in the future. New technologies such as intelligent grids that use software and virtual power plants to coordinate transmissions and distribution more effectively will fundamentally transform the whole electricity system. In a combination of centralized and decentralized units, nuclear power might function as a braking technology for the transformation of the electricity system. Others, however, argue that nuclear power must be extended as a bridging technology until the system is fit enough to accept a large share of renewables.

There is common consensus in Germany and the EU that the major concern will be increasing energy efficiency.[26] Besides, widening the energy mix, and in particular the trajectory to explore solar power for a large scale use, are important steps for sustainable energy security.

Germany is traveling an unexplored path and there are many signs that the structured nuclear phase-out is exerting the necessary pressure to move forward toward a more sustainable energy system. In other words, without the phase-out there would no immediate need to advance reform, innovation and modernization. The big unknown that remains concerns energy efficiency, but it is certain that the solution will require huge financial investments in electricity grids.

Section 4

NUCLEAR ENERGY IN THE REGION: A COMPARATIVE OUTLOOK

9

Nuclear Energy in India: In Retrospect and Prospect

R. Chidambaram and *R.K. Sinha**

India is the fifth largest producer of electricity in the world. In the year 2007–08, total generation in the country was 704.47 billion units. Coal is the main source of electricity generation in the country. The share of different energy resources in the Indian electrical energy mix is shown in Table 9.1.

Table 9.1
Total Installed Capacity in India (October 31, 2008)[1]

Fuel	Installed Capacity (MWe)	Percentage share
Total Thermal	92,892.64	64.6
Coal	76,988.88	53.3
Gas	14,704.01	10.5
Oil	1,199.75	0.9
Hydro (Renewable)	36,497.76	24.7
Nuclear	4,120.00	2.9
RES** (MNRE)	13,242.41	7.7
Total	1,46,752.81	

* The authors acknowledge, with gratitude, the assistance provided by I.V. Dulera and J. Aparna in the preparation of this paper.
**Renewable Energy Sources (RES) include Small Hydro Power, Bio-Gas, Urban & Industrial wastes, and Wind Energy.

[263]

India's emissions, at 1,342 million tons of CO_2 per year, amount to about 4.6 percent of global emissions. Per-capita emissions are among the lowest in the world, at 1.2 tons CO_2/person/year.[2] Energy security considerations demand that a large developing country like India makes the maximum use of its domestic energy resources.

Three Stages of the Indian Nuclear Power Program

Domestic reserves of uranium in India are limited, whilst those of thorium are quite large. In order to fully exploit these resources, India has adopted a three-stage nuclear power program, based on a closed fuel cycle, requiring reprocessing of spent fuel from every reactor so as to judiciously utilize the available fissile material for peaceful purposes. In the first stage of the indigenous program, natural uranium-fuelled pressurized heavy water reactors (PHWRs), which produce electricity efficiently, also generate plutonium to launch the second stage.

The second stage, based on fast breeder reactors (FBRs), initially using plutonium recovered from the spent fuel of the first stage reactors, multiplies the installed capacity several times by converting the fertile material (reprocessed uranium, and at a later stage thorium) to fissile material whilst producing electricity.[3] The third stage envisages utilization of thorium to provide a sustainable supply of energy to meet the country's needs. A multiplication of India's fissile inventory is also needed to establish a higher power base for using thorium in the third stage of the program. These three stages are depicted in Figure 9.1.

Historical Growth and Current Status of Nuclear Power in India

Dr. Homi Jehangir Bhabha formulated the strategy for setting up a nuclear research program in India even before India had attained independence. He wanted India to be self-reliant in this newly emerging area. The Indian nuclear program began in 1945 with the establishment of the Tata Institute of Fundamental Research (TIFR).

Figure 9.1

Phases of the Indian Nuclear Program

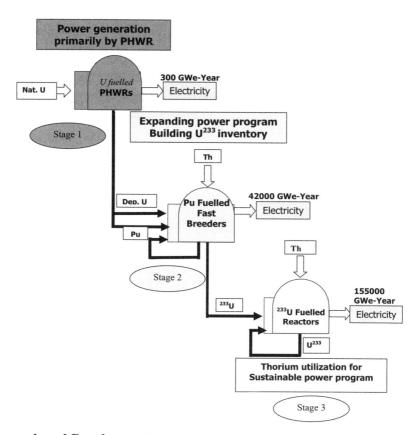

Research and Development

Bhabha Atomic Research Centre

In 1957, research and development work specific to nuclear energy was shifted to the newly established Atomic Energy Establishment Trombay (AEET), renamed the Bhabha Atomic Research Centre (BARC) in 1967. Over the past several decades, a multidisciplinary infrastructure for conducting research and development in nuclear sciences and engineering has been set up at BARC. These include several research reactors, and a large number of laboratories and other research facilities dealing with basic as well as applied sciences and engineering development. Most of the other units of the Indian Department of Atomic Energy (DAE), dealing with various aspects of the nuclear power program, also had their origins in BARC.

RESEARCH REACTORS

One of the earliest decisions Dr. Bhabha made was to build research reactors of different types to develop an intimate understanding of the complex issues involved in the control of the nuclear chain reaction. The design of reactors involves optimization of geometry, fuel design, safety, materials selection, irradiation behavior of fuel and structural materials. The first of these research reactors was a swimming pool-type reactor (see Figure 9.2), aptly christened "APSARA" – after the celestial water nymph – by the first Prime Minister of India, Pandit Jawaharlal Nehru.[4] With APSARA, India became the first Asian country outside the erstwhile Soviet Union to have designed and built its own nuclear reactor. It is useful to recall how close the date of criticality of APSARA (August 4, 1956) was to the first electricity-generating reactor, the AM-1, a 5 MW graphite-moderated and water-cooled reactor that began operation at Obninsk in Russia on June 27, 1954. The next crucial step materialized in 1960 with the building of CIRUS, a high power (40 MWth) research reactor. This reactor, then known as the Canada India Reactor and now as CIRUS, was built in collaboration with Canada. Experiments carried out with CIRUS and APSARA have provided the necessary confidence and expertise for the design and safe operation of many other nuclear power reactors in the country. In early 1961, a zero energy critical facility named ZERLINA (Zero Energy Reactor for Lattice Investigations and New Assemblies) was built to study various geometrical aspects (lattice parameters) of a reactor fuelled with natural uranium and moderated with heavy water. The next logical step was to build a critical facility, which used plutonium as fuel. Such a test reactor was built in 1972 and was named

PURNIMA (Plutonium Reactor for Neutron Investigations in Multiplying Assemblies). This reactor was intended for studying the behavior of plutonium fuel in a pulsed fast reactor (which was contemplated as an experimental facility for neutron scattering experiments but ultimately was not taken up for construction). Following this, a critical facility called PURNIMA-2 was also designed, with a solution containing 400g of uranyl (based on U-233) nitrate serving as the fuel for this facility. It attained criticality in 1984. The need was then felt for a research reactor with even larger neutron flux and irradiation volumes than CIRUS, to meet the growing requirements for radioisotopes and research. This culminated in building of a totally indigenous 100 MW research reactor, boasting the highest flux in Asia at that time. It attained criticality in August 1985 and was named DHRUVA (see Figure 9.3).[5]

Figure 9.2

The APSARA Reactor

Figure 9.3
The DHRUVA and CIRUS Reactors

Indira Gandhi Center for Atomic Research

Initial R&D work on fast reactors, including on sodium technology, was started at BARC. The work on setting up a Reactor Research Centre at Kalpakkam was begun in 1971, essentially to pursue a program for developing fast breeder technology. RRC was renamed the Indira Gandhi Center for Atomic Research (IGCAR) in December 1985. This Center has modern facilities for developing and testing fast breeder reactor materials, components and systems including those required to work with high temperature sodium.

A fast breeder test reactor (FBTR) was commissioned in 1985, with indigenous plutonium–uranium mixed carbide fuel, providing valuable design and operational experience. Based on the successful operation of the FBTR, it was decided to embark on the commercial phase of the fast reactor program, in terms of a 500 MWe FBR at Kalpakkam, in 2003.

As part of the studies with U-233 fuel, a 30 kWt pool-type research reactor, KAMINI (Kalpakkam MINI), was designed and built (Figure 9.4).[6] Prior to this, a mock-up of the core of this reactor was built and became critical in April 1992. It was given the name PURNIMA-3. KAMINI was became operational in 1996. This reactor is being extensively used as a neutron source for research applications such as neutron radiography of irradiated nuclear fuel and pyro devices for the Indian space program.

Figure 9.4

The KAMINI Reactor

Commercial Nuclear Power Plants

Nuclear Power Corporation of India Limited (NPCIL)

Initially, work on the PHWR program was conducted at BARC. In 1967, the Department of Atomic Energy constituted a Power Projects Engineering Division (PPED) with the responsibility for design, engineering, procurement, construction, commissioning, operation and maintenance of atomic power plants. With the proposed expansion of the nuclear power program, a Nuclear Power Board was constituted in 1984 to implement the program. The Nuclear Power Board was converted into a

Corporation and the Nuclear Power Corporation of India (NPCIL) was registered as a Public Limited Company in 1987.

The setting up of PHWRs and associated fuel cycle facilities is already in the industrial domain. NPCIL is presently operating seventeen nuclear power units (including two BWRs) at six locations, is implementing the construction of six (including two PWRs and one FBR) on-going nuclear power projects, and is handling other related activities. The existing operating power stations are: Tarapur Atomic Power Station (TAPS) Units 1, 2, 3, and 4 in Maharashtra; Rajasthan Atomic Power Station (RAPS) Units 1, 2, 3, and 4 in Rajasthan; Madras Atomic Power Station (MAPS) Units 1 and 2 in Tamil Nadu; Narora Atomic Power Station (NAPS) Units 1 and 2 in Uttar Pradesh; Kakrapar Atomic Power Station (KAPS) Units 1 and 2 in Gujarat; and Kaiga Atomic Power Station Units (Kaiga) 1, 2, and 3 in Karnataka.

The Tarapur Atomic Power Station (TAPS)

The Tarapur Atomic Power Station (TAPS), based on a BWR, was an exception to the three-stage nuclear power program drawn up by the AEC. This deviation was also based on the fact that the power shortage in the western region could be most economically reduced by building a nuclear power plant. This station, built by the General Electric (GE) Company, USA on a turnkey basis, began commercial operation in 1969. The construction of TAPS was very useful in acquiring initial experience in the construction and operation of nuclear power plants.

Pressurized Heavy Water Reactors (PHWRs)

In parallel with the construction of TAPS-1 and -2, work started on a second nuclear power station near Kota, Rajasthan, as a joint Indo-Canadian venture. The first unit of this station began commercial operation in 1972. This was followed by the construction of one more reactor at Rajasthan and two more reactors at Kalpakkam. The emphasis at this stage was on increasing local participation in design, equipment

[270]

manufacture and construction. A standardized Indian version was designed at this point, with several important improvements such as two fast-acting shutdown systems, high pressure as well as low pressure Emergency Core Cooling System (ECCS) injection, full double containment, new ball-filled end-shield design, etc. This design was used from Narora onwards. A chronological account of the construction of currently operating Indian nuclear power plants is available elsewhere.[7,8,9]

Nuclear power in India has passed through various stages of development. TAPS-1 and -2, and RAPS-1 served to demonstrate the technology through international co-operation. Subsequent efforts led to the indigenization of the technology (RAPS-2, MAPS-1 and -2), followed by standardization and consolidation of the knowledge base (commissioning of NAPS-1 and -2, KAPS-1 and -2, RAPS-3 and -4). India has now successfully achieved commercialization of the PHWR technology and is engaged in the design and deployment of enhanced capability PHWRs.

On the route from the Rajasthan-1 to Kakrapar-2, the designs of the PHWRs were progressively upgraded to further improve their safety, economics and reliability. At present, three units of 220 MWe PHWRs are in an advanced stage of construction at Kaiga and Rajasthan. While TAPS-3 and -4 are 500 MWe PHWRs, a program of construction of additional PHWRs of 700 MWe capacity has been chalked out. In addition, two VVER type PWRs, each of 1,000 MWe capacity, being built with Russian collaboration, are in the advanced stages of construction in Kudankulam, Tamil Nadu.

Gestation Period and Construction

The construction period of Indian nuclear power plants is comparable with the best practices in other plants in the world (about five years). The completion cost of an indigenous Indian nuclear power plant (PHWR) is substantially lower than international costs.

Fast Reactor Program

Studies on the content of the FBR program, and type of test reactor to be built, were undertaken in the early 1960s. A collaboration agreement was signed in 1969 with France for technical know-how to build a test reactor in India similar to the French RAPSODIE reactor. In order to gain experience with the steam generators and power plants in the context of FBRs, it was decided to add these facilities to the FBTR.

The construction of the FBTR was completed in 1984. Critical components of the FBTR, such as the reactor vessel, rotating plugs, control-rod-drive mechanisms, sodium pumps, steam generators and component-handling machines were manufactured in India with know-how from France. Only 20 percent of the total cost of the reactor was in foreign exchange, paid mainly for know-how and raw materials. Sodium for the reactor was procured from local suppliers and purified in IGCAR. After commissioning of various systems in 1984–85, the FBTR was made critical in 1985. The reactor produced nuclear steam in January 1993 and reached a milestone when the power level was increased to 10.5 MWt in December 1993. The rolling of the turbine using nuclear steam was achieved in 1996 and the reactor was connected to the grid on July 11, 1997. A highlight of the operation of the FBTR is the excellent performance of the sodium pumps, intermediate heat exchangers and the steam generator.

After completing the design, associated R&D and the development of manufacturing technology for a 500 MWe prototype fast breeder reactor (PFBR), whose characteristics[10] are given in Table 9.2, the construction of the PFBR plant at Kalpakkam was started in 2003 and is now in an advanced stage. An independent corporation called BHAVINI has been set-up for this purpose. This new venture exemplifies the synergy between the research and development strengths of Indira Gandhi Centre for Atomic Research and the project planning and construction expertise of NPCIL.

Table 9.2

Main Characteristics of the PFBR

Thermal power	1250 MWt
Electrical power	500 MWe
Primary circuit concept	Pool
Reactor coolant	Sodium (Na)
Fuel	$PuO_2 - UO_2$
Core height/diameter	1/2
Fuel pin diameter/no. of pins per Sub Assembly	6.6 mm / 217
Number of Primary Sodium Pumps	2
Number of IHX	4
Number of secondary loops	2
Number of Steam Generator (SG) per loop	4
Number of TG	1
Primary sodium temp. at reactor inlet/outlet	670/820 K
Secondary sodium inlet/outlet temp.	798/628 K
Water temp. at SG inlet	508 K
Steam condition at SG TSV	763 K at 16.7 MPa
In vessel fuel handling	2 Rotatable plugs + 1 Transfer arm
Containment building	RCC rectangular shape
Reactor site	Kalpakkam
Design life	40 Years

In this context, it is instructive to examine the current status of fast reactors in the world. Currently, France (Phenix) and Russia (Beloyarsky-3) have one operating fast reactor each. France (Super-Phenix) has one reactor under permanent shutdown, as do the USA (Enrico Fermi-1), Germany (KNK-II) and Kazakhstan (BN350). The United Kingdom (Dounreay) has two reactors under permanent shutdown. Japan (Monju) has one fast reactor under long-term shutdown. India, which is building the PFBR, is one of two countries which are building new fast reactors; the other being Russia (Beloyarsky–4).

[273]

In an open fuel cycle, the available world uranium resources cannot meet – in the foreseeable future – the lifetime fuel demand for new reactors. Thus, use of the open fuel cycle will limit the contribution of nuclear energy as a clean option to address the climate change threat, and may drive an enhanced use of fossil fuels to make up for the energy deficit. In other words, for nuclear energy to be a *sustainable* mitigation technology in the context of the climate change threat, it is essential to close the nuclear fuel cycle with fast breeder reactors.[11]

Infrastructure for Nuclear Power: Establishment and Upgrades

Nuclear Fuel

The first nuclear-grade purity ingot of uranium was produced in India on January 19, 1959, and over half the initial core of metallic uranium fuel for the Canada India Reactor (CIRUS) was fabricated in 1960 indigenously, its performance being as good as that supplied by Canada. The Uranium Metal Plant (UMP) was operated continuously from 1959 to 1980 and later underwent an expansion to meet fuel demands of the research reactor DHRUVA.[12]

Work on the manufacture of uranium dioxide as a power reactor fuel began in the early sixties—about the same time as in other countries. In fact, when ZERLINA needed a new core in 1965, two tons of uranium dioxide fuel elements of 19-rod cluster type were fabricated for this reactor. The confidence gained led to a decision to make fuel bundles for half of the initial core for the first PHWR at Rajasthan. This technology development led to the creation and establishment of a facility for the production of uranium dioxide-based fuels as well as reactor components at the Nuclear Fuel Complex (NFC) at Hyderabad in the early seventies. The oxide fuel for PHWRs is manufactured at NFC from the yellow cake (uranium concentrate) obtained from the Uranium Corporation of India

Ltd. An enriched fuel fabrication plant for the fabrication of fuel for TAPS reactors has also been established at NFC.[13]

Figure 9.5
FBTR Fuel Sub-assemblies

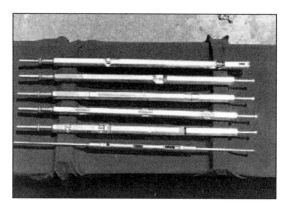

Figure 9.6
MOX Fabrication Facility at AFFF, Tarapur

Simultaneously, efforts were in progress during the early sixties and seventies to develop and fabricate the more complex next generation fuel required for the second stage of India's nuclear program—plutonium-based fuels. The facility to develop and fabricate plutonium-based ceramic, alloy and dispersion fuels was set up and used for the first time in 1972 for PURNIMA-1, which used pure plutonium oxide fuel. With this experience, a facility to fabricate uranium–plutonium mixed oxide fuel for both thermal and fast reactors was set up at Trombay. An important achievement was the fabrication of mixed carbide fuel of high plutonium content. Earlier, mixed oxide fuel with high plutonium content was found to be chemically incompatible with sodium, with which the fuel would be in contact in case of a breach in the cladding tube. This indigenously-designed and -developed fuel set a record for mixed carbide fuel when it was used as a driver fuel for the first time in the world (Figure 9.5). This unique fuel was successfully fabricated for the entire core of the FBTR and has successfully seen a burn-up of over 154 GWd/t, without any fuel failure.

Another high-point has been the successful fabrication of uranium–plutonium mixed oxide fuels as an alternate for enriched uranium fuel for the BWR at Tarapur (Figure 9.6) and as a resource conservation measure for the natural uranium fuelled PHWRs. The in-pile irradiation experiments carried out at BARC during the 1980s gave greater confidence in the capability to design and fabricate plutonium-based fuels. This led to the establishment of the Advanced Fuel Fabrication Facility (AFFF), which fabricated the MOX fuel for loading at TAPS reactors.[14] In the experimental MOX-7 fuel bundle of the PHWR, the inner seven elements are replaced by MOX fuel elements.

The technology for fabrication of fuels for reactors such as the advanced heavy water reactor (AHWR), incorporating the fissile isotope U-233, is complicated by the associated isotope U-232, whose products emit high-energy gamma radiation on decay. However, a good amount of development has already been achieved in the manufacture of these fuels.

Heavy Water

The PHWR uses heavy water as a moderator and coolant. The construction of the first heavy water plants in India began in 1975 with imported equipment from France and Germany. Subsequently, these plants have been designed and constructed with Indian technology and equipment. The Heavy Water Board operates eight heavy water plants in India. The plants at Kota and Manuguru are based on the indigenously-developed hydrogen sulfide–water exchange process, while those at Baroda, Hazira, Tuticorin, Talcher and Thal are based on an ammonia–hydrogen exchange process. The Nangal plant uses hydrogen distillation process.

Heavy water production capacity is adequate for meeting the planned program of nuclear power generation in the foreseeable future, and has also enabled export of heavy water to other countries.

Special Materials and Structural Products

Apart from nuclear fuel, the NFC also produces core structural components for the entire nuclear power program of India. The NFC processes zircon sand through a series of chemical and metallurgical operations using indigenously-developed flow sheets to finally produce zirconium alloy pressure tubes, calandria tubes, fuel cladding and several other products used in nuclear power and research reactors. The Nuclear Fuel Complex has also diversified its activities by making seamless tubes of stainless steel, carbon steel, titanium and many other special alloys of nickel, magnesium, etc. by employing hot extrusion and cold pilgering processes. Apart from the Department of Atomic Energy, the customers of the NFC include several segments of Indian industry. Indian Rare Earths Limited (IREL) has been set up for the separation of nuclear materials from indigenous ores.

Spent Fuel Reprocessing

India's reprocessing program was launched with the design, construction and commissioning of the demonstration plant at Trombay. The plant was commissioned in 1964 to reprocess the spent fuel from the 40 MWt research reactor CIRUS.

The Trombay plant gave sufficient impetus to continue R&D in the domain of reprocessing. The results of these efforts were integrated into the design of the second plant constructed at Tarapur (commissioned in 1975) for reprocessing of Zircaloy-clad oxide spent fuel from the Tarapur and Rajasthan Atomic Power Stations. Several campaigns of reprocessing were carried out under international safeguards.

The Tarapur plant also provided experience in the design of appropriate packages (as per the guidelines of the Atomic Energy Regulatory Board, AERB) for safe in-land transportation of spent fuels, which is a vital consideration in locating reprocessing facilities. Since 1975 there have been safe shipments of spent fuel involving thousands of kilometers by rail and road.

With the expansion of the nuclear program, the need arose to augment the reprocessing capacity to treat the spent fuel from increased nuclear power generation. To cater to the needs of reprocessing Zircaloy-clad natural uranium oxide spent fuel from Madras Atomic Power Station, a new plant was designed and commissioned (with the involvement of Indian industries) near the power station at Kalpakkam. The design aims to ensure availability of the plant capacity throughout the operation of the power station.[15]

To meet the challenges of the thorium-based fuel cycle, R&D efforts are directed towards extractive metallurgy of thorium, fuel fabrication and utilization in reactors, reprocessing of irradiated thorium for U-233 recovery and studies on U-233 based reactor systems. The THOREX (thorium uranium extraction) process has been developed for treating irradiated thorium. An engineering-scale facility is in operation at Trombay for the processing and recovery of U-233 from the irradiated thorium fuel rods from CIRUS and DHRUVA.

Nuclear Waste Management

From the very beginning, special emphasis has been placed on the development of technologies for both high- and low-level wastes. For the processing of low- and intermediate-level wastes, the Effluent Treatment Plant (ETP) was commissioned at Trombay in 1967, which demonstrated efficient treatment and management of radioactive waste from the CIRUS reactor. This plant continues to be operational and has gone a long way toward helping in the design and establishment of effluent treatment facilities for power reactor wastes and low-level effluents from reprocessing plants.[16] Presently, all the Indian PHWR base plants have dedicated waste management systems.

R&D activities concerning the management of high-level waste (HLW) were directed to develop and characterize a number of alternative glass matrices suitable for the immobilization of HLW on the one hand and also to develop, evaluate and perfect conditioning processes and techniques on the other. The first waste immobilization plant (WIP) at Tarapur was set up by the early 1980s. It used a pot-glass process for vitrification. During the 1980s, ceramic melters using Joule heating emerged as an attractive alternative to metallic melters, and an industrial-scale ceramic melter-based vitrification facility has been built at Tarapur. The WIP being set up at Kalpakkam is designed to cater to both PHWR and FBR high-level waste requirements. After the establishment of the WIP at Tarapur, the first solid storage and surveillance facility (SSSF) co-located with a vitrification plant was also constructed.

For solid waste disposal (SWD), preliminary work began in 1962 at BARC with the first laboratories for radiochemistry and isotope production.[17] This was followed by the establishment of a near-surface disposal facility at BARC, known as the Radioactive Waste Storage and Management Site. Over the years, considerable experience and expertise has been committed to refining and improving the design and construction of these disposal modules.

For the treatment of gaseous waste, very efficient gas cleaning techniques have been developed indigenously and are being successfully employed in all the nuclear installations. Combined particulate and iodine filters, charcoal-impregnated sampling filters and filter banks have also been developed through sustained R&D efforts and are currently in routine use in nuclear power plants.

Ongoing R&D work includes the development of alternate vitrification technology based on the cold crucible induction melting technique, and the development of synthetic rock[18] (Synroc) as a futuristic alternative to borosilicate glass—especially suited for fast reactor HLW. Remote handling and robotics play a very important role in waste management. All the equipment required for these functions is manufactured and available on a regular basis indigenously and has been standardized for such applications.[19]

Human Resource Development

While pursuing a comprehensive nuclear power program, Dr. Bhabha and his colleague Dr. Raja Ramanna were convinced that basic research – including what has been called "directed basic research"[20] – and the development of the right kind of manpower were crucial for the success of the program,[21] considering the fact that topics related to nuclear technology did not form part of the regular courses in most Indian universities at the time. Therefore, the BARC Training School was launched in 1957. This School, with a current annual intake of about 150 MSc, B.Tech and M.Tech students, is now training its 52nd batch of entrants. In high technology areas, "hiring and training" is more cost-effective than "training and selecting." The BARC Training School has played a major role in the development of nuclear science and technology in the country. The curriculum for every discipline is designed under three headings – foundation courses, core courses and advanced courses – and are continuously monitored.[22] The faculty for teaching comprises working scientists and engineers, a majority of whom are past Training School graduates themselves. This provides an excellent link for continuity and synergy. For some select advanced courses, faculty from the Indian Institute of Technologies (IIT) and other premier institutes are also invited. The program ensures a good balance between theoretical and practical skills. Having realized the need for highly-specialized professionals in specific areas, the department has initiated the DAE Graduate Fellowship Scheme (DGFS) for engineers in collaboration with the six IITs at Mumbai, Kanpur, Delhi, Kharagpur, Roorkee and Chennai. The scheme also strengthens the research links with these premier institutes of the country. Furthermore, the Department has launched four affiliated training schools:[23]

- The Raja Ramanna Center for Advanced Technology (RRCAT), Indore: for specialization in areas related to lasers, accelerators, cryogenics, superconductors, materials science, power electronics and microwaves.

- The NFC, Hyderabad: for specialization in areas related to nuclear fuel fabrication and heavy water production to support the nuclear power program.

- The NPCIL with training centers at Tarapur, Rawatbhata, Kaiga and Kalpakkam: for operation and maintenance of nuclear power plants.

- The IGCAR, Kalpakkam: for specialization in core engineering disciplines with emphasis on specific fast reactor technologies, nuclear reactor physics and nuclear fuel cycle chemistry.

Over 7,700 young men and women have graduated so far from these training schools, and overall a total of 300–350 people are now being trained annually. Considering India's expanding nuclear program, this number is likely to rise.

Nuclear Program as a Catalyst for Development

The nuclear program in India has been a catalyst for the development of various technologies ranging from welding and water desalination to precision manufacturing and super-computing.

Indian Industrial Support to the Nuclear Power Program

Indian Industries have played a vital role in the development of nuclear energy by supplying nuclear components and equipment and by carrying out execution-related work, as well as commercializing technologies developed at the R & D institutions of DAE.

At the time of the country's independence in 1947 and for several years thereafter, the industry's capability was limited, necessitating large efforts by DAE and NPCIL[24] to develop the Indian industry and achieve high standards in manufacturing of equipment for developing nuclear power technology. In addition to transfer of technologies, measures like investments in development of workshops and facilities for testing were initiated to enable the industry to develop its capabilities. Over the years, Indian industry has developed its capabilities in design, engineering and manufacturing of equipment, which is comparable with international standards. It has also grown to take up large package contracts for the construction of a number of nuclear power plants simultaneously. Indian industry, as a result of its interaction with the nuclear establishment, has also developed a range of capabilities valuable in the production of other technologies.

The Atomic Energy Regulatory Board (AERB)

The Atomic Energy Regulatory Board (AERB) was constituted on November 15, 1983, to carry out regulatory and safety functions[25] independently as a structure outside the Department of Atomic Energy. The mission of the Board is to ensure that the use of ionizing radiation and nuclear energy in India does not cause undue risk to health and the environment. The AERB is supported by other bodies such as the Safety Review Committee for Operating Plants (SARCOP), Safety Review Committee for Applications of Radiation (SARCAR) and the Advisory Committees for Project Safety Review (ACPSR)—for nuclear power projects, light water reactor projects, waste management projects, etc. The

ACPSR recommend to the AERB issuance of authorizations at different stages of plant construction, after reviewing the submissions made by the plant authorities based on the recommendations of the associated Design Safety Committees. The SARCOP carries out safety surveillance and enforces safety regulations in the operating units of the Department of Atomic Energy (DAE). The SARCAR recommends measures to enforce radiation safety in medical, industrial, and research institutions, which use radiation and radioactive sources.

The AERB also receives advice from the Advisory Committee on Nuclear Safety (ACNS). This Committee is composed of experts from the AERB, DAE and other external institutions. The ACNS provides recommendations on safety codes, guides and manuals prepared for siting, design, construction, operation, quality assurance and decommissioning/life extension of nuclear power plants, which have been prepared by the respective advisory committees for each of these areas. It also advises the Board on generic safety issues.

The administrative and regulatory mechanisms that are in place ensure multi-tier review by experts nation-wide. These experts come from reputed academic institutions and governmental agencies. The AERB also has a mechanism to check its effectiveness and quality assurance in its activities, and a process by which it improves its systems through its own experience, feedback and international regulatory practices.

Institutional Setup of the Indian Atomic Energy Program

All activities pertaining to the utilization of nuclear energy in India come under the purview of the Indian Atomic Energy Commission. A number of industrial units, research organizations, corporations, and state-aided institutions provide the required inputs ranging from R&D support, materials and components and other specialized inputs to the corporate organizations to further the design, construction, operation and maintenance of nuclear power plants in the country. Many other organizations are involved in high-end research and development in fields pertaining to nuclear energy and allied services. A non-comprehensive institutional setup of the Indian atomic energy program is shown below in Table 9.3.

Table 9.3
Institutional Setup of the Indian Atomic Energy Program

Nuclear Fuel and special materials	Atomic Minerals Division, Uranium Corporation of India Limited, Nuclear Fuel Complex, Indian Rare Earths Ltd., Heavy Water Board.
Research Organizations	Bhabha Atomic Research Center, Indira Gandhi Centre for Atomic Research, RR Centre for Advanced Technology, Variable Energy Cyclotron Centre.
Corporates	Nuclear Power Corporation of India Ltd., Electronics Corporation of India Ltd., Bharatiya Nabhikiya Vidyut Nigam Ltd.
Aided Research Institutes	Tata Institute for Fundamental Research, Saha Institute of Nuclear Physics, Institute of Plasma Research, Institutes of Mathematical Sciences, etc.
Regulatory Body	Atomic Energy Regulatory Board.
Supporting Indian Corporates	BHEL, L&T, others, joint ventures considered.

Major Components of Indian R&D Program for the Next Generation of Nuclear Energy Systems

Advanced Heavy Water Reactor (AHWR)

The Bhabha Atomic Research Centre (BARC) has completed the design and development of an Advanced Heavy Water Reactor (AHWR) to achieve large-scale use of thorium for the generation of commercial nuclear power. This reactor will produce most of its power from thorium, with no external input of uranium-233, in the equilibrium cycle.

The AHWR (see Figure 9.7) is a vertical, pressure tube-type, boiling, light water-cooled, and heavy water moderated reactor.[26] It is a land-based nuclear power station designed to produce 300 MWe of power (gross), along with 500 m^3/day of desalinated water. The reactor incorporates a number of passive safety features and is associated with a fuel cycle with reduced environmental impact. At the same time, the reactor possesses several features that are likely to reduce its capital and operating costs. The AHWR employs natural circulation for cooling the reactor core under operating and shutdown conditions. All event scenarios initiating from non-availability of main pumps are, therefore, excluded. The Main Heat Transport (MHT) system transports heat from fuel pins to the steam drums using boiling light water as the coolant. Steam is separated from the steam–water mixture in steam drums and is supplied to the turbine. The condensate is heated in moderator heat exchangers and feed heaters and is returned to steam drums by feed pumps. Four down comers connect each steam drum to the inlet header. Several major experimental facilities have been set up and used with this objective, and some others are under construction for generation of additional data. The latter include a Critical Facility, with a capability to simulate AHWR core lattice and fuel configurations, and a full height Integral Test Loop to simulate the Main Heat Transport (MHT) System of AHWR. Major design parameters of the AHWR are shown in Table 9.4.

Important Safety Features of the AHWR include:[27]

- Slightly negative void coefficient of reactivity.

- Passive safety systems working on natural laws.

- Large heat sink in the form of gravity-driven water pool with an inventory of 6,000 m^3 of water, located near the top of the Reactor Building.

- Removal of heat from core by natural circulation.

- Emergency Core Cooling System injection directly inside the fuel.

- Two independent shutdown systems.

- Passive poison injection in moderator in the event of non-availability of both primary as well as secondary shut down systems owing to failure or malevolent insider action.

- Consistent with the approach used in standardized Indian PHWRs, the AHWR is provided with double containment. For containment isolation, a passive system has been provided in the AHWR.

Table 9.4
Major Design Characteristics of the AHWR

Attribute	Design Particulars
Core configuration	Vertical, pressure tube-type
Fuel	$(Th-Pu)O_2$ and $(Th-^{233}U)O_2$
Coolant	Boiling Light Water
Moderator	Heavy Water
Cover Gas	Helium
Number of coolant channels	452
Pressure tube ID	120 mm
Pressure tube material	20% Cold Worked Zr-2.5% Nb Alloy
Lattice pitch	245 mm
Active fuel length	3.5 m
Calandria diameter	7.4 m
Calandria material	Stainless Steel - Grade 304L
Steam drum pressure	7 MPa
Mode of core heat removal	Natural circulation
MHT loop height	39 m
Primary shut-down system	40 mechanical shut-off rods
Secondary shut-down System	Liquid poison injection in moderator

Figure 9.7
Advanced Heavy Water Reactor (AHWR)

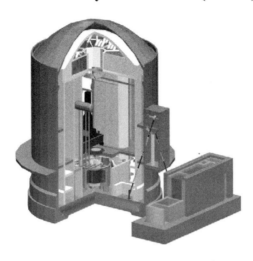

Figure 9.8
Compact High-Temperature Reactor

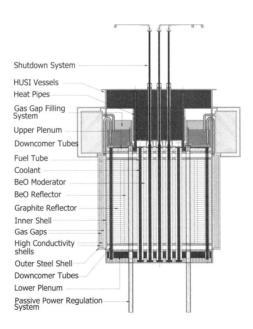

The AHWR fuel cycle will be self-sufficient in U-233 after initial loading. The spent fuel streams will be reprocessed and thorium and U-233 will then be recycled and reused. The AHWR fuel cycle has enough flexibility to accommodate a large variety of fuelling options. Incidentally, the thorium fuel cycle also presents low proliferation risks, a factor considered significant by several nations for export of nuclear technology.

The Indian regulatory body AERB has already completed a pre-licensing safety appraisal of the reactor. In April 2008, the AHWR critical facility was commissioned to carry out reactor physics-related experiments. This reactor has been internationally recognized as an innovative design. The IAEA's International Project on Innovative Nuclear Reactors and Fuel Cycles (INPRO) addresses several aspects of future sustainable nuclear energy systems. A case study of the AHWR was carried out as a part of INPRO. The study established the conformance of the AHWR design features with the futuristic INPRO requirements in the areas of sustainability, economics, safety of nuclear installations, environment, waste management, and proliferation resistance.

High Temperature Reactors

BARC is currently developing the concept of a high temperature nuclear reactor[28] capable of supplying process heat at a temperature in the range of 600-1,000 degrees Celsius. These nuclear reactors are being developed with the objective of providing energy to facilitate combined production of hydrogen, electricity, and drinking water. Under the program, India is currently developing a compact high-temperature reactor (CHTR) (see Figure 9.8) as a demonstrator for associated technologies. The CHTR is a mainly U-233–thorium fuelled, lead–bismuth-cooled and beryllium oxide-moderated reactor. This reactor, initially being developed to generate about 100 kWth of power, will have a core life of fifteen years and will have several advanced passive safety features to enable its operation as a compact power pack in remote areas not connected to the electric grid. The reactor is being designed to operate at 1,000 degrees Celsius, to facilitate demonstration of technologies for high-temperature process heat applications such as hydrogen production by splitting water through a high efficiency thermo-chemical process. Molten lead-based coolant has been selected for the reactor in order to achieve a higher level of safety. For this reactor, developmental work in the areas of fuel, structural materials, coolant technologies, and passive systems are being conducted in the BARC. Experimental facilities are also being set up to demonstrate associated technologies. In parallel, design work has been initiated for the development of a 600 MWth high temperature reactor for commercial hydrogen production by high temperature thermo-chemical water splitting processes. Technologies being developed for the CHTR would be utilized for the development of this reactor. Various analytical studies have been carried out in order to compare different options regarding fuel configuration and coolants. Initial studies carried out indicate the selection of a pebble bed reactor configuration with either lead or molten salt-based cooling by natural circulation. In addition, a 5 MWth reactor is also being designed, operating at a temperature of 600 degrees Celsius, to function as a compact nuclear power pack to supply electricity in regions not connected to the electricity grid of the country.

Accelerator-Driven Systems

Accelerator-driven sub-critical reactor systems (ADS) have received considerable attention from the global nuclear community owing to their capability to incinerate the minor actinides and long-lived fission products, radiotoxic waste, and also for their ability to convert fertile material into fissile material. Since India has vast resources of thorium, an ADS offers a potential route for accelerated thorium utilization.[29] The ADS consists of a sub-critical reactor coupled to a high power proton accelerator through a spallation target. For the practical realization of ADS, challenges include development of a high energy (>1 GeV) and high current (>20 mA) proton accelerator to produce the intense spallation neutron source needed to drive the sub-critical reactor assembly.[30] It is necessary that the accelerator is reliable, rugged, and stable in order to provide uninterrupted beam power to the spallation target over long periods of time. It should also have a high efficiency of conversion of electric power to beam power. R&D organizations in the DAE are working toward the development of the different technologies associated with ADS-based systems. These include development of a low-energy (20 MeV) high-intensity proton accelerator as the front-end injector for the 1 GeV accelerator for the ADS program, spallation targets, heat removal technologies, and other special materials required for such systems.

Fusion-Based Systems

Research on thermo-nuclear fusion and related technologies is being carried out at the Institute of Plasma Research in Gandhinagar, Gujarat State. The research institute is equipped with ADITYA, the first indigenously designed and built tokamak of the country. It was commissioned in 1989. ADITYA is a medium size tokamak and has a major radius of 0.75 m and minor radius of 0.25 m.[31] During this period, experiments on edge plasma fluctuations, turbulence and other related areas have been conducted. In addition, a steady-state superconducting tokamak SST-1 is being designed and fabricated at the institute. The objectives of the SST-1 include studying the physics of the plasma processes in the tokamak under steady-state conditions and mastering related technologies. These studies are expected to contribute to the tokamak physics database for very long pulse operations.[32]

In the field of nuclear fusion, internationally, a major effort is in hand under the umbrella of the International Thermonuclear Experimental Reactor (ITER) program. India has joined a select group of seven partners including the USA, the European Union, Japan, Russia, China and South Korea to jointly provide special hardware items and expertise in selected areas of this program.[33] The Indian contribution to this program would be worth over half a billion US dollars.

Applications of Radioisotopes

Applications of Radioisotopes in Healthcare and Industry

Various radioisotopes are being produced for healthcare, agricultural and industrial applications. Among them Iodine-131, Iodine-125, and Technetium-99m are used for diagnostics and Cobalt-60, Cesium-137 and Iridium-192 are used for therapy. In the field of agriculture, Cobalt-60 is used for crop improvement and food preservation and Phosphorus-32 is used for plant uptake studies. In industrial applications, Iridium-192 is used in radiography and Bromine-82, Mercury-197 and Tritium are used as radiotracers. Cesium-137 and Cobalt-60 are utilized in nucleonic gauges and Tritium is used in hydrology.

Radioisotopes and their formulations have found wide applications in nuclear medicine and healthcare. The diagnostic applications consist of organ/function imaging and Radioimmunoassay (RIA). Tele-therapy, brachytherapy and internal administration are some of the therapy applications. There are over 500 RIA labs, about 100 nuclear medicine centers and many cancer hospitals functioning in India. The Tata Memorial Center, a state-aided institution under the DAE, was recognized as an outstanding cancer organization for its excellence in cancer control within and beyond India's borders by the International Union for Cancer Control, Washington DC, in 2006. The Board of Radiation and Isotope Technology (BRIT) has been operating the radiation sterilization plant ISOMED – now ISO-9002 accredited – at Trombay. ISOMED has been in operation for close to three decades and the availability factor in recent years has been more than 90 percent. Also, radiation sterilization plants are operating in Delhi and Bangalore.

In the area of radiation medicine, Bhabhatron – an Indian telecobalt machine – has been developed and ten machines have been successfully put into service in cancer treatment. Bhabhatron has many advanced features – including safety and battery backups suitable for semi urban and rural conditions – and is cheaper than its alternatives. In 2007, the IAEA, the government of India and the government of Vietnam signed a memorandum of understanding (MoU) to donate one Bhabhatron-II to Vietnam under the Program of Action for Cancer Therapy (PACT) program of the IAEA, to make radiation therapy more widely available in developing countries.

Applications in Agriculture

In the agriculture sector, 35 high-yield varieties of seeds (groundnut 12; mustard 3; soybean 2; greengram 7; blackgram 4; pigeon pea 3; cowpea 1; sunflower 1; rice 1; and jute 1) have been developed and released for cultivation.[34] The mutant groundnut seeds developed at the BARC contribute to nearly 25 percent of total the groundnut cultivation and 22 percent of blackgram (*urad*) production in the country. In the state of Maharashtra, this percentage is as high as 95 percent.

Applications in Food Preservation

The Indian government has approved the radiation processing of certain food items both for export and domestic consumption, such as onion, potato, ginger, garlic, shallots, mango, rice, semolina, wheat flour, raisins, figs, dried dates, meat and meat products including chicken and spices. A radiation processing plant with a capacity of 12,000 tons per year has been operating at Navi Mumbai since January 2000 for high dose radiation processing of spices. The Krushi Utpadan Sanrakshan Kendra (KRUSHAK) irradiator at Lasalgaon near Nashik[35] has a capacity of 10 tons per hour and this plant processes onion, pulses, rawa and turmeric. Considerable progress has been achieved in the setting-up of radiation processing plants in the private sector.

Applications in Environmental Care

Nisargaruna, an ongoing project of the DAE, utilizes biodegradable wastes to generate energy as well as organic manure.

Environmental isotope techniques have been employed to identify the recharge springs in India, in order to construct artificial recharge structures for rainwater harvesting and ground water augmentation for their rejuvenation.[36]

Under the Rural Technology Action Group (RuTAG) program of the Principal Scientific Adviser to the Government of India, a model isotope hydrology project had been taken up as part of the Open Platform Innovation Strategy of the Himalayan Environment Studies and Conservation Organization (HESCO) and BARC in the mountainous region of Gaucher, Chamoli District, Uttarakhand. The rates of discharge of many springs in this region have increased by three to nine times and two new springs have been accessed. Almost all of the springs have become perennial. This holds great promise for the hilly regions of India, as well as outside.

Strategies for Long Term Energy Supply in India and the Emerging Nuclear Renaissance

Figure 9.9 shows a variation of human development index (HDI) that corresponds to per capita electricity consumption (PCEC) for different countries. In order to reach developed country status, a desirable goal should be to reach a PCEC of about 5,000 kWh/capita/year. According to a DAE study, India needs to reach this PCEC by the year 2050.

[293]

Figure 9.9

Variation of Human Development Index with Per Capita Electricity Consumption

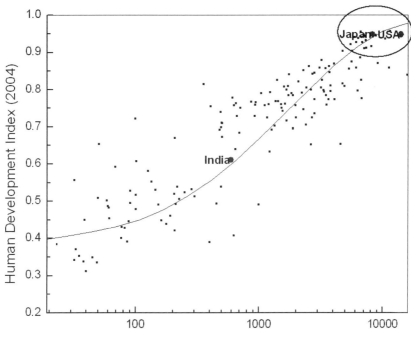

Per Capita Electricity Consumption in kWh/Year

Source: United Nations Development Program (UNDP), "Human Development Report" 2004–2006.

It has been estimated that even after utilizing all currently known domestic energy resources in India, there would be a deficit of 412 GWe generation capacity in the year 2050.

To fill this gap, the alternatives are the importation of either nuclear reactors and nuclear fuel or huge quantities of coal (about 1.6 billion tons in 2050). The importation of 40 GWe LWRs during the period 2012–2020, along with domestic resources, can satisfy the energy demand in 2050.[37]

Conclusion

The enhanced use of nuclear energy as a substitute for fossil fuels is essential in addressing climate change. Such an enhanced role cannot be implemented without closing the nuclear fuel cycle.[38] Furthermore, it is important to consider natural hydrocarbons as an important resource for industrial applications in a large number of sectors, and to conserve it for such applications in the future, rather than burning these away for heat generation.

We are now on the threshold of a global nuclear renaissance. Even so, it is noted that since 1985 there has not been any additions to the group of 33 countries that had initiated construction of nuclear power plants.[39] This situation needs to be corrected through the implementation of new technological and institutional mechanisms which address the concerns that have hampered the extension of the global reach of nuclear power. India, as a developing country, with a full nuclear capability in terms of the exploitation of nuclear energy for power generation, is fully-equipped to contribute in this scenario with the technological and institutional solutions needed for the global expansion of nuclear power. India is in a strong position to interact with other developing countries and address the specific needs of such countries, with an understanding acquired through its own experience as a developing country. There are definite cultural advantages to achieving such co-operation as a component of a wider partnership to secure global growth of nuclear power. In particular, India can provide support during project evaluation, technical consultancy during the construction phase, training, technical support for regulatory organizations, and assistance in the commissioning and operation phases. India would also co-operate with other nuclear countries in providing nuclear power plants for countries new to this technology area.

Nuclear energy is recognized as an important mitigating technology in the control of global climate change. For it to be sustainable, however, it is necessary to close the nuclear fuel cycle. The technologies that are considered important in the context of depleting global fossil fuel resources – renewables, nuclear and increasingly efficiency technologies – are the same as those that are relevant to mitigating climate change.

10

Status and Prospects of Nuclear Energy in Pakistan

Ansar Parvez and *G.R. Athar*

With the commissioning of the Karachi Nuclear Power Plant (KANUPP) in 1971, Pakistan became a member of the elite club of countries using nuclear power for electricity generation. The plant was built by Canadian General Electric (CGE) under a turn-key contract. However, very early in KANUPP's life, international embargoes were imposed on the supply of fuel, spare parts and technical services. The scientists and engineers of the Pakistan Atomic Energy Commission (PAEC) took on this challenge and within a few years were successful in manufacturing nuclear fuel and establishing a technical support system for the plant. Facilities were also developed to manufacture some necessary spare parts for KANUPP. Consequently, it was possible to keep the plant operating without any vendor support. Toward the end of its design life, KANUPP underwent various refurbishments and safety retrofits to extend plant's life by at least fifteen years. The plant has now been provisionally re-licensed by the Pakistan Nuclear Regulatory Authority (PNRA).

The construction of Pakistan's second nuclear power plant, the Chashma Nuclear Power Plant Unit-1 (C-1), began in 1992 with the help of the People's Republic of China (PRC). The plant began commercial operation in September 2000 and has since maintained a very good performance record. Construction of the third nuclear power plant, the

Chashma Nuclear Power Plant Unit-2 (C-2), is now underway, and the plant is expected to start commercial operation in 2011.

To secure a sustainable and affordable electricity supply, Pakistan requires an optimal mix of all cost-effective electricity generation technologies. The government of Pakistan has chalked out a plan which supports the expansion of the country's nuclear power program; targeting a nuclear power generation capacity of 8,800 MW by the year 2030. Considering the current international scenario under which Pakistan's nuclear power program remains subject to embargoes, and the large capital requirements of nuclear power plants, this will be a formidable challenge.

To put the country's nuclear power program in perspective, a power sector profile is presented in this paper. This is followed by a summary of the current nuclear power program and an examination of the future prospects for nuclear energy in the country. The next section reviews Pakistan's indigenous capability to support its nuclear power program, which is followed by a description of the national regulatory infrastructure for nuclear power. Next, the paper presents an overview of the international cooperation and commitments made by Pakistan. The penultimate section examines the benefits of nuclear power to Pakistani society, which is followed by a conclusion.

National Power Sector Profile

The Current Status of Pakistan's Power Sector

Electricity consumption in Pakistan has been growing at a pace greater than that of overall energy consumption and economic growth as a result of urbanization, industrialization and rural electrification. Table 10.1 presents selected economic and energy data for the past three decades. During this period, grid electricity supply increased by an average of 6.9 percent per annum, which is 1.27 times higher than the growth in Gross

Domestic Product (GDP) and 1.1 times higher than the growth of primary energy supply. This ratio would increase further if electricity supply from captive power[1] is added. Despite this rather large increase in electricity supply, the level of electricity consumption in Pakistan is still very low compared to global norms; i.e., approximately one fifth of the world average.[2]

Table 10.1
Evolution of the Power Sector in Pakistan

	1979/80	1999/2000	2007/08	Average annual growth during 1979/80–2007/08
Population (million)	80.2	137.5	161.0	2.5 %
Installed electricity generation capacity (MW)	3,510	17,400	19,420	6.3 %
Total grid electricity supply (million kWh)	14,974	65,752	95,860	6.9 %
Per capita electricity supply (kWh)	186	478	595	4.2 %
Total primary commercial energy supply (million TOE)	11.63	38.38	62.92	6.2 %
Primary commercial energy supply per capita (TOE)	0.14	0.28	0.39	3.7 %
Total Gross Domestic Product (million US$ in 2008 prices)	37,776	106,801	164,692	5.4 %
Gross Domestic Product per capita (US$ in 2008 prices)	471	777	1,023	2.8 %

Notes: (i) population estimates are as of January 1 in mentioned financial years, i.e., for 2007/08, the data is for January 1, 2008; (ii) one ton of oil equivalent (TOE) is equal to 44.2 gigajoules.
Sources: HDIP (2009)[3] and GOP (2008).[4]

Figure 10.1
Pakistan's Electricity Generation Mix

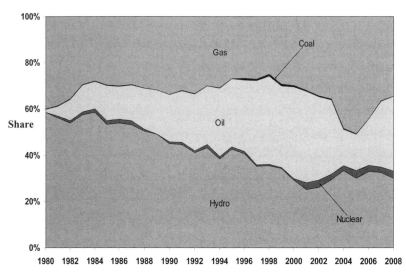

Source: Based on data from HDIP (2009).[5]

Figure 10.1 displays the electricity generation mix in Pakistan. It shows that electricity generated from oil-fired plants accounted for only a few percent in 1980/81.[6] However, it jumped to more than 20 percent by 1989/90 and then further increased to 40 percent by 2000/01 owing to increased electricity demand, a government policy directive on the limited use of natural gas in the power sector, and a softening of oil prices in international market. However, following a policy change which promoted the allocation of natural gas for power generation, the share of gas-based electricity generation increased to fifty percent by 2004/05. However, this much load could not be sustained by natural gas, and the relative share of gas declined owing to supply limitations. The current electricity generation mix and sectoral consumption are given in Table 10.2.

Owing to an inadequate supply of electricity, load shedding has at various times been undertaken in the country. Lately, however, load shedding has had to be increased considerably because of the supply–demand mismatch. Table 10.3 shows the extent of the shortfall in electricity supply in the area supplied by the National Transmission and Dispatch Company (NTDC) in 2007/08.[7] NTDC covers the entire country except the large metropolitan city of Karachi where Karachi Electric Supply Corporation (KESC) is responsible for supply of electricity. The two systems can buy or sell electricity from each other, but like NTDC, the KESC system is also passing through capacity shortages. Therefore, the entire country had to resort to load shedding.

Table 10.2

Electricity Supply and Consumption in Pakistan (2007/08)

Electricity generation	**95,661**	**100%**
Hydro	28,707	30.0%
Gas	32,923	34.4%
Oil	30,818	32.2%
Nuclear	3,077	3.2%
Coal	136	0.2%
Auxiliary consumption	**3,688**	**3.9% of total generation**
Net generation	**91,973**	
Imported	**199**	**0.2% of net supply**
Net supply	**92,172**	
T & D losses	**18,742**	**20.3% of net supply**
Sectoral consumption	**73,400**	**100%**
Residential	33,704	45.9%
Industrial	20,729	28.2%
Agricultural	8,472	11.6%
Commercial/Services	6,145	8.4%
Others	4,350	5.9%

Source: HDIP (2009).[8]

Table 10.3
Peak Demand in the System of NTDC in 2007/2008 against Available Capability

Month	Computed peak demand profile		
	System peak (MW)	Available capability (MW)	Surplus(+) Deficit(-) (MW)
July	15,941	13,706	-2,235
August	15,862	13,666	-2,196
September	16,056	13,644	-2,412
October	13,737	14,092	355
November	12,401	11,590	-811
December	12,154	9,679	-2,475
January	12,255	9,104	-3,151
February	12,123	10,122	-2,001
March	13,682	9,845	-3,837
April	15,124	11,568	-3,556
May	16,649	11,195	-5,454
June	17,398	12,442	-4,956

Source: Data acquired from the National Transmission and Dispatch Company (NTDC).

Future Requirements of Electric Power

In 2004, the Planning and Development Division of the government of Pakistan, together with the NTDC, carried out a study under the Energy Security Plan (ESP) 2005–30 that projected the future electricity demand of the country.[9] These projections were based on the then prevailing high economic growth scenario corresponding to an average annual growth rate of 8.7 percent during the planning period 2004/05 through 2024/25. However, during the financial year 2007/08, economic growth slowed to 5.8 percent, mainly as a result of the unprecedented increase in global oil prices, lower domestic food production and higher international food prices, all of which resulted in an increase in inflation, a drop in foreign exchange reserves and the depreciation of the national currency. The current worldwide recession is also likely to have a negative impact on the growth of the national economy in the near future. Analysts foresee real GDP growth slipping further in the financial year 2008/09, and it may take a few years for the economy to rebound.[10]

An in-house assessment of electricity demand was conducted by the PAEC, assuming that the economy would gradually regain 6 percent annual growth in 4–5 years' time.[11] The peak load projections based on different growth scenarios are listed in Table 10.4. Allowing a margin of fifteen percent to account for the reserves required to meet planned and unplanned outages, the installed capacity requirements are given in Table 10.5. These projections, though lower than those made for the ESP 2005–30, still require a 2.8-fold increase for the 'low' scenario and a 4.5-fold increase for the 'high' scenario in the installed electricity generation capacity over the next 23 years.

Table 10.4

Projection for Peak Load in Pakistan

Year	High scenario*		Medium scenario*		Low scenario*	
	Generation (million kWh)	Peak Load (MW)	Generation (million kWh)	Peak Load (MW)	Generation (million kWh)	Peak Load (MW)
2006/07 (actual data)	98,213	15,999	98,213	15,999	98,213	15,999
2009/10	113,927	18,632	112,794	18,447	112,044	18,324
2014/15	147,797	24,452	140,661	23,271	135,246	22,375
2019/20	202,677	33,531	182,789	30,241	167,522	27,715
2024/25	301,299	49,848	255,467	42,265	218,481	36,146
2029/30	458,118	75,792	371,888	61,526	289,727	47,933

* Scenarios have been based on the assumed values of average annual economic growth during 2007/08–2029/30 as; high growth (7.3 percent), medium growth (6.5 percent) and low growth (5.0 percent).
Source: Athar et al. (2008).[12]

Table 10.5
Estimated Capacity Requirements (MW)

Year	Energy Security Plan	PAEC assessment		
		High scenario	Medium scenario	Low scenario
2007/08	19,420 (actual)			
2009/10	27,420	21,427	21,214	21,073
2014/15	47,540	28,120	26,762	25,731
2019/20	72,270	38,561	34,777	31,872
2024/25	110,760	57,325	48,605	41,568
2029/30	162,590	87,161	70,755	55,123

Sources: (i) GOP (2005),[13] (ii) HDIP (2009).[14]

Supply Potential of Domestic Fossil Fuel, Hydro and Renewable Resources

It is pertinent here to provide an overview of the available indigenous resources and their role in meeting the energy requirements above.

Natural Gas

The balance of recoverable reserves of natural gas as of June 30, 2008 was 29.8 trillion cubic feet (0.8 trillion cubic meters). If used exclusively for electricity generation, these reserves can support 14,000 MW of capacity for thirty years. However, demand for natural gas is continuously increasing in non-power sectors. During 2007/08, fertilizer feedstock, industry, houses, transport, and commercial sectors contributed to 66 percent of gas consumption.[15]

Gas reserves are being depleted faster than the rate of additional resources being added as a result of new discoveries. Projections show that gas production, after reaching a level of 4,424 million cubic feet per day (mmcfd) in 2009/10, will decline to 3,001 mmcfd by 2019/20. A shortfall of about 141 mmcfd of natural gas is projected by 2009/10, which could increase to 6,113 mmcfd by 2019/20.[16]

These projections imply that the future availability of indigenous gas for electricity generation will remain restricted if no substantial new reserves are discovered.

Crude Oil

The balance of recoverable reserves of crude oil as of June 30, 2008 was 326.7 million barrels. The production during 2007/08 was 70,000 barrels per day (bpd) which accounted for only 19 percent of the total oil consumed in the country.

As with natural gas, projections of oil discoveries are also not encouraging. The Ministry of Petroleum and Natural Resources has projected that the daily production of crude oil could shrink to around 13,000 bpd by 2017 if major oil discoveries are not made.

Coal

The total coal resources of Pakistan are reported to be 186 billion tonnes,[17] of which the measured reserves are 3.4 billion tonnes.[18] These measured reserves alone can support 24,000 MW of power generation capacity for thirty years. However, 95 percent of the coal reserves of Pakistan lie in the Thar coal field in the Sindh province. Thar coal is of lignite quality with high (30–56 percent) moisture content and is of low heating value. Its average heating value of 8,645 British thermal units (Btu) per pound is about 50 percent of that of standard coal.[19] Efforts to exploit the Thar coal field for electricity generation have largely been limited to undertaking feasibility studies and developing a modest amount of infrastructure. Lately, the government has renewed its efforts to invite private enterprise to develop the Thar coal mines and build power plants around these mines.

Indigenous coal production during 2007/08 was 4.1 million tons from various small coal fields. At present, the largest productive coal field is

Lakhra, which produced 1.0 million tons of coal during 2007/08. The national Energy Security Plan projects that coal production must increase to 153 million tons by 2029/30 in order to achieve an increase in coal-based electricity generation capacity from the present level of 150 MW to 19,910 MW.[20]

Hydro

The identified hydro power potential of Pakistan is about 46,000 MW,[21] of which only 6,480 MW (14 percent) has been exploited so far. The ESP envisages 26,000 MW of additional hydro-based capacity by 2030. However, sociopolitical and environmental issues coupled with project financing requirements remain the main barriers to large-scale development of hydro power.

Renewables

The coastal area of Pakistan, mainly in the Sindh province, has a considerable potential for wind energy—the gross wind power potential being around 50,000 MW. However, in view of the constraints of area utilization, the exploitable electric power potential of this area is more realistically estimated to be around 11,000 MW.[22] Current government policies encourage utilization of renewable energy resources. A policy for the development of renewable energy for power generation, launched by the government in 2006, provides large incentives to prospective investors. Groundbreaking ceremonies of the first two wind farms (50 MW each) have already been performed.

The Alternative Energy Development Board (AEDB), in collaboration with international organizations, is assessing the potential of solar power. Thus far, the only notable development is the preparation of solar maps.

To summarize, the country has limited oil and gas resources; hydro resources are associated with sociopolitical, environmental and project financing issues; coal of lignite quality, though abundant, is hampered by the daunting requirement of large-scale mining infrastructure; and renewable resources are in their initial stages of development and in any event are not expected to meet base-load electricity requirements. Hence, nuclear power has a viable role to play in meeting the base-load electricity generation requirements of the country.

Nuclear Power Development

The Pakistan Atomic Energy Council was established in 1956 with the aim of executing development projects involving nuclear power plants, scientific research and the promotion of peaceful uses of nuclear technology in agriculture, medicine and industry. In 1965, the Council became the Pakistan Atomic Energy Commission (PAEC).

Besides the nuclear power program, the PAEC has over the last five decades grown to incorporate the following:

- the Pakistan Institute of Nuclear Science and Technology (PINSTECH)—a large Research and Development (R&D) center with two research reactors;

- radiopharmaceutical production facilities;

- nuclear medical centers throughout the country;

- agriculture, biotechnology and genetic engineering research centers;

- food irradiation facilities;

- mechanical manufacturing complexes;

- instrumentation and control workshops; and

- human resource development centers.

The nuclear power program of the PAEC involved setting up and operating nuclear power plants as well as the development of the front-end fuel cycle. Pakistan now has two nuclear power plants in operation and one under construction. A brief overview of these plants is given below.

[307]

The Karachi Nuclear Power Plant

The contract for the first nuclear power plant was signed in 1965 with CGE of Canada to construct the Karachi Nuclear Power Plant (KANUPP). It is a Pressurized Heavy Water Reactor (PHWR) of 137 MWe gross capacity. At the time, Pakistan did not have many professionals trained in the field of nuclear power technology. Therefore, a team of scientists and engineers was sent to Canada to acquire skills and expertise in relevant areas, and to interact with the Canadian teams working on the KANUPP project.

KANUPP began commercial operation in 1972. For the smooth operation of KANUPP, Pakistan relied on Canada for the supply of fuel, spares and technical services. However, embargoes in 1976 caused vendor support to be withdrawn from KANUPP. In response, the PAEC engineers and scientists were ultimately successful in producing fuel and providing technical support for the plant. The PAEC also developed the capability to manufacture some critical spare parts for KANUPP. Consequently, despite the restrictions and embargoes, the plant was able to operate with no incidents leading to safety concerns for the plant, its personnel, the general public or the environment. Nevertheless, the embargoes did hamper the performance of KANUPP, resulting in lower lifetime availability and capacity factors.

Before the completion of the design life of KANUPP in 2002, the PAEC had initiated work on its life extension as plant monitoring and periodic inspections had indicated that major plant equipment including fuel channels, steam generators, steam condensers, turbine generators, primary heat transport piping, feeders, etc., were in good condition.

After a request from the PAEC for technical advice and safety assessments, the IAEA arranged expert missions and initiated technical cooperation projects to support KANUPP. Under the umbrella of the IAEA, the vendor country also showed a willingness to assist in certain safety related issues. The missions/reviews recommended actions for safety enhancement and measures to overcome issues of ageing and obsolescence. An IAEA technical project, the Safe Operation of KANUPP

[308]

(SOK), was initiated to support the safe operation of the plant by reassessing design safety, averting plant degradation owing to ageing, introducing modern operational practices and improving design safety. SOK, declared a model project by the IAEA, was later extended to become the Improving Safety Features of KANUPP (ISF) project.

Under a comprehensive balancing, modernization and rehabilitation (BMR) project, KANUPP's conventional equipment was also upgraded. The replacement of its obsolete control, instrumentation and regulating computers was carried out under a technological up-gradation project (TUP).

After various inspections and reviews carried out by international experts, KANUPP approached PNRA for license to operate the plant for another 15 years. Various documents including safety review and analysis reports were developed and provided to PNRA which ultimately granted interim re-license to KANUPP, but put a cap on thermal power that limited the electrical rating to a maximum of 90 MWe. On fulfilling the regulatory requirements of PNRA, National Electric Power Regulatory Authority (NEPRA) and the concerned provincial Environmental Protection Agency, KANUPP was restarted in January 2004 after a re-licensing outage of about two years, and has been operating safely since then.

The Chashma Nuclear Power Plant Unit-1 (C-1)

After the commissioning of KANUPP, the PAEC began work on site selection for the second nuclear power plant of the country. Chashma, located in the center of Punjab province and close to electricity demand centers, was selected in early 1970s as the location of the plant. However, despite the keen interest of Pakistan in building additional plants, it took some two decades to begin construction of the second nuclear power plant owing to the non-responsive international environment as well as a lack of indigenous technological and industrial capability for the construction of nuclear power plants. The development of the nuclear power industry in the People's Republic of China opened a window for Pakistan. A contract was signed with the China National Nuclear Corporation (CNNC) on

December 31, 1991, for a 325 MWe Pressurized Water Reactor (PWR). This plant is named the Chashma Nuclear Power Plant (CHASNUPP-1 or C-1).

C-1 was built under a turn-key contract with the main contractor, CNNC, who awarded subcontracts to various Chinese companies as well as to the PAEC. The main designer was the Shanghai Nuclear Engineering Research and Design Institute (SNERDI). C-1 is an example of south–south cooperation and its construction has been a great learning experience for both China and Pakistan. It was the very first nuclear power plant exported by China, and was also Pakistan's first PWR.

At the request of the PAEC, the IAEA reviewed the basic design of C-1 and provided technical assistance during construction and operation through expert missions involving:

- review of site evaluation;
- review of reactor internals modifications;
- pre-operational safety review (OSART Mission);
- support in developing maintenance training;
- symptom-based emergency operating procedures;
- development of full scope training simulator;
- severe accident management guidelines; and
- a human resource development program.

Licensing of C-1 was carried out by the Directorate of Nuclear Safety and Radiation Protection (DNSRP), then functioning under the PAEC. The Chinese National Nuclear Safety Authority (NNSA) provided extensive support to the DNSRP on regulatory matters. Regulatory requirements in China at that time were based on the standards of the IAEA Safety Series No. 50 C-D,[23] whereas the PAEC evaluated C-1 on the basis of standards of IAEA Safety Series No. 50 C-D Rev. 1.[24]

C-1 was connected to the national grid on June 13, 2000 and has since been operating well. After some early teething problems, related mainly to performance of equipment in the conventional island, the availability factor of the plant improved from 70.2 percent in Cycle 1 to 79.2 percent in Cycle 2 and increased to well above 90 percent in subsequent cycles (see Table 10.6). By August 3, 2008, when C-1 was shut down for its fifth refueling outage, the plant had generated 16,458 million kWh and had lifetime availability and capacity factors of 76.0 percent and 71.7 percent, respectively (Figure 10.2). The plant is now back in service after completing the outage.

Although C-1 was a turnkey project, the PAEC performed several wide-ranging tasks which provided its personnel with experience and enhanced their confidence. These tasks included:

- siting studies;

- preparation of feasibility study;

- contract negotiations;

- establishment of project organization;

- carrying out owner's role;

- reviewing plant design and modifications;

- development and implementation of Quality Assurance (QA) program;

- verifying control points;

- fabrication of certain process equipment;

- construction of a few buildings outside the nuclear and conventional islands; and

- development of a full-scope training simulator.

Table 10.6
Operational History of C-1

Status	Date	Availability Factor	Capacity Factor
Cycle-1	15-09-2000 to 29-09-2002	70.2%	64.3%
RFO-1	30-09-2002 to 18-01-2003		
Cycle-2	19-01-2003 to 14-04-2004	79.2%	76.8%
RFO-2	15-04-2004 to 24-07-2004		
Cycle-3	25-07-2004 to 30-09-2005	96.4%	93.0%
RFO-3	01-10-2005 to 17-11-2005		
Cycle-4	18-11-2005 to 14-02-2007	97.3%	95.5%
RFO-4	15-02-2007 to 02-05-2007		
Cycle-5	03-05-2007 to 02-08-2008	95.6%	91.5%
RFO-5	03-08-2008 to 23-01-2009		

Notes: RFO = refueling outage.
Source: data acquired from C-1.

Figure 10.2
Availability and Capacity Factors of C-1

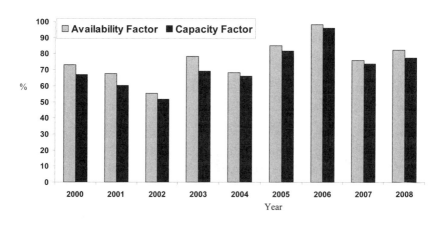

Source: Data acquired from C-1.

[312]

Chashma Nuclear Power Plant Unit-2 (C-2)

The successful functioning of C-1 gave the country the confidence and direction to construct more nuclear power plants in collaboration with China. In May 2004, Pakistan and China signed another contract to construct a second unit, namely CHASNUPP-2 (C-2), adjacent to C-1. The gross generation capacity of C-2 is expected to be 340 MWe.

C-1 is the reference plant for C-2, but the new plant will have additional safety features added to meet the current regulatory requirements of the PNRA such as complementing a deterministic safety analysis with a probabilistic safety assessment (PSA) and incorporating several countermeasures against severe accidents. Human factor engineering was also considered in the design. Owing to the introduction of enhanced safety features and its licensing on the basis of PNRA regulations based on the latest IAEA safety standards, (Safety Series NS-R-1),[25] C-2 can be called a Generation II+ plant.

The First Concrete Pour for the nuclear island was carried out on December 28, 2005 and the plant is scheduled to begin commercial operation in 2011 (see Table 10.7).

Table 10.7
Status of Existing Nuclear Power Plants and those
Under Construction

	KANUPP	C-1	C-2
Type	PHWR	PWR	PWR
Gross Design Capacity (MWe)	137	325	340
Net Design Capacity (MWe)	125	~ 300	~ 315
Operational Capacity (MWe)	90*	~ 300	~ 315
Status	Operational	Operational	Under construction
Reactor Supplier	CGE (Canada)	CNNC (China)	CNNC (China)
First Concrete Pour	Aug. 1966	Aug. 1, 1993	Dec. 28, 2005
Criticality	Aug. 1, 1971	May 3, 2000	Apr., 2011**
Grid Connection	Oct. 18, 1971	Jun. 13, 2000	May, 2011**
Commercial Operation	Dec. 7, 1972	Sep. 15, 2000	Aug., 2011**

Notes: * re-licensed at de-rated power of 90 Mwe; ** projected.
Sources: data acquired from KANUPP, C-1 and C-2.

[313]

Future Prospects for Nuclear Energy in the Country

Nuclear power in Pakistan has now been underway for 36 years. With two plants operating, the PAEC has more than 44 reactor years of operation experience to its credit. During this period, safety has been the hallmark of the country's nuclear industry.

The use of nuclear power would help the country in enhancing the availability and security of electricity supply through diversification of its power sector. The scarcity of conventional energy resources in the country makes utilization of nuclear energy potently attractive. Nuclear power development is capital-intensive, requires broad industrial infrastructure and long lead-times. Therefore, the expansion of the nuclear power program is likely to be a gradual and slow process for a developing country like Pakistan. The PAEC, which is responsible for nuclear power development in Pakistan, has formulated a nuclear power program that considers the increasing demand for electricity, limited indigenous energy resources and nearly four decades of its experience in the nuclear power industry. The government of Pakistan approved the plan as a part of its ESP.[26] The program envisages a nuclear capacity of 8,800 MW by the year 2030.

Indigenous Capability for Support of Nuclear Power Program

Beginning with self-reliance in the operation and maintenance of its existing nuclear power plants, Pakistan aims to gradually increase its contribution to future nuclear power projects. The objective is to conserve foreign exchange and reduce overall costs in the long run, while raising the level of the nation's industrial and technological base.

Nuclear Power Planning

The PAEC began nuclear power planning activities in the early 1960s and has developed expertise in energy forecasting, power plant economics and power system expansion, leading to a capability to develop feasibility studies for nuclear power projects.

In collaboration with the IAEA, the PAEC has assisted various developing countries in their power system expansion studies and planning for the introduction of nuclear power through expert missions and fellowship programs. The PAEC also provides human resources for the national, regional and inter-regional training courses of the IAEA on various aspects of nuclear power planning.

Site Selection and Evaluation

Since the early 1970s, the PAEC has conducted siting studies for its nuclear power projects. Although the existing sites at Karachi and Chashma can accommodate additional units, more sites for future nuclear power plants are being investigated. Through IAEA expert missions, professionals from the PAEC have also been assisting other developing countries in their studies for nuclear power project sites.

Design, Engineering and Project Management

At the global level, nuclear power projects have been contracted in many different ways. At one extreme, a single contractor or a consortium of contractors takes overall responsibility for the entire task as a turn-key project, while at the other extreme the owner buys only the basic hardware of the nuclear steam supply system (NSSS) from a supplier, and handles the rest of the power plant. In the latter case, the owner also buys all other necessary equipment by itself or through a hired architect-engineer.

The PAEC has been working to develop project management capabilities to foster its own architectural engineering skills. The design and engineering phase of the nuclear power project in Pakistan was initiated in 1980s. Although KANUPP, C-1 and C-2 have been turn-key projects, the PAEC has been involved in various project management activities. This experience will help PAEC to manage the future nuclear power projects.

An Engineering Design Organization (EDO) has been established to ultimately act as architect engineer for the future indigenous nuclear power projects. EDO will also continue to provide technical support to operational nuclear power plants.

Manufacturing

Some capabilities exist in the public and private sectors to manufacture thermal power plant boiler components, heat exchangers and electrical equipment. In the PAEC, efforts have been made to develop expertise in instrumentation and control, materials and manufacturing of some critical spares. Already, the PAEC's manufacturing facilities are providing services to local industry in both the public and private domains. The PAEC has also provided mechanical equipment and has contributed to design, development, engineering and installation services in various projects of the European Organization of Nuclear Research (CERN).[27]

Since technical know-how, production capability and various inspection and testing facilities exist within the country, there is the potential to undertake the manufacture of mechanical equipment, components and heavy steel structures for the conventional island and balance of plant (BOP) of nuclear power plants.

Fuel Cycle

The PAEC initiated its nuclear fuel cycle activities with a modest prospecting program in the early 1960s. A number of promising areas were identified, some of which are presently being exploited. The PAEC is capable of uranium prospecting, mining and ore processing. It also has the indigenous capability to fabricate KANUPP fuel bundles.

Waste Management

Nuclear power plants generate radioactive waste of two types:

[316]

- operational radioactive wastes in solid, liquid and gaseous forms, which are by and large low and medium level radioactive wastes; and

- spent fuel discharged from the reactor which is high level radioactive waste.

Generally, nuclear industry manages radioactive wastes using three principles: (i) delay and decay, (ii) dilute and disperse, and (iii) concentrate and contain. PAEC also follows these principles for waste management.

As part of the plant design, appropriate radioactive waste management systems are installed in the nuclear power plants to remove radioactive liquid, gaseous and solid wastes arising from the operation of these plants. Releases during operation are maintained well below the permissible levels.

For high-level radioactive waste, the necessary storage capacities have been installed at plant sites. This waste is packed in standard drums for storage and subsequent removal to interim storage facilities for 60–70 years prior to their permanent disposal.

Decommissioning of Nuclear Facilities

Decommissioning costs are included in the operation and maintenance costs of nuclear power plants and are recovered as a part of tariffs. Decommissioning funds have been established for both operating power plants (KANUPP and C-1), where a specified proportion of generation receipts is set aside. The amounts allocated to decommissioning are invested in medium- to long-term bonds to account for inflation.

Human Resource Development

The PAEC has set up a number of institutes at different locations to provide higher education and job-specific training for developing the human resources required for its programs. Every year, scientists and engineers graduating[28] from universities in various disciplines are selected at the national level. Some of them are given one year of special training; others are admitted to two-year postgraduate programs. The trained personnel then move on to work in their relevant fields. These training/educational centers include:

- The Pakistan Institute of Engineering and Applied Sciences (PIEAS) located adjacent to PINSTECH. Established in 1967, PIEAS offers programs at Masters and Ph.D level in nuclear engineering, system engineering, process engineering, material engineering, mechanical engineering, medical physics, laser, plasma and computational physics, computer science, and nuclear medicine. Besides the regular academic programs, PIEAS also conducts training courses in various specialized topics, i.e., reactor supervision and operation, health physics, laser technology, computer applications and advanced reactor safety, etc. It also runs courses in management.

- The KANUPP Institute of Nuclear Power Engineering (KINPOE), located near the Karachi Nuclear Power Plant. KINPOE offers Masters Degrees in nuclear power engineering and a one-year diploma in nuclear technology to engineering and science graduates. It also offers a post-diploma program in nuclear technology for technicians who have earned a three-year diploma after high school.

- The CHASNUPP Centre for Nuclear Training (CHASCENT), located adjacent to the Chashma Nuclear Power Plant. CHASCENT provides one-year training in nuclear power plant technology to engineers and technicians.

- The Computer Training Centre (CTC) was established in Islamabad in 1982 to meet the requirements of computer professionals in the PAEC, and in universities and other R&D

organizations. CTC offers post-graduate training courses to scientists and engineers. It also conducts short training courses in computer applications.

- The National Center for Non-Destructive Testing (NCNDT), established in Islamabad in 1985. The NCNDT imparts training to engineers and technicians of the PAEC and industry in non-destructive testing techniques.

- The Pakistan Welding Institute (PWI), established in 1995 to provide training in welding technology to professionals of the PAEC and the wider industry.

Regulatory Infrastructure

The regulatory infrastructure in the country aims to ensure that a framework of design and operational safety is in place that will:

- keep the public, radiation workers and the environment protected against accidents or malicious acts that may lead to harmful effects of ionizing radiations; and

- minimize the risk of a major economic loss owing to any accident or malicious act, thereby protecting investment in the nuclear industry.

At the time of the construction of KANUPP in 1965, a regulatory setup was established within the PAEC which gradually evolved into an independent regulatory body. Established in 2001, the Pakistan Nuclear Regulatory Authority (PNRA) is mandated to control, regulate and supervise all matters related to nuclear safety and radiation protection in the country. The timing of the establishment of the PNRA coincided with that of the commissioning of C-1.

The PNRA issues licenses for the production, storage, disposal, trade and use of nuclear substances and radioactive materials. The licensing procedure for nuclear power plants is multistage, i.e.: (i) site registration, (ii) construction license, (iii) fuel loading permit, and (iv) operation license. The licensing requirements include:

- site evaluation report;

- preliminary and final safety analysis reports;

- Probabilistic Safety Assessment (PSA);

- Quality Assurance (QA) program;

- pre-service and in-service inspections programs;

- commissioning reports;

- radiation protection program;

- radioactive waste management program;

- environmental monitoring program;

- decommissioning strategy; and

- Non-Objection Certificates (NOCs) from the provincial Environmental Protection Agency and other concerned agencies.

International Cooperation and Commitments

At the international level, the development of nuclear power was initially restricted to only a few industrialized countries. Later, certain developing countries established nuclear power programs with the assistance of those countries which had already developed their own programs. The list has been very select, and although a number of countries have attempted to use nuclear power to meet their growing electricity demand, only three

(Mexico in 1989, China in 1991 and Romania in 1996) have been able to join the group of nuclear power producers during the last twenty years.

Pakistan needs nuclear power to continue its economic growth, ensure the security of energy supplies and to preserve its environment. It was one of the first few developing countries to set up a nuclear power plant. However, the development of its nuclear power program has been compromised since as early as 1976 by the embargoes imposed on the transfer of nuclear technology to Pakistan, notwithstanding the fact that the entire nuclear power program of the country is under comprehensive IAEA safeguards.

Though Pakistan is continuing to develop its nuclear power program through indigenous efforts and with the help of People's Republic of China, the country would welcome international cooperation to achieve its nuclear development targets. In fact, Pakistan has proposed that 'Nuclear Parks' may be set up in the country where foreign companies could build and operate nuclear power plants as an economic venture on a build–own–operate (BOO) basis.

Pakistan has been actively pursuing international cooperation for the safe operation of its nuclear power program. In addition to being an active member state of the IAEA, Pakistan is a member of the World Association of Nuclear Operators (WANO) and the CANDU Owners Group (COG). Missions from the IAEA and these organizations are routinely invited to visit Pakistan's nuclear power plants to advise and suggest improvements. Furthermore, Pakistan is already a party to conventions on:

- Nuclear Safety.
- Physical Protection of Nuclear Materials.
- Early Notification of a Nuclear Accident.
- Assistance in case of a Nuclear Accident or Radiological Emergency.

[321]

Pakistan is ready to share its experience of nuclear power with the developing nations and can assist in initiating their nuclear power programs under the umbrella of the IAEA, particularly in the areas of nuclear power planning, siting studies, establishment of regulatory authorities and human resource development.

Benefits from Use of Nuclear Power

Economics

Nuclear power plants are more capital-intensive than fossil fuel-powered plants. However, the fuel cost of a nuclear power plant is much lower and, therefore, the overall cost of electricity generation is generally in favour of nuclear. Another advantage of nuclear power is that once a nuclear power plant has been built, the cost of nuclear electricity generation is relatively impervious to fluctuation in the price of fuel since its fuel costs are only around 10–20 percent of the total generation cost. Favorable cost economics, coupled with freedom from dependence on fossil fuels, make nuclear power attractive, particularly for the energy deficient, oil-importing countries.

Environment

Nuclear power plants are environment-friendly compared to fossil fuel-fired power plants. In fact, nuclear power is the only technology used for electricity generation which from the very beginning of its development has laid great emphasis on possible environmental impacts. The radioactivity released into the environment by a nuclear power plant is continuously monitored and is maintained well below permissible limits. A nuclear power plant does not produce carbon dioxide or oxides of sulfur or nitrogen, which lead to adverse environmental degradation through the greenhouse effect and acid rain. Emissions avoided to-date by KANUPP and C-1 are listed in Table 10.8.

Energy Security

Nuclear power helps to enhance security of energy supply as it reduces dependence on imported fuels. Furthermore, owing to its round-the-year availability, nuclear power enhances power system reliability in Pakistan through reducing the negative seasonal impacts of hydro (in low water months) and gas plants (in cold months when natural gas is diverted to the residential sector for heating purposes).

Table 10.8

Emissions Avoided by Nuclear Power in Pakistan (thousand tons)

Pollutant	If electricity was generated by:		
	Oil-fired steam plant	Gas-fired steam plant	Gas-fired Combined Cycle plant
Carbon dioxide	20,056	15,617	11,869
Oxides of sulfur	130	Negligible	Negligible
Nitrogen oxide	62	70	40

Desalination

Another potentially important application of nuclear power is the desalination of seawater. The PAEC is already working to couple a desalination plant of 400,000 gallon per day capacity to KANUPP.[29] The plant is based on multi-effect distillation (MED) technology, where steam tapped from KANUPP will provide thermal energy to the distillation plant.

Concluding Remarks

In view of the resource limitations and constraints being faced in the development of fossil fuel-based power and hydropower, Pakistan needs to exercise the option of making increasingly larger use of nuclear power to meet its electricity requirements. The government of Pakistan has set the target of producing 8,800 MW of nuclear power by the year 2030. The PAEC has been safely operating two nuclear power plants and has around four decades of experience in the relevant fields. It also possesses qualified manpower and is working to expand the infrastructure support that will play a key role in achieving this target.

Pakistan requires international cooperation from industrialized countries for its nuclear power program, the entirety of which is under IAEA safeguards. At the same time, Pakistan can assist aspiring countries in initiating their own nuclear power programs under the IAEA umbrella.

11

The Israeli Nuclear Weapons Program

John Steinbach

Our aim should be to create a security environment, and you can't do that
if you don't recognize publicly that Israel has nuclear weapons …
George Perkovich[1]

Should war break out in the Middle East again, or should any Arab nation
fire missiles against Israel, as the Iraqis did, a nuclear escalation, once
unthinkable except as a last resort, would now be a strong probability.
Seymour Hersh[2]

With several hundred weapons and a robust delivery system, Israel
has quietly supplanted Britain as the world's fifth largest nuclear
power, and now rivals France and China in terms of the size of its nuclear
arsenal. Although it maintains an official policy of nuclear ambiguity –
neither acknowledging nor denying possession of nuclear weapons –
Israel is universally recognized as a major nuclear power. As former UN
Chief Weapons Inspector Hans Blix has noted, "The whole world is fairly
sure that they have about 200 weapons, and beating around the bush I
think doesn't change very much—they are part of the nuclear landscape"[3];
and according to the authoritative Center for Defense Information, "the
Israeli nuclear weapon infrastructure is probably quite large, including the
full range of strategic and tactical battlefield weapons."[4]

While much attention has recently been lavished on the potential
threat posed by Iranian weapons of mass destruction, the major nuclear
power in the region, Israel, has been largely ignored. Possessing a

sophisticated nuclear arsenal with an integrated strategy for its use in combat, Israel's nuclear monopoly provides the major regional impetus for the proliferation of weapons of mass destruction. With India and Pakistan, the other nuclear-armed non-signatories to the Nuclear Nonproliferation Treaty (NPT), the Israeli nuclear program imperils future nuclear nonproliferation efforts. Israel's nuclear arsenal reinforces the prospect that future conflicts in the region could rapidly escalate into a regional or global nuclear cataclysm.

In 1963, Shimon Peres enunciated Israel's policy of nuclear ambiguity or opacity, neither confirming nor denying its nuclear program; "Israel will not be the first country to introduce nuclear weapons in [the Middle East]."[5] In 2001, Aluf Benn writing in the *Bulletin of the Atomic Scientists* discussed the policy of Israel's Nuclear opacity; "although everyone knows what capabilities Israel has, it remains silent about them."[6] Because of its draconian military censorship, the world has derived most of its knowledge about the Israeli nuclear program from whistle-blowers, unguarded comments by Israeli political leaders, and analysis of evidence by scientists and arms control experts. Information contained in this paper was collected from the historical record and from contemporary authoritative sources and press accounts. Where possible, direct quotes from Israeli officials, commentators and nuclear experts are used to illustrate points, and primary sources are referenced. Careful analysis and cautious skepticism are prerequisites for presenting an accurate overview of Israel's nuclear program.

Israel's Civilian Nuclear Energy Program

The official position of Israel is that it is committed to the development of a civilian nuclear power program. "Since the 70's the Government of Israel decided that an option to produce electricity using nuclear reactors should be prepared and maintained. This option requires promoting nuclear knowledge and research, preparing sites suitable for building nuclear power plants and continuously examining the economical benefit of operating such plants."[7] A description of the mission of the Department for Nuclear Engineering includes the following statement:

> The Department for Nuclear Engineering operates in a range of areas, mainly: promoting the topic of reactor power in Israel and the development of the necessary infrastructure (nuclear fuel and clearing nuclear waste); long-term planning, including testing advanced reactor technologies and following-up with their future implementation; researching desalinization of sea-water combined with nuclear energy.[8]

Despite this claim, an exhaustive search of publicly available sources indicates the existence of no meaningful Israeli civilian nuclear energy program, past or present.

The reason for this apparent discrepancy is inherent in the Israeli policy of nuclear opacity.

> In its public identity, the IAEC was presented as a coordinating and advisory scientific agency in the prime minister's office, composed of some of the nation's nuclear scientists, with a mission to guide and to coordinate Israel's commitment to be part of the global nuclear age. But this civil-scientific identity was largely a façade to shield the IAEC's true identity: a defense-related research unit.[9]

From its inception, the Israeli nuclear program has centered on developing a nuclear weapons program, with any other nuclear development being incidental.

Since the 1950s, Israel has repeatedly advanced the idea of developing a civilian nuclear power program. After first embracing the idea in the 1950s, in the 1960s Israel entered negotiations with the United States to develop a 300 MW desalination/electrical generation nuclear facility. "This cooperative project proceeded well and a demonstration desalination unit powered by a conventional energy source was built at the Ashdod power station. But in the meantime the retreat from nuclear energy had begun, and so in the early 1970's the project was scrapped."[10] Israel's commitment to a nuclear weapons program had effectively negated any realistic possibility of developing civilian nuclear energy.

[327]

Israeli public statements about plans to build a nuclear power generating station have continued up to the present. However, this is unlikely to happen unless and until Israel abandons its policy of nuclear ambiguity and begins to negotiate in good faith a Middle East Nuclear Weapons Free Zone.[11] To develop a nuclear power program without international cooperation from the International Atomic Energy Agency (IAEA) and the Nuclear Suppliers Group (NSG) would be enormously complex and expensive—far beyond the capability of a relatively small nation like Israel. Apparently the latest proposal for a nuclear power station was predicated upon Israel receiving a special dispensation to maintain its nuclear weapons program while enjoying access to civilian nuclear technology, similar in nature to the recent agreement between India and the United States. According to the Director of the Galili Center for Strategy and National Security, Dr. Reuven Pedatzur:

> It won't be open nuclear development even if Israel will decide to go in the direction of a reactor for electricity. If there will be a real and practical plan for building a reactor, Israel will make a condition that if there will be international supervision it will only be of the civilian reactors and not of Dimona ...

> The United States agreed that India will separate its nuclear program into two different [programs], one civilian the other military. And regarding the military part there will be no international supervision and no UN supervision. I believe that Israel will try the same idea.

The United States quickly discouraged such speculation, maintaining that no similar deals would be forthcoming for the other two non-signatories to the Nuclear Non-Proliferation Treaty (NPT), Israel and Pakistan. Under Secretary of State Nicholas Burns stated emphatically, "We've always felt of India as an exception [sic] ... But we're not anticipating, in any way, shape or form, a similar deal for any other country."[12]

According to the Swiss International Relations and Security's (ISN) Security Watch, Israel has again "shelved" plans for the most recent proposed nuclear power station.[13]

Background Analysis

Meaningful discussion of Israel's nuclear policy requires critical examination of the roots of the Israeli attitudes that led to the development of its nuclear weapons program. First and foremost, the shadow of the Holocaust weighed heavily on the minds of David Ben-Gurion and Ernst David Bergman as they became preoccupied with the idea of an Israeli bomb. Avner Cohen writes in *Israel and the Bomb*, "Israel's project was conceived in the shadow of the Holocaust, and the lessons of the Holocaust provided justification and motivation for the project."[14] According to Shimon Peres, Bergman stated, "I am convinced ... that the State of Israel needs a defense research program of its own, so that we shall never again be as lambs led to the slaughter."[15] In 1966, Bergman wrote to Meir Ya'ari, the leader of the leftist political party MAPAM; "I cannot forget that the Holocaust came upon the Jewish people as a surprise. The Jewish people cannot allow themselves such an illusion for a second time."[16] David Ben-Gurion in his farewell address to the Israeli Armaments Development Authority (RAFAEL), defended the nuclear project saying, "I am confident, based not only on what I heard today, that our science can provide us with the weapons that are needed to deter our enemies from waging war against us."[17] Speculating about a meeting between Israeli Prime Minister Golda Meir and US President Richard Nixon that laid the basis for Israel's policy of nuclear opacity, Cohen wrote, "Meir may have assured Nixon that Israel thought of nuclear weapons as a last-resort option, a way to provide her Holocaust-haunted nation with a psychological sense of existential deterrence."[18]

A second major factor influencing Israel's strategic policies is its close identification with Europe and the West. Historically, Israel's leadership has identified with Europe rather than the Middle East, exacerbating already great regional tensions.[19] Ashkenazim (Jews of European ancestry), although a minority of the Israeli Jewish population, disproportionately occupy positions of leadership, and heavily influence opinion and policy.[20] (Mizrahim and Sephardim, Asian and Mediterranean Jews, comprise 39 percent of Israel's population, compared to 37 percent Ashkenazim.[21]) Writing for Gush Shalom, Uri Avnery, a leader of Israel's Peace Bloc, wrote, "In his book *The Jewish State*, the founding document of the Zionist movement, Theodor Herzl famously wrote: 'For Europe we shall be (in Palestine) a part of the wall against Asia ... the vanguard of culture against barbarism...' This attitude is typical for the whole history of Zionism and the State of Israel up to the present day."[22] In 2007 Naftali Tamir, Israeli ambassador to Australia, is quoted in *Haaretz* as saying, "Israel and Australia are like sisters in Asia. We are in Asia without the characteristics of Asians. We don't have yellow skin and slanted eyes. Asia is basically the yellow race. Australia and Israel are not – we are basically the white race."[23] While the Israeli Foreign Office quickly repudiated Tamir, the fact that a high-ranking Israeli diplomat would assume license to make such transparently racist comments underscores Avnery's argument.

This century-long history of Israeli–Western chauvinism and anti-Arab racism in turn generates Arab anti-Semitism, thus creating a vicious cycle of mutual suspicion and mistrust. A related complicating factor in negotiating a just and lasting Arab/Israeli peace is the emergence of radical religious movements in Israel, and the Arab world, injecting religion into a conflict historically based on a colonial/anti-colonial struggle.[24, 25]

Despite these stumbling blocks, majority Arab public support for "a just and lasting peace with Israel" has increased significantly since 2006,[26] while a 2007 survey indicated a substantial majority of Israelis support a comprehensive peace agreement with the Arabs.[27] It should be noted, however, that while majority public opinion in Israel, Palestine and the Arab nations supports a comprehensive peace agreement, there remains deep skepticism within both communities about prospects for peace.

The History of the Israeli Bomb

The Israeli nuclear program began in the late 1940s under the direction of Ernst David Bergmann, who established the Israeli Atomic Energy Commission in 1952. Convinced that nuclear weapons would solve Israel's security problems, Israel's first Prime Minister, David Ben-Gurion, and a young Shimon Peres became the principal architects and the driving force behind the nascent Israeli program.[28]

For a variety of reasons including revulsion over the Holocaust, guilt about manipulating Israel into attacking Egypt to justify the 1956 Suez War, and mutual interest in developing a nuclear weapons program, France provided the bulk of early nuclear cooperation with Israel, culminating in construction of a heavy water-moderated, natural uranium reactor and plutonium-reprocessing facility situated near Beersheba in the Negev Desert.[29, 30] Originally designed as a 24 Megawatt (MW) facility, the Dimona reactor was built with a cooling system three times larger than needed and numerous sources suggest that in fact Israel did enlarge its capacity significantly.[31] Because of the lack of any electrical generation capacity and the inclusion of a plutonium-reprocessing facility, it is clear that from the beginning that the French understood that they were providing Israel with a nuclear weapons option.[32]

The aftermath of the 1956 Suez Crisis precipitated top-secret talks that cemented and accelerated the Israeli–French nuclear collaboration. From its inception in the early 1950s, Israeli scientists were active partners in the French nuclear weapons program, providing critical nuclear expertise and actively participating in French bomb tests in Algeria in the early 1960s. According to Seymour Hersh, "The French would become dependent for the next decade – the first French nuclear test took place in 1960 – on Israeli computer skills."[33] The current Israeli nuclear program should be understood largely as an extension of this earlier collaboration.

The United States became aware of the Dimona reactor construction in late 1960 and the outgoing Eisenhower administration demanded an explanation. The official Israeli response was that Dimona was for "peaceful purposes," including scientific, industrial and medical applications.[34] In reality, the sole purpose of the Dimona plant was to produce nuclear bombs.[35] After several years of delay – caused by Charles de Gaulle's decision in 1960 to end official French involvement in Dimona's construction – the reactor finally went on line in 1964 and plutonium reprocessing began shortly thereafter.[36] Despite various Israeli claims that Dimona was a manganese plant or a textile factory, the extreme security measures employed told a very different story. In 1967, Israel shot down one of their own Mirage fighters that strayed too close to Dimona, and in 1973 the Israelis destroyed a Libyan civilian airliner which strayed off course, killing 104.[37]

According to Michael Karpin, author of *The Bomb in the Basement*, on November 16, 1966, Israel, which had by then separated enough plutonium for a primitive nuclear weapon, successfully carried out a sophisticated laboratory test that established the viability of its nuclear bomb.[38] There is substantial reporting from several credible sources that Israel participated as a full partner in the Algerian–French nuclear tests, and thus had no need to actually test a nuclear weapon.[39] Some historians suggest that Israel's fear that Egypt would preemptively attack Dimona was a factor in Israel's decision to attack Syria leading to the 1967 war.[40] According to Avner Cohen, during the 1967 war Israel assembled "two deliverable nuclear explosive devices."[41] By the time of the 1973 war, Israel possessed an arsenal of perhaps several dozen deliverable atomic bombs, and reportedly went on full nuclear alert.[42]

Possessing advanced nuclear technology and 'world class' nuclear scientists, Israel was confronted early with a major problem—how to obtain the necessary uranium and heavy water to operate the Dimona reactor. Israel's own uranium source – phosphate deposits in the Negev – was inadequate to meet the needs of its rapidly expanding program. During the early 1960s, France had supplied Israel with relatively small quantities of uranium but as Israeli/French relations cooled, a larger source was needed. The short-term answer in 1968 was to collaborate with West Germany in the 'Plumbat affair,' successfully diverting 200 tons of yellowcake (uranium oxide).[43] The West German authorities subsequently covered up their role in this clandestine operation.[44] Seymour Hersh disputes US Central Intelligence Agency (CIA) allegations that a US corporation called the Nuclear Materials and Equipment Corporation (NUMEC) diverted hundreds of pounds of enriched uranium to Israel from the mid-50s to the mid-60s.[45] In the late 1960s, Israel solved the uranium problem by developing close ties with South Africa leading to a quid pro quo arrangement whereby Israel supplied the technology and tritium for the 'apartheid bomb,' while South Africa provided as much as 600 tons of uranium.[46]

Heavy water was required as a moderator for the natural uranium reactor at Dimona and Israel solved this problem by purchasing 20 tons from Norway. With the assurance that it would be used strictly for peaceful purposes, the agreement gave Norway the right to inspect the heavy water for thirty-two years but Israel refused to permit meaningful inspections. Israel agreed in 1990 to return half of the heavy water to Norway.[47] As of 1990 Israel had reportedly used two tons of heavy water and retained approximately eight tons more for future use.[48]

In 1977, the Soviet Union warned the United States that satellite photos indicated South Africa was planning a nuclear test in the Kalahari Desert; in response, the Apartheid regime appeared to back down under pressure. But on September 22, 1979, a US satellite detected an atmospheric test of a small nuclear device in the Indian Ocean off South Africa. Apparently because of Israel's possible involvement, a carefully selected scientific panel – kept in the dark about important details – issued a report questioning the accuracy of the Vela satellite. Later it was learned through Israeli sources that the detonation detected in the Indian Ocean was actually the last of three carefully guarded tests of miniaturized Israeli 155mm nuclear artillery shells.[49, 50]

The Israeli/South African collaboration continued until the fall of Apartheid in 1990, and was especially active in developing and testing of ballistic missiles during the 1970s and 1980s.[51] The RSA series of South African ballistic missiles appear to be virtually identical to Israel's Jericho series.[52] Israel and South Africa conducted numerous joint missile tests at the Overberg Test Range. In addition to uranium and test facilities, South Africa provided Israel with large amounts of investment capital, while Israel enabled the Apartheid state to avoid international economic sanctions.[53] The Israeli–South African nuclear collaboration officially ended in 1989.

Although the French and South Africans were principal collaborators with the Israeli leadership in developing the Israeli nuclear program, the United States also shares some responsibility. Investigative journalist Mark Gaffney wrote, "[the Israeli nuclear program] was possible only because of calculated deception on the part of Israel, and willing complicity on the part of the US."[54] Israel was the second nation to sign up to Eisenhower's 'Atoms for Peace' program, and became the recipient of a 5 MW highly enriched uranium research reactor at Nahal Soreq which went online in 1960. This reactor later became the centerpiece of much of Israel's basic nuclear research, including the training of nuclear scientists and technicians.

President Kennedy, concerned about the Israeli nuclear program, insisted that US scientists be allowed to inspect the Dimona reactor (at the time under construction) to ensure that it was, as Israel claimed, strictly for peaceful purposes.[55] The Israelis went to extremes to prevent the inspectors from discovering the existence of a nuclear weapons program. Kennedy's successor, Lyndon Johnson, had a more ambivalent attitude toward nuclear proliferation and a more pro-Israeli viewpoint. Shortly after his election, President Richard Nixon and Israeli President Golda Meir met in 1969. Nixon agreed to end the Dimona inspections and remove US pressure on Israel to join the Nuclear Nonproliferation Treaty (NPT).[56] In retrospect, the Meir–Nixon understanding set the stage for Israel's ongoing policy of nuclear ambiguity. Nixon formally ended the Dimona inspections in 1970.[57]

[334]

From its inception, the United States was heavily involved in the Israeli nuclear program. Israeli scientists were largely trained at US universities and were generally welcomed at nuclear weapons labs.[58] In the early 1960s, the controls for the Dimona reactor were obtained clandestinely from a company called Tracer Lab – the main supplier of US military reactor control panels – purchased through a Belgian subsidiary, apparently with the acquiescence of the US National Security Agency (NSA) and the CIA.[59] In 1971, the Nixon administration approved the sale to Israel of hundreds of krytons, a type of high-speed switch vital to the development of sophisticated nuclear bombs.[60] In 1979, President Carter provided ultra high-resolution photos from a KH-11 spy satellite that were used two years later to bomb the Iraqi Osirak Reactor.[61]

Throughout the Nixon and Carter administrations and accelerating dramatically under Ronald Reagan, US advanced technology transfers to Israel have continued unabated to the present day, most recently including 'supercomputers' capable of being used to design advanced nuclear weapons and missiles.[62] It has been widely argued that illegal acquisition of US technology essential to nuclear weapons production facilitated the Israeli nuclear program.[63]

Following the 1973 war, Israel intensified its nuclear efforts while continuing its policy of deliberate nuclear ambiguity. By 1976, the CIA estimated that Israel possessed an arsenal of 10–20 plutonium bombs.[64] According to Seymour Hersh, in 1981, five years prior to the Vanunu revelations, an Israeli scientific defector provided Washington with photographic evidence that Israel possessed an arsenal of more than one hundred thermonuclear weapons. A senior intelligence official said, "Why do they need a thermonuclear device? Israel was more advanced and better than any of our people had presumed it to be—clean bombs, better warheads."[65] Despite this information, until the mid-1980s most US intelligence estimates of the Israeli nuclear arsenal remained of the order of two-dozen weapons. In 1986, the explosive revelations of Mordechai Vanunu, a nuclear technician in the Dimona plutonium reprocessing plant, changed everything overnight.

[335]

The Vanunu Revelations

A leftist supporter of Palestine, Mordechai Vanunu believed that it was his moral obligation to expose Israel's nuclear program. He managed to smuggle dozens of photos and valuable scientific data concerning the operation out of Israel and in 1986 his story was published in the London *Sunday Times*.[66] Rigorous scientific scrutiny of the Vanunu revelations by nuclear scientists including bomb designers Theodore Taylor and Frank Barnaby, led to the disclosure that Israel possessed as many as 200 highly sophisticated nuclear bombs. The revelations indicated that the Dimona reactor's capacity had been expanded several fold and that Israel was producing enough plutonium to make ten to twelve bombs per year. Vanunu proved unequivocally that Israel operated a large nuclear bomb production project that included plutonium reprocessing, uranium enrichment, fuel rod fabrication, depleted uranium munitions fabrication and lithium 6, tritium and deuterium production (used in advanced nuclear weapon design).[67] After interrogating Vanunu for several days, Barnaby concluded, "The acquisition by Israel of lithium deuteride implies that it has become a thermonuclear-weapon power – a manufacturer of hydrogen bombs … Israel has the ability to turn out the weapons with a yield of 200–250 kilotons."[68] Upon examining the Vanunu evidence, a 'senior U.S. intelligence analyst' said of the Vanunu data, "The scope of this is much more extensive than we thought. This is an enormous operation."[69]

Just prior to publication of his information Vanunu, lured to Rome by a female American Mossad agent with CIA connections, was beaten, drugged, kidnapped and transported to Israel.[70] Following a campaign of disinformation and vilification in the Israeli press, Vanunu was convicted of treason by a secret security court and sentenced to 18 years in prison. He served over 11 years in solitary confinement in a six-by-nine-foot cell. Vanunu was released from prison in 2004, but has since been held under virtual house arrest under draconian 1945 British Mandate Emergency Regulations.[71] The world press, especially in the United States, has largely ignored the Vanunu revelations and Israel continues to affect an ambiguous nuclear posture.[72]

Israel's Nuclear Arsenal

Today, estimates of the Israeli nuclear arsenal range from 100 to over 400 bombs. Given the magnitude of destruction caused by even the smallest nuclear weapon, the size of Israel's nuclear arsenal – whether 100 or 500 bombs – is irrelevant. The Hiroshima and Nagasaki bombs, primitive and small by modern standards, utterly destroyed two major cities. Within a two kilometer radius of the Hiroshima epicenter there was total destruction of all buildings and massive mortality.[73, 74] There is little doubt that Israeli nuclear weapons are among the world's most sophisticated, and largely designed for war fighting. According to various sources, the Israeli arsenal includes boosted fission weapons and small neutron bombs, designed to maximize deadly gamma radiation while minimizing blast effects and long-term radiation—in essence designed to kill people while leaving property intact.[75] Other weapons include ballistic missiles capable of reaching Moscow, nuclear-capable fighter aircraft, cruise missiles, land mines,[76] and artillery shells with a range of 45 miles.[77] In June 2000 an Israeli submarine launched a cruise missile that hit a target 950 miles away, making Israel only the third nation after the US and Russia with that capability. Israel currently deploys three of these virtually impregnable submarines, each carrying at least four cruise missiles.[78] The Israeli nuclear arsenal clearly dwarfs the actual or potential arsenal of all other Middle Eastern states combined and is much greater than any conceivable need for defensive deterrence.

Like the major nuclear powers, Israel bases its strategic nuclear threat on a 'triad' of delivery systems – aircraft, land-based ballistic missiles and submarine-based cruise missiles – with which it can threaten the entire Middle East and beyond. While numerous Israeli aircraft have nuclear capability,[79] three primary strategic aircraft are designed specifically to deliver nuclear weapons. Additionally, Israel's tactical arsenal is widely understood to include nuclear artillery shells, nuclear-capable short-range and cruise missiles, and nuclear land mines.[80] Israel's arsenal includes the following devices:

[337]

Jericho-1

The Jericho-1 short-range ballistic missile (SRBM) was developed during the 1960s with assistance from France. Deployed in 1973, the Jericho 1-is designed to carry nuclear, chemical and conventional warheads, and reportedly has a 500 kg payload with a range of 480–750 km. Approximately 100 Jericho missiles are deployed roughly 20 km east of Jerusalem at the Sedot Mikha launch site near the Tel Nof airbase. The Jericho-1 can reach Cairo and Damascus. There are also reports that the Jericho-1 can be deployed using mobile launchers.[81] Some reports claim that the Jericho-1 is no longer operational.[82]

Jericho-2

The Jericho-2 is a two-stage nuclear-capable intermediate range ballistic missile (IRBM) with a generally reported range of 1,500-3,500 km.[83] According to the respected *Jane's Intelligence Review*, the extended range of the Jericho 2 is 5,000 km, with a payload of 2,500 kg.[84] Developed and flight-tested in collaboration with South Africa, the Jericho-2 – apparently identical to the South African RSA-2 – was deployed sometime in the late 1980s or early 1990s.[85] With its extended range and sophisticated inertial and terminal guidance system, the Jericho-2 can target virtually the entire Middle East. Approximately fifty Jericho-2 missiles are reportedly deployed at a facility named Zachariah (Hebrew for God remembers with a vengeance.)[86]

Jericho-3 "Shavit"

On January 17, 2008, Israel launched a long-range ballistic missile (LRBM) from the Palmahim Air Base, Israel's satellite launch site.[87] There is speculation that this was a test launch of the Jericho-3 missile. The Jericho-3, which reportedly closely resembles the South African RSA-3 ballistic missile, is designed with a range of up to 4,000 km, carrying a payload of 1,000 kg.[88] According to the Israeli Defense

Ministry, the missile is capable of carrying "an unconventional" payload.[89] Although apparently not yet currently deployed, cities like Moscow, Islamabad and far beyond will be within range of the Jericho-3.

Other Potential Nuclear-Capable Missiles

Other potential nuclear-capable missiles include: Lance MGM-52 short-range missiles (130 km range); Gabriel-4 cruise missiles (200 km range); Harpoon Cruise Missiles (120 km range); Popeye Turbo air/sea launched cruise missiles (300 km range); and Popeye-1 and -3 cruise missiles (100 and 300 km ranges).[90]

F-4E 2000 Phantom (Kurnas-2000 Sledge-Hammer)

In 1968 the United States agreed to sell Israel the F-4E, at that time considered to be "the most advanced airplane in service, anywhere."[91] Approximately fifty planes were upgraded and deployed beginning in 1989. Modifications include reinforced skin and fuel cells, rewiring, advanced APG-76 radar, new J-79-GE-17 turbojet engines, air-to-air Sparrow, Sidewinder or Python missiles, and guided bombs.[92, 93, 94] With a carrying capacity of 7,200 kg and a relatively long range of 1,600 km, the F-4E was reportedly placed on nuclear alert during the 1973 war and may still be allocated to a nuclear role.[95]

F-15I Raam (Thunder)

The F-15I, the largest plane in the Israeli Air force, has a range of 4,450 km (enough to reach Tehran and return without refueling). Detachable conformal fuel tanks such as those found jettisoned in Turkey following the 2007 raid on Syria permit the F-15I a greatly expanded range.[96] In 2003, Israel demonstrated its range by flying three F-15s one-way from Israel to Poland without refueling, a distance of 1,600 km, approximately the same as Tehran.[97] According to John Pike, "The 25 F-15Is operational since 1999 [and the 100 F-16Is] were procured first and foremost to deal with the Iranian threat"[98]

F-16I Sufa (Storm)

The F-16I is the most recent and advanced addition to the Israeli nuclear arsenal. A total of 102 aircraft have been purchased. The F-16I has a comparable range and capability to the F-15E. Both planes took part in the massive military exercise that took place in June 2008 over the Mediterranean and Greece, widely seen as a direct threat to Iran. Over 100 F-15s and F-16s participated.[99] Both the F-15I and F-16I are nuclear-capable.

Submarines

Three Dolphin class submarines, with "the most advanced sailing and combat systems in the world," complete the third and potentially most important leg of Israel's nuclear 'triad.'[100] The first three Dolphin class submarines were delivered and deployed in the late 1990s, and have since undergone retrofitting. The Dolphin has six 533 mm and four 650 mm torpedo tubes. Built by the German shipyard HDW, it has been widely reported that the submarines have been modified to carry nuclear-tipped cruise missiles.[101]

According to press reports, in 2000 Israel conducted a successful test of a submarine-launched cruise missile that hit a target 900 miles away.[102] Among the cruise missiles suspected of being modified to carry a nuclear warhead are the Harpoon, with a range of 130 km, a modified Popeye Turbo with a range of 320 km, a modified Gabriel-4LR with a range of 200 km, as well as a new indigenously designed missile with a longer range.

Dolphin class submarines are designed to patrol the Mediterranean Sea, the Red Sea and the Persian Gulf. The Dolphin has a range of 4,500 miles and can remain on patrol for more than a month at a time.[103, 104] In addition to the current submarine fleet, in 2005 Israel purchased two more advanced Dolphin class submarines from Germany. The new submarines feature a larger fuel capacity, extending the range to over 10,000 km and operational endurance to approximately fifty days, thus permitting deployment to the Persian Gulf without refueling.[105] The new submarines, scheduled for delivery in 2010, will be equipped with a super quiet new air-independent, fuel-cell-based propulsion system (AIP) which will enable them to stay submerged for extended periods.[106] Israel is also in the process of retrofitting two of its older Gall Class submarines to accommodate new weapons. With a fleet of seven near invulnerable nuclear-capable submarines, Israel will soon have the capability to target the whole of Europe, the Middle East, and most of Africa and Asia.

Chemical and Biological Weapons

Israel reportedly also possesses and can readily deploy a comprehensive arsenal of chemical and biological weapons.[107] Like its nuclear program, much of what is known is based on conjecture and analysis.[108] According to the London *Sunday Times*, Israel has produced both chemical and biological weapons with a sophisticated delivery system. The *Sunday Times* quoted a senior Israeli intelligence official as saying: "There is hardly a single known or unknown form of chemical or biological weapon … which is not manufactured at the Nes Tziyona Biological Institute."[109] The same report described F-16 fighter jets specially designed for chemical and biological payloads, with crews trained to load the weapons at a moment's notice. In an effort to recruit East European Jewish Scientists, David Ben-Gurion was quoted as saying in 1948: "… either increase the capacity to kill the masses or to cure the masses; both are important."[110]

Israel signed the Chemical Weapons Convention (CWC) in 1993, but has since refused to ratify it. While chemical weapons research probably continues, "It is unlikely that this offensive CW program exists today."[111] Israel has never signed the 1972 Biological Weapons Convention (BWC). The current status of Israel's biological weapons status is unclear, however, "For all practical purposes, Israel acts as if it maintains a policy of biological ambiguity."[112]

[341]

Nuclear Weapons Facilities[113]

Israel's major nuclear weapons facilities are the Nahal Soreq Nuclear Research Center and the Negev Nuclear Research Center (Dimona), both of which are discussed below. Other facilities include:

- Eliabun storage facility for tactical nuclear artillery shells, nuclear landmines and other tactical nuclear weapons;

- Haifa Rafael-Israel Armament Development Authority, "Reportedly the location of a nuclear weapons design laboratory (Division 20), a missile design development laboratory (Division 48) and a weapons assembly plant;"[114]

- Haifa Port, the main Israeli Navy base and homeport of Israel's submarines;

- Mishor Rotern: Negar Phosphates Chemical Company (uranium mining);[115, 116]

- Yodefat nuclear weapons assembly facility. According to Vanunu, plutonium is transported from Dimona to Yodefat;[117]

- Triosh strategic weapons storage facility;

- Beit Zachariah Jericho IRBM base missile launch facility;

- Palmachim Airbase missile test range and space launch facility;

- Be'er Yaakov research and missile assembly facility;

- Rehovot Heavy Water Production Plant;

- Tel Nof air base and reportedly location of a strategic nuclear weapons storage bunker;

- Nevatim Air Base and site of an underground strategic air command center.[118]

Nahal Soreq Nuclear Research Center

The heart of the Nahal Soreq Research Center is a five-megawatt, highly-enriched uranium research reactor provided by the United States as part of the Atoms for Peace program. Despite IAEA oversight of the research reactor to ensure that it operates only for peaceful purposes, Nahal Soreq is widely reported to be a major nuclear weapons research and production facility involved in plutonium reprocessing and nuclear weapons research and design, and should be considered analogous to Lawrence Livermore and Los Alamos Labs in the United States.[119, 120] According to John Pike, the Soreq facility is "the functional equivalent of the US Livermore or Los Alamos national weapons laboratories. It is responsible for nuclear weapons research, design and fabrication."[121] Nahal Soreq is also involved in commercial nuclear research.[122] The reactor, scheduled to be decommissioned in eight years, will be replaced with a super-fast particle accelerator purchased from Germany.[123]

Negev Nuclear Research Center (Dimona)

Located in the Negev Desert, the Dimona complex is the largest nuclear site in Israel comprising a wide array of nuclear facilities. Mordechai Vanunu, the Dimona whistle-blower provided much of the information available about Dimona. The Dimona complex is an enormous enterprise buried six stories deep below the Negev Desert. According to Vanunu, when US scientists inspected Dimona in the 1960s, the elevator shaft was bricked over so the inspectors were completely unaware of the underground operation.[124] Dimona's centerpiece, the heavy water moderated natural uranium / plutonium production reactor was originally rated at 24 MW, but is understood to have been expanded to between 70 and 150 MW.[125] In the mid-1960s, a plutonium reprocessing plant was completed. Reportedly, Dimona can produce 40–60 kg of plutonium each year, enough for approximately 5–10 nuclear bombs.[126] The reactor has been operated nearly continually for almost 45 years and is at the end of its operational life. According to *Jane's Intelligence Review*, "However, after about 35 years in operation, there is growing concern even within the Israeli Government that Dimona is no longer able to meet the needs of an expanding nuclear arsenal. According to internal Dimona reports, the nuclear reactor there is suffering severe damage from 35 years of operation."[127]

The operations of Dimona are broken down into individual facilities called Machons. Machon-1 is the plutonium production reactor and also produces tritium for 'boosted' fission bombs. Machon-2 is the top-secret lab, six floors deep, where plutonium is reprocessed and where lithium-6 deuteride, used to boost fission bombs, is separated. Machon-3 processes natural uranium for reactor use and converts lithium deuteride into solids. Machon-4 treats high- and low-level wastes. High-level waste is stored and low-level waste is mixed with tar and buried. Machon-5 is a fuel rod assembly plant. Machon-6 is an operations service center for the other facilities. Machon-8 is a uranium enrichment facility and experimental laboratory. Machon-9 is a laser uranium separation and plutonium enrichment facility. Machon-10 manufactures depleted uranium munitions. In addition to the nine Machons, satellite photos show dozens of buildings and suggest a radioactive waste burial site nearby.[128]

Radiological, Environmental and Health Concerns

The plutonium isotope 239 used in a nuclear bomb is produced when neutrons in the core of a nuclear reactor bombard Uranium-238. Weapons grade plutonium generally consists of approximately 93 percent Plutonium-239 and 7 percent Plutonium-240. Dedicated plutonium reactors, such as Dimona, require the irradiated uranium to be removed every few weeks to prevent the buildup of undesirable isotopes.[129] The process of producing, separating, processing and weaponizing plutonium is extremely complex, generating large amounts of extremely high-level radioactive waste and even more low-level radioactive waste. Given the well-documented history of the failure of the world's other nuclear nations to manage their own nuclear waste issues, despite its secrecy blanket, it is unlikely that Israel has fared any better.[130] Writing in the *Washington Post*, Jonathon Broder pointed out, "It took the end of the Cold War for the United States to begin addressing environmental disasters like the Hanford nuclear waste site in Washington state. In their tiny, crowded country, Israelis don't have the luxury of waiting until peace permits such environmental issues to be discussed."[131]

The Dimona reactor has been operating nearly continually for 45 years, making it one of the oldest reactors in the world.[132] Reports in the Israeli and international media suggest that the Dimona reactor, which is nearing or at the end of its operational life, has suffered significant deterioration and represents a serious radiological threat.[133] Employees at Dimona have complained about workplace radiation exposure and ensuing cancer and other health effects, and several have filed and won lawsuits against the Israeli government.[134, 135] In 2002, Israel Television broadcast a special report detailing first hand the dangers Dimona poses to plant workers and the surrounding environment.[136]

The Palestinian Authority has expressed concerns about radiation exposure from Dimona, and documents increased cancer rates among nearby populations.[137] Jordan also has expressed concern about radiation dangers posed by the Dimona facility. Sufian al-Tell, a Jordanian environmental specialist, calculates that the Dimona reactor may have produced as much as 4,000 tons of nuclear waste.[138] Others claim that low-level waste is buried in the nearby Negev. Israel, to the extent that it addresses these concerns at all, claims that Dimona poses no threat to workers or the environment. According to Elhanan Abramov, Deputy CEO of the Negev Nuclear Research Center, "We stop the reactor, check the systems and renovate them ... Israeli citizens, apart from being assured that the reactor is safe, can sleep more soundly because this reactor is working."[139] In 2004, the Israeli military distributed potassium iodide pills to some Dimona residents, apparently to protect them in case of radiation leakage.[140]

After 45 years of operation, it becomes questionable whether even the most comprehensive overhaul of a nuclear reactor can address inherent safety issues, particularly brittleness in the reactor vessel itself caused by intense neutron bombardment. Jane's reported, "According to internal Dimona reports, the nuclear reactor there is suffering severe damage from 35 years of operation. This damage has come from the neutron radiation from the reactor core, and this bombardment has changed the reactor structure at the atomic level. Metal supports have become brittle and warped as the neutrons have created gas bubbles in the metal itself."[141] Israeli nuclear scientists have called on Israel to shut down Dimona, citing environmental concerns and pointing out that Israel has no need of further plutonium production. The London *Sunday Times* reported that Professor Uzi Eben, a former senior official in Israel's nuclear program, called for Dimona's shutdown to "avert a catastrophe."[142]

Radiological Concerns and Censorship

Like every other aspect of Israel's nuclear policy, the issue of health and environmental concerns posed by Dimona and the research reactor at Nahal Soreq is tightly censored and seldom mentioned by either the government or by the Israeli press. Owing to censorship concerns, much about what is known about Israel's nuclear program has been gleaned indirectly. Avner Cohen published his groundbreaking work, *Israel and the Bomb*, in the United States to evade censorship, but was prevented for several years from returning to Israel because of arrest threats.[143]

An example dealing specifically with censorship about Dimona's radiological hazards concerned Israeli retaliation against the British Broadcasting Corporation (BBC) for broadcasting a documentary in March 2003.[144] In the program, the BBC featured five Dimona workers who appeared on Israeli Television Channel 2, discussing the health and environmental consequences of their work. The BBC attempted to get them to testify again on camera but was told about threats to punish them like Vanunu if they cooperated. Ariel Speiler, one of the Dimona workers interviewed on Israeli TV, refused to talk about his experience working at Dimona saying, "The Secret Service silenced me. They've silenced me completely. They told me not to say one word. What can I do? What can I do? They told me; 'You'll end up like Vanunu.' How long has he been in prison? 15 years? Do you want me to go to jail? I really wanted to talk. I asked the others but they refused. Nobody wants to talk."[145]

The BBC went on to interview Israeli investigative journalist Ronen Bergman about Yehil Horev, the Israeli Ministry of Defense chief censor. Bergman recounted the story of how he interviewed Brigadier-General Yitzhak Yaakov, former chief weapons scientist of the IDF, about his fictional memoir. When Bergman submitted his story to Horev's censors, Yaakov, was publicly vilified as a traitor and forced to spend two years in prison. On camera Bergman stated, "Horev was afraid that veterans of the Israeli army, the Israeli intelligence, the Israeli nuclear effort, would try to maintain their footprint in the history of Israel and tell their story. And he wanted to frighten them. In this sense he was successful." After the documentary was aired, Israel announced that it was cutting off all ties and ending cooperation with the BBC.[146]

Israeli Nuclear Strategy

The question of the 'existential threat' posed by Israel's small size and lack of territorial depth, in juxtaposition with the size, resources and population of its Arab neighbors has been posited as a primary justification for Israel's nuclear program.[147] This theme of defensive nuclear deterrence still resonates within Israel, but the size and sophistication of its nuclear arsenal and statements by public officials

[347]

strongly suggest that deterrence is only one aspect of a much broader and far-reaching nuclear strategy. In the popular imagination, the Israeli bomb is a "weapon of last resort," to be used only at the last minute to avoid annihilation. At least in its early years, this formulation was motivated by fresh memories of the Holocaust.[148] Whatever truth this idea may have had in the minds of the early Israeli nuclear strategists, today the Israeli nuclear arsenal is inextricably linked to and integrated with overall Israeli military and political policy.

Israeli nuclear expert Oded Brosh said in 1992, "… we need not be ashamed that the nuclear option is a major instrumentality of our defense as a deterrent against those who attack us."[149] Seymour Hersh points out in *The Sampson Option*; "The Samson Option is no longer the only nuclear option available to Israel."[150] Over the decades, Israel has made numerous veiled nuclear threats against the Arab nations and against the Soviet Union (by extension Russia since the end of the Cold War). For example Ariel Sharon, former Israeli Prime Minister is quoted as saying, "Arabs may have the oil, but we have the matches."[151] In 1983 and again in 2003, Sharon suggested that India consider joining with Israel to attack Pakistani nuclear facilities.[152]

Possessing an overwhelming nuclear superiority allows Israel to act with impunity even in the face of worldwide opposition. For example, during the 1982 invasion of Lebanon Israel destroyed Beirut, resulting in 20,000 deaths, mostly civilian.[153] Despite the near destruction of a neighboring state, not to mention the utter destruction of the Syrian Air Force, Israel was able to carry out the war for months and an occupation for many years, at least in part owing to its nuclear threat.

In late 2008, despite worldwide condemnation, Israel implemented a blockade of Gaza characterized by Richard Falk, United Nations Special Rapporteur for Palestinian human rights, as "a crime against humanity, a flagrant and massive violation of international humanitarian law as laid down in Article 33 of the Fourth Geneva Convention."[154]

The implicit US nuclear umbrella enjoyed by Israel may also be a factor in encouraging Israeli adventurism. Writing in the fall of 2004, James Russell pointed out:

> It might be equally argued that the strategic umbrella provided by US forces could in fact encourage Israel to act more aggressively than it otherwise would, since its actions would be backed not just by its own nuclear force but also by the thousands of warheads in the US arsenal and the array of standoff conventional munitions used to great effect in Afghanistan and Iraq. The NPR [US Nuclear Posture Review] implies that the defense of Israel represents a core mission for the strategic deterrent by identifying several near-term contingencies involving an attack on Israel that could lead to the use of nuclear weapons by the United States.[155]

Israel uses its nuclear arsenal not just in the context of deterrence or direct war fighting, but in other subtler but no less important ways. For example, the possession of weapons of mass destruction can be a powerful lever to maintain the status quo, or to influence events to Israel's perceived advantage, such as to protect so-called moderate Arab states from internal insurrection, or to intervene in inter-Arab warfare.[156] In Israeli strategic jargon this concept is called 'non-conventional compellence,' as illustrated by the following quote from Shimon Peres, "acquiring a superior weapons system [nuclear] would mean the possibility of using it for compellent purposes—that is forcing the other side to accept Israeli political demands, which presumably include a demand that the traditional status quo be accepted and a peace treaty signed."[157] Cohen quotes Peres as saying, "Israeli nuclear weapons were important in encouraging Arab realism … [They were] instrumental in bringing Egyptian President Anwar Sadat to Jerusalem in 1977 and may have been even more important in convincing other Arabs, particularly the Palestinians, to recognize that the Arab-Israeli conflict could not be resolved by the sword."[158] Neo-conservative writer Robert Tucker raised the question; "What would prevent Israel … from pursuing a hawkish policy employing a nuclear deterrent to freeze the status quo?"[159]

Discussing Israeli nuclear compellence, dissident historian Israel Shahak further observes, "Israel is preparing for a war, nuclear if need be, and for the sake of averting domestic change not to its liking, if it occurs in some or any Middle Eastern states."[160]

Another major use of the Israeli bomb is to compel the US to act in Israel's favor, even when it runs counter to its own strategic interests. As early as 1956 Francis Perrin, head of the French A-bomb project wrote: "We thought the Israeli Bomb was aimed at the Americans, not to launch it at the Americans, but to say, 'If you don't want to help us in a critical situation we will [compel] you to help us; otherwise we will use our nuclear bombs'."[161] During the 1973 war, Israel used the nuclear threat to pressure Kissinger and Nixon to airlift massive amounts of military hardware to Israel. The Israeli Ambassador, Simha Dinitz, is quoted as saying at the time, "If a massive airlift to Israel does not start immediately, then I will know that the US is reneging on its promises and we will have to draw very serious conclusions ..."[162] Another example was spelled out explicitly by Amos Rubin, economic adviser to then Prime Minister Yitzhak Shamir; "If left to its own Israel will have no choice but to fall back on a riskier defense which will endanger itself and the world at large ... To enable Israel to abstain from dependence on nuclear arms calls for $2 to 3 billion per year in US aid."[163] Since then Israel's nuclear arsenal has expanded exponentially, both quantitatively and qualitatively, while the US currently provides Israel with approximately $3 billion in annual military aid.[164]

Regional and International Implications

In the event of a future Middle Eastern war, the possible Israeli use of nuclear weapons should not be discounted. According to Shahak, "In Israeli terminology, the launching of missiles on to Israeli territory is regarded as 'nonconventional' regardless of whether they are equipped with explosives or poison gas."[165] Israeli nuclear doctrine dictates that an unconventional attack requires a nonconventional (nuclear) response; a perhaps unique exception being the Iraqi SCUD attacks during the Gulf War.[166] Seymour Hersh warns, "Should war break out in the Middle East again, or should any Arab nation fire missiles against Israel, as the Iraqis did, a nuclear escalation, once unthinkable except as a last resort, would now be a strong probability."[167] Ezer Weizman, former Israeli President, said, "The nuclear issue is gaining momentum [and the] next war will not be conventional."[168] Jonathan Schell and Martin Sherwin appeal, "Israel and the entire Middle East are approaching a stark existential choice: a nuclear holocaust or a nuclear-free Middle East ... In a desperate effort to assure its local nuclear monopoly, Israel is in danger of courting national suicide."[169]

The Israeli nuclear arsenal has profound implications for the Middle East, and the global community. Israel Shahak has argued, "Israel's insistence on the independent use of its nuclear weapons can be seen as the foundation on which Israeli grand strategy rests."[170] According to Seymour Hersh, "the size and sophistication of Israel's nuclear arsenal allows men such as Ariel Sharon (and Benjamin Netanyahu)[171] to dream of redrawing the map of the Middle East aided by the implicit threat of nuclear force."[172] General Amnon Shahak-Lipkin, former Israeli Chief of Staff is quoted in the Hebrew language newspaper *Maariv*; "It is never possible to talk to Iraq about no matter what; it is never possible to talk to Iran about no matter what. Certainly about nuclearization. With Syria we cannot really talk either." Munya Mardoch, Director of the Israeli Institute for the Development of Weaponry, said in 1994, "The moral and political meaning of nuclear weapons is that states which renounce their use are acquiescing to the status of vassal states. All those states which feel satisfied with possessing conventional weapons alone are fated to become vassal states."[173]

Russia – and before it the Soviet Union – has long been an implied target of Israeli nuclear weapons. It is widely reported that the principal purpose of Jonathan Pollard's spying for Israel was to furnish satellite images of Soviet targets and other super sensitive data relating to US nuclear targeting strategy.[174] According to the widely respected security analyst John Pike, "The USSR (by extension Russia) has always been one of the primary targets of Israel's nuclear force, as Israeli assumptions hold that no Arab nation would attack Israel without Soviet support."[175] Since it began launching its own spy satellites in 1988, Israel no longer requires US spy secrets. Israeli nuclear weapons aimed at the Russian heartland seriously complicate disarmament and arms control negotiations and lower the threshold for their actual use. Investigative journalist Mark Gaffney cautions, "... if the familiar pattern [Israel refining its weapons of mass destruction with US complicity] is not reversed soon – for whatever reason – the deepening Middle East conflict could trigger a world conflagration."[176]

Prospects for a Nuclear Free Middle East

Despite the fact that all states in the region, including Israel, are on record as supporting in principle a Middle East Nuclear Weapons-Free Zone (MENWFZ), prospects for an agreement are discouraging. For its part, Israel conditions any discussions about eliminating its nuclear program upon the prior implementation of a comprehensive and lasting Israeli/Arab peace; "Israel insists that even *discussions* regarding the possibilities of limiting sensitive arms – primarily in the nuclear realm – should begin only after comprehensive peace has been achieved and minimal degrees of confidence and mutual trust have been established."[177]

Proposals for a MENWFZ began in earnest in 1974 when Iran, under the Shah, introduced a resolution in the UN General Assembly. This proposal and others that followed require all states in the region to foreswear nuclear weapons, accept IAEA safeguards and inspections, and agree not to accept or transfer nuclear technology to other states or parties. Israel, which steadfastly refuses to sign the Nuclear Non-Proliferation

Treaty or a Fissile Material Cutoff Treaty (FMCT), has offered conditional support for a MENWFZ, and participates actively in the Nonproliferation Treaty Organization. However, as Avner Cohen and Marvin Miller caution:

> Israel's attitude towards an FMCT has now evolved into strong opposition. At the same time, Israel is attempting to "balance" this opposition and its purely rhetorical support for the establishment of a Middle East Nuclear Weapons Free Zone (NWFZ) by emphasizing various actions it has taken in recent years in support of the global nonproliferation regime, such as its active participation in the Comprehensive Test Ban Organization and its adherence to international norms with regard to the export of nuclear and other military technology. In this manner, it seeks to make the case that Israel is a "responsible" albeit opaque nuclear state in contrast to "rogue" states such as Iran.[178]

David Albright, the director of the Institute for Science and International Security urged President-Elect Barack Obama to, "make a key priority of persuading Israel to join the negotiations for a universal, verified treaty that bans the production of plutonium and highly enriched uranium for nuclear explosives, commonly called the Fissile Material Cutoff Treaty (FMCT),"[179]

Israel, along with India and Pakistan, is not a signatory to the NPT. Ze'ev Shiff, dean of Israeli military commentators declared, "Whoever believes that Israel will ever sign the UN Convention prohibiting the proliferation of nuclear weapons ... is day dreaming,"[180] Other statements by Israeli leaders suggest that this position may ultimately change, albeit only in the context of a long-term peace agreement.[181] The position of the Arab states is that Israel's nuclear capabilities are destabilizing and must be "addressed as a precondition to peace and security in the region."[182] (It should be pointed out, however, that the Saudi/Arab League Peace Proposal of 2002 contains no such ultimatum.) These polarized viewpoints give rise to an intractable problem, providing little room for negotiations. To resolve this conundrum, a small minority of Israelis recognize that Israel must be prepared to renounce nuclear weapons in return for a regional peace agreement. Zeev Maoz suggests, "Should Israel renounce nuclear weapons? The answer I suggest is yes, but for a price. In return for greater regional security, Israel must give up its nuclear weapons."[183]

Israel has clearly indicated that it will not permit any other state in the Middle East to acquire nuclear weapons.[184] General Shahak-Lipkin declared "I believe that the state of Israel should from now on use all its power and direct all its effort to preventing nuclear developments in any Arab state whatsoever. In my opinion, all or most [available] means serving that purpose are legitimate."[185] After the 1981 raid on the Osirak reactor, the UN Security Council passed UN resolution 487 condemning Israel's attack on a International Atomic Energy Agency (IAEA) safeguarded facility and calling on Israel, "urgently to place its nuclear facilities under IAEA safeguards."[186] For the past 27 years, Israel, with the complicity of the West, has ignored the resolution. This insistence on retaining a nuclear monopoly in the region provides motivation for the Arab states to develop their own deterrence, destabilizing non-proliferation efforts in the region.

While publicly saying it has no objections to other Middle Eastern state's plans to develop civilian nuclear energy under IAEA safeguards, there is little question that Israel has serious concerns. Richard Beeston, *The Times* Diplomatic Correspondent reported, "The sudden rush [by Arab States] to nuclear power has raised suspicions that the real intention is to acquire nuclear technology which could be used for the first Arab atomic bomb."[187] The attack on the IAEA-safeguarded Osirak reactor and the ongoing threats against Iran's nuclear program suggest that Israel may be reluctant to allow development of civilian nuclear programs by neighboring states. [The head of Haifa University's National Security Studies Center, Professor Gabriel Ben-Dor said, "The fear is that various other countries in the region will follow in Israel's footsteps and also develop nuclear energy programs of their own. And we all know that once a nuclear energy program gets underway it is difficult to make the

[354]

distinction between the peaceful uses of nuclear energy and military developments and ramifications."[188]] If it desires, there is little doubt about Israel's military capability to prevent the other Middle Eastern states from going nuclear.

The Arab states, long aware of Israel's nuclear program, bitterly resent its coercive implications, perceiving it as the paramount threat to peace in the region. In May 2008, the Arab League declared that if Israel were to announce that it has nuclear weapons, the Arab states would withdraw from the NPT.[189] Mohamed Elbaradei, Secretary General of the International Atomic Energy Agency vehemently repudiated the Arab League statement:

> Arab countries' walkout is not the solution. A walkout by Arab countries will create a great deal of tension in the region that may lead to more pressures and even the use of military force against some Arab countries … We should understand that the use of nuclear weapons will be the beginning of the end of humanity. The use of nuclear weapons by any region like the Middle East—this means the destruction of the entire Middle East. The solution is that we build a strategy as to how we can reach a region free from weapons, including the weapon of the Israeli nuclear programme.

The International Pugwash organization concludes, "Even after a comprehensive peace is reached, Israel will probably decide to retain a strategic deterrent capability for some period as an insurance policy against bellicose regimes coming to power in the neighboring states. Until this perception is reversed, it is difficult to foresee dramatic steps in the direction of a WMDFZ."[190]

Unless Israel decides to negotiate openly over its covert nuclear program, it is doubtful that a near-term resolution of the Israeli/Arab conflict can be achieved. Joseph Cirincione, former director for nonproliferation at the Carnegie Endowment for International Peace points out:

> The world does well to remember that most Middle East weapons programs began as a response to Israel's nuclear weapons. Everyone

> already knows about Israel's bombs in the closet, bringing them out into the open and putting them on the table as part of a regional deal may be the only way to prevent others from building their own bombs in their basements. It should be obvious that Israelis are better off in a region where no one has nuclear weapons than in one where many nations have them.[191]

Israel's nuclear monopoly in such a historically politically unstable region as the Middle East jeopardizes future arms control and disarmament agreements such as a MENWFZ, and could threaten nuclear escalation.

Changes in the Middle East Political Environment

The political environment in the Middle East has changed dramatically since the 1973 war. Ironically, Israel attributes the Arab states' newfound willingness to negotiate to the existence of its nuclear weapons program.[192] However, Zeev Maoz disagrees with this analysis; "I argue that the balance sheet of Israel's nuclear policy is decidedly negative: not only did the policy fail to deter Arab attacks in 1973 and 1991, but it has been unrelated or only marginally related to Arab decisions to make peace with the Jewish state"[193] The Palestine Liberation Organization (PLO) represented the Palestinian people in direct negotiations with Israel, leading to formal recognition of Israel. The Arab states have expressed the desire to normalize their relationship with Israel based on Israel's return to 1967 borders with Jerusalem becoming the capital of both Israel and Palestine, a position most recently articulated (and recently renewed) in the 2002 Saudi/Arab League proposal. Egypt and Jordan have signed separate peace agreements with Israel. In 2003, with the support of President Mohammad Khatami and supreme religious leader Ali Khamenei, Iran proposed opening a broad dialogue with the United States, "including full cooperation on nuclear programs, acceptance of Israel and the termination of Iranian support for Palestinian militant groups."[194] The proposal, which would have aligned Iranian policies with those of Egypt, Saudi Arabia and others in the region, was summarily rejected by the Bush administration.[195]

The Palestinian Authority and Hamas have accepted in principle a two-state solution. In April, 2008, Khaled Meshaal, the head of Hamas' Political Bureau was quoted in the Fatah-controlled Palestinian newspaper *Al-Ayyam*, "In my heart, of course I believe all of Palestine belongs to the Palestinians. But practically speaking, our political position is a de facto two-state solution."[196] A February 2008 *Haaretz* poll showed that 64 percent of the Israeli public supports direct talks with Hamas.[197] Following the 2007 Annapolis Conference, where the Palestinian Authority, Israel and the United States agreed in principle to a Two-State solution, Ehud Barak, Israel's Prime Minister, was quoted in *Haaretz*, "I believe that there is no path other than the path of peace. I believe that there is no just solution other than the solution of two national states for two peoples."[198]

International pressure has been mounting on Israel to acknowledge its nuclear arsenal in the context of broader non-proliferation negotiations. Mohammed Elbaredi declared, "This is not really sustainable that you have Israel sitting with nuclear weapons capability there while everyone else is part of the non-proliferation regime."[199] Peter Kuznick, Director of the Nuclear Studies Institute at the American University asks:

> Countries like Iran look at what's happened to India and Pakistan – countries that have tested nuclear weapons recently – and they see that there's really very little sanctions against them, and [countries like Iran ask], 'Why is Israel allowed to have nuclear weapons without even any serious discussion about eliminating their nuclear weapons, and why can't Iran [have the same weapons]?' So it seems to them that there's a lot of double standards being imposed at this point.[200]

Although prospects for a negotiated peace in the Middle East show some promise, especially widespread hope that the election of Barack Obama will herald a more even-handed treatment, the window may swiftly close.[201] Benjamin Netanyahu, the hard-line Likud leader who opposes the current peace process and supports turning the occupied territories into economic Bantustans, is poised to form a hard-right government, with extremist Avigdor Lieberman becoming Israeli Foreign Minister.[202] In addition, the current global economic crisis and plummeting oil prices may destabilize Middle East governments, making negotiations more problematic.[203] After the debacle of the 2005 Nuclear Non-Proliferation Treaty Review Conference, progress toward resolving the current standoff over Israel's nuclear arsenal may prove critical to the success of the upcoming 2010 Conference. Referring to former nuclear proponents turned disarmament advocates such as former US Secretary of State Henry Kissinger, Rebecca Johnson warns, "The terrifying prospects of an eroded NPT and potential nuclear free-for-all, starting in the Middle East, have undoubtedly contributed to the new found enthusiasm of many born-again nuclear abolitionists."[204]

Conclusion

Placing the issue of Israeli nuclear weapons directly and honestly on the table would achieve several salutary effects. First, it would highlight a primary destabilizing dynamic compelling the region's states to each seek a deterrent of their own. Second, it would expose a perceived double standard, which sees the US and Europe on the one hand condemning Iran and Syria for developing weapons of mass destruction, while simultaneously protecting and enabling Israel. Third, acknowledging Israel's nuclear program would focus international public attention on the necessity of a MENWFZ agreement. Finally, a Nuclear Free Middle East would make a comprehensive regional peace agreement much more likely. George Perkovich, writing for the Carnegie Endowment for International Peace, stated: "Our aim should be to create a security environment, and you can't do that if you don't recognize publicly that Israel has nuclear weapons."[205]

[358]

Israel's nuclear arsenal is symptomatic of another, larger problem – the global proliferation of nuclear weapons – universally acknowledged as a paramount threat to human survival. Speaking before the UN Disarmament Conference in New York, UAE Ambassador to the UN, Abdulaziz Nasser Al-Shamsi, called for the abolition of nuclear weapons, declaring nuclear weapons could lead to, "the destruction of people and the threatening of the natural, environment and civil legacy of our world, which is the major source of our strength and resources, as well as being the major factor of our survival and stability."[206]

According to Hans Blix, "Israel is not likely to give up its nuclear weapons until you have a peaceful settlement in the Middle East, and let's hope that that comes sooner rather than later."[207] For more than 30 years the UN General assembly has passed annual resolutions by overwhelming majorities calling for the establishment of a Middle East Nuclear Weapons-Free Zone. Can the world afford the luxury of waiting another 30 years?

A Sustainable Nuclear Vision for the Gulf

Abdelghani Melaibari

The world is witnessing increasing energy demand owing to the progressive growth in global population size, rising standards of living in many parts of the world, and the mismanagement of energy resources through excessive consumption, which necessitates a culture of rational energy use. If we are to achieve a sustainable energy future, it will involve a paradigm shift in terms of national energy security policies and the regional energy mix.

Nuclear Success Stories

Those who have followed the global development of electric power reactors – from 1951, when electricity was first produced by the experimental EBR-1 nuclear reactor in Idaho, USA, to 1954 when the electricity produced by the Russian APS-1 Obninsk reactor was linked to the electricity grid south of Moscow – recognize that national nuclear power programs have unique features tailored to the circumstances of the countries in which they operate.

The United States

In the United States, nuclear power began with two experimental reactors – EBR-1 and S1W – built for the benefit of the US Navy. The first nuclear submarine – the USS *Nautilus* SSN-571, inaugurated in 1954 – relied on

Pressurized Water Reactor (PWR) technology. The submarine was designed and produced by the Westinghouse company. This was followed by operation of the Shippingport Atomic Power Station in late 1957; the first nuclear reactor to produce electric power in the United States. This technology formed the basis of more than half of the nuclear reactors operating in the world today. Westinghouse was later joined by Other US companies in the field of nuclear reactor production such as Combustion Engineering, Babcock & Wilcox, General Electric, General Automatic, and Allis-Chalmers.

In 2008, there were 104 nuclear reactors operating in the United States, with a total production capacity of 100,582 MWe. In 2007, they produced 806.55 terawatt hours (TWh) of electricity, equivalent to 19.39 percent of the electricity needs of the United States. Development in the field of US nuclear power not only depends on building new reactors, but also on reactor lifetime-extension of 60 years or more, achieving the maximum possible capacity and power uprate. During the last fifteen years the United States has increased its electricity production from existing nuclear reactors by the equivalent of that produced by an additional ten, large 1,500 MW reactors.[1]

Canada

In Canada, owing to a desire to rely on enriched uranium, Atomic Energy of Canada Ltd. (AECL) has developed CANDU (Canada–Deuterium–Uranium) reactors that use natural uranium as fuel and heavy water as a neutron moderator. Canada has also adopted pressure tubes instead of pressure vessels, allowing fuel provision during operation, which places these reactors at the forefront of operating performance. Although Canadian experimental reactors relying on natural fuel and heavy water began operating in 1947 – the NRX reactor and NRU reactor in 1957 – and benefited from cooperation with America, Britain, Italy and Japan, the first CANDU reactor for producing electricity was only connected to the grid in 1962.

[362]

In 2008 there were 14 CANDU-6 nuclear reactors operating in Canada, producing 935 MW of the country's total production capacity of 12,610 MWe. In 2007, Canada produced 88.19 TWh of electricity, i.e. 14.17 percent of its electricity needs. AECL now offers its advanced third generation reactors – ACR-1000s – whose production capacity is equivalent to 1,200 MW. Reactors based on the CANDU design have been constructed in Argentina, India, Korea, Pakistan, Romania and China.

France

The French nuclear program began with gas-cooled reactors, but swiftly made the radical shift to PWRs after a number of French companies allied themselves with the US company Westinghouse to form the *Franco-Américaine de Constructions Atomiques* (FRAMATOME). This decision by the French electricity company *Èlectricité de France* (EDF) resulted in the most ambitious electric power production program in the world. It introduced new principles to the nuclear field such as large reactors with a capacity of more than 1,400 MW using mixed oxide fuel (MOX).

In 2008, there were 59 operating nuclear reactors in France with a total generating capacity of 63,260 MW. In 2007, France generated 42.13 TWh, i.e. 76.86 percent of its electricity needs. All nuclear reactors in France are presently of the PWR type except for the Phenix reactor, which has been in operation since 1973 and is a fast-breeder reactor, generating 130 MW.

Japan

The Japanese nuclear program provides an example of a gradually perfected industrial productive capacity. The program began with the construction of the Tokai-1 gas-cooled reactor (GCR) in 1965, but then shifted to reliance on light water reactors (LWRs). The construction of nuclear reactors in Japan flowed from those built by US companies; the 320 MW Mihama-1 built by Westinghouse was followed by successive reactors of this type, starting with Mihama-2, constructed by Mitsubishi

Heavy Industries Ltd. (MHI); Westinghouse built the Takahama-1 reactor (780 MW), with MHI building all successive reactors of this type beginning with the Takahama-2; this was followed by the Westinghouse-built Ohi-1 and -2 reactors (1,120 MW each), which were followed by Mitsubishi's Tsuruga-2 and successive reactors. This model was also used in the construction of boiling water reactors (BWRs). The US company General Electric built the Tsuruga-1 and Fukushima Daiichi-1 reactors (440 MW each), with Hitachi and Toshiba building all successive reactors of this type beginning with Shimane-1 and Hamaoka-1. Then General Electric built Fukushima Daiichi-2 (760 MW), with Hitachi and Toshiba building Fukushima Daiichi-3 and -4 and others. Finally, General Electric built the Tokai-2 and Fukushima Daiichi-6 reactors (1,067 MW each), which were followed by successive reactors of this type beginning with Fukushima Daini-1 and -2, built by Hitachi and Toshiba respectively. Advanced boiling water reactors (ABWRs), generating 1,315 MW each, were then built by Toshiba and Hitachi, beginning with Kashiwazaki Kariwa-6 and -7.

In 2008, there were 55 operating nuclear reactors in Japan with a total generating capacity equivalent to 47,587 MW, and Japan generated 267.34 TWh in 2007, i.e. 27.54 percent of its total electricity needs.

Great Britain, Russia and Others

The British nuclear program relied on gas-cooled reactors (GCRs and AGRs) with one last experiment in PWR named Sizewell-B, whilst the Russians experimented with light water-cooled graphite-moderated reactors (LWGRs) as well as PWRs. There are also other unique experiences to be found in the programs of each country that has entered the field of nuclear energy.[2]

It thus becomes obvious from the above that peaceful nuclear programs for electricity generation largely depend on three main factors: availability of industrial and research capabilities; possible alliances and partnerships; and fuel supply security.

Table 12.1
Nuclear Power Reactors in the World

Country	Reactors in Operation		Reactors Under Construction		Nuclear Electricity Supplied in 2007	
	No. of Units	Total (MWe)	No. of Units	Total (MWe)	Total (MWe)	% of Total
Argentina	2	935	1	692	6.72	6.20
Armenia	1	376			2.35	43.48
Belgium	7	5,824			45.85	54.05
Brazil	2	1,795			11.65	2.84
Bulgaria	2	1,906	2	1,906	13.69	32.10
Canada	18	12,610			88.19	14.70
China	11	8,572	5	4,220	59.30	1.92
Czech Rep.	6	3,619			24.64	30.25
Finland	4	2,696	1	1,600	22.51	28.94
France	59	63,260	1	1,600	420.13	76.85
Germany	17	20,430			133.21	27.28
Hungary	4	1,829			13.86	36.81
India	17	3,782	6	2,910	15.76	2.52
Iran			1	915	NA	NA
Japan	55	47,587	1	866	267.34	27.54
Korea (Rep. of)	20	17,451	3	2,880	136.60	35.34
Lithuania	1	1,185			9.07	64.36
Mexico	2	1,360			9.95	4.56
Netherlands	1	482			3.99	4.10
Pakistan	2	425	1	300	2.31	2.34
Romania	2	1,305			7.08	13.02
Russian Fed.	31	21,743	6	3,639	147.99	15.97
Slovakia	5	2,034			14.16	54.30
Slovenia	1	666			5.43	41.57
South Africa	2	1,800			12.60	5.45
Spain	8	7,450			52.71	17.44
Sweden	10	9,034			64.31	46.12
Switzerland	5	3,220			26.49	40.03
Taiwan	6	4,921	2	2,600	38.96	19.30
UK	19	10,222			57.52	15.12
Ukraine	15	13,107	2	1,900	87.22	48.09
USA	104	100,582	1	1,165	806.55	19.39
Total	439	372,208	33	27,193	2,608.14	NA

Source: IAEA, Reference Data Series No. 2, *Nuclear Power Reactors in the World* (Vienna: IAEA, 2008).

Patterns of Energy Consumption in the Gulf

Oil, Gas and Electricity

The GCC countries are similar in terms of their culture, economic conditions, and social circumstances. The total population of these countries was nearly 42 million in 2008. Table 12.2 shows that the GCC countries are among the largest consumers of oil, gas and electricity. Together, they rank 5th in terms of oil consumption after the United States, China, Japan, and Russia.

As for natural gas consumption, they rank third after Russia and the United States. In terms of electricity generation, they occupy the 13th rank globally.

Table 12.2
GCC Consumption of Oil, Gas and Electricity (2007)

Country	Electricity Produced (KWh)	Rank	Oil Consumption (bpd)	Rank	Natural Gas Consumption (cu m)	Rank
Saudi Arabia	165,600,000,000	22	2,000,000	13	68,320,000,000	12
UAE	57,060,000,000	43	372,000	34	39,560,000,000	18
Kuwait	41,110,000,000	55	333,000	38	11,800,000,000	42
Qatar	13,540,000,000	78	95,000	78	17,930,000,000	34
Oman	11,890,000,000	87	66,000	89	8,795,000,000	49
Bahrain	8,187,000,000	98	31,000	105	10,270,000,000	44
Total GCC Pop. 42M	297,000,000,000	13	2,897,000	5	157,000,000,000	3

Source: The 2008 world fact Book, CIA. https://www.cia.gov/library/publications/the-world-factbook/

There is also a progressive growth in demand for electricity in all the GCC countries. This is illustrated by a study carried out by the Electricity and Co-Generation Regulatory Authority in the Kingdom of Saudi Arabia (see Table 12.3).

A further example is the projected growth in demand for electricity in the UAE, where it is expected to rise to nearly 15,000 MW in 2008 and to 42,000 MW by 2020.[3]

Table 12.3

Projected Growth in Electricity Generation Capacity (2008–2023)

Year	Most Likely Case		High Growth Case		Low Growth Case	
	Peak Load (MW)	Energy (GWh)	Peak Load (MW)	Energy (GWh)	Peak Load (MW)	Energy (GWh)
2008	33,930	198,766	33,930	198,767	32,816	192,244
2013	41,940	248,930	45,253	268,589	39,468	234,254
2018	50,218	298,058	63,415	376,386	46,371	275,229
2023	57,808	343,110	93,779	556,607	52,588	312,127

Source: Electricity and Co-Generation Regulatory Authority, Saudi Arabia (http://www. ecra.gov.sa/arabic/studies. htm).

Water Desalination

The second major source of energy consumption in the GCC countries is water desalination. In view of the scarcity of water in the region, there is a heavy dependence on large water desalination plants on the eastern and western coasts of the Arabian Peninsula, accounting for more than half the world's total fresh water production, estimated at 32.4 million m^3 daily.[4]

The *Desalination Year Book 2008–2009* shows that water desalination plants are being built with huge capacities, with the largest desalination plant in the world in Fujairah, UAE, which has a daily production capacity of 456,000 m^3. Also, five plants are currently under construction; the production capacity of each is in excess of 500,000 m^3. The largest of these is the Shoaiba-3 unit, at the Al Shoaiba plant to the west of Mecca, whose production capacity is 880,000 m^3 daily. The first water desalination plant to produce 1,000,000 m^3 daily is being constructed at the Ras Azzuor plant in northeastern Saudi Arabia; construction works are scheduled to begin in early 2009. By way of comparison, the largest water desalination plant in the western hemisphere is the Carlsbas plant being constructed near San Diego, with a production capacity of nearly 190,000 m^3 of water daily.[5] The Saudi Water and Electricity Company estimates that demand for water will rise to 10 million m^3 daily in 2025.[6]

[367]

Table 12.4

Independent Water and Power Projects being Executed by the Water and Electricity Company in Saudi Arabia

Project	Water (m3/ day)	Power (MW)	Water & Power Technology	Expected Commercial Operation Date	Remarks
Shuaibah: Phase 3	880,000 (194MIGD)	900	Desalination (MSF) Turbine (back-pressure).	Feb 2009	Cost: US$2,434 million. Construction: 86% completed (July 2008)
Shuqaiq: Phase 2	212,000	850	Desalination (RO) Turbine (condensing)	May 2010	Cost: US$ 1,850 million. Construction: 26% completed (July 2008)
Ras Azzuor: Phase 1	1,000,000	850–1,100		Feb 2012	Bid Submission: June 2008

Notes: Shoaiba, Phase 3: West of Holy Mecca on the Red Sea Coast, KSA; Shuqaiq, Phase 2: Southwestern Saudi Arabia on the Red Sea Coast; Ras Azzuor, Phase 1: Northeastern Saudi Arabia on the coast of the Gulf.

Sources: Andejani, Farah Imam, *Review of WEC IWPPs in Saudi Arabia*, IFAT, July 2008.

Transport

The third major source of energy consumption in the GCC countries is the transport sector. The region lacks developed public transport systems within its major cities and it still largely depends on personal means of transport. The region also lacks railways for commercial transport of raw materials and goods, continuing to rely on trucks.

Conclusion #1: National Policies

Every state must have a national policy that suits its particular situation and which reflects:

- its resources such as oil, gas, metals, and sunlight;
- its energy needs for electricity generation, water desalination, transport, and industry;

- the energy mix necessary to suit its requirements;

- the direction of its economy;

- the desired impact on the environment;

- prevailing social conditions;

- the security of energy sources;

- sustainability guarantees; and

- benefits for coming generations.

National Energy Policies

National energy policies encompass a number of different approaches and methodologies to deal with various energy-related challenges, including: developing and diversifying energy sources; improving the production and distribution of energy; curbing energy consumption; devising regimes and laws relating to energy; achieving bilateral agreements with other countries and global agreements in the energy field; providing investment incentives; building general frameworks of consumption rationalization; implementing taxes, charges, and government subsidization; determining the minimum limit of energy to be provided by the state to every citizen; and fostering public opinion supportive of the national interest.

They also comprise arrangements made by the state in the energy field, such as regimes, agreements, and directives including:

- Specific formulas for planning, generating, transmitting, distributing and using energy, to which all state sectors must be both subject, and committed to.

- Legislation relevant to the activity of the private sector in the field of energy, such as trade, transmission, and storage.

[369]

- Legislation affecting energy use, such as efficiency and emission standards.

- Organizational entities with specified responsibilities and authority; also with clearly defined relations between them.

- Regulations specific to geological surveys, energy exploration, and energy-specific research.

- Energy security, especially when parts of the generation process rely on imports.

- Measures to ensure the balance between supply and demand, and imports and exports,

- Means to balance development and related impacts on the environment.

- Diversifying dependence between depletable energy sources and sources of renewable energy.

Conclusion #2: Rational Use of Energy

Per capita energy consumption in the GCC countries is very high, even when compared with advanced countries. This includes the consumption of electricity, desalinated water and transport, and illustrates patterns of consumption that can be modified or improved via raising awareness, providing better education, organization and incentives, improving infrastructures, and fostering a culture of responsible energy use.

We must realize that our excessive consumption of fossil and non-renewable energy sources will be to the detriment of future generations, and that the depletion of oil and gas is a reality that has already become more tangible in certain regions of the world. It must also be noted that investing time and funds in programs promoting rational energy use, and improving the efficiency of production and consumption, can contribute to meeting the increasing demand for energy.

[370]

Figure 12.1
Number of Operating Reactors Worldwide

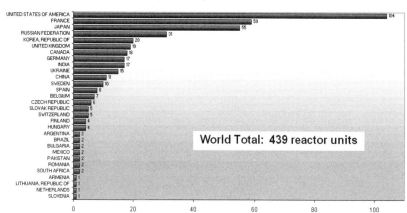

Source: IAEA, Reference Data Series No.2, *Nuclear Power Reactors in the World* (Vienna: IAEA, 2008).

Is Nuclear Energy an Option to the GCC Countries?

An initial answer to this question requires the following:

- determining the amount of energy which can be generated by using nuclear reactors, and hence the approximate number of reactors required;

- specifying the period over which the nuclear energy plan extends, the timetable for completion, and the obligations associated with the plan; and

- the future features of the electric grid connecting the GCC countries.

Nuclear electricity-generation is a mature field that has proven its security and safety for more than half a century. The cumulative hours of reactor operation is equal to 13,036 years, all based on the fact that the interaction of one fissile material (Uranium 335 or Plutonium 239) can produce 50 million times that produced by burning one carbon atom.[7]

Figure 12.2
Electricity Generation from Nuclear Reactors in 2007 (%)

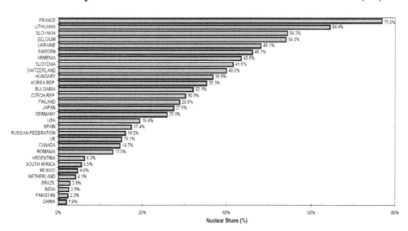

Source: IAEA, Reference Data Series No.2, *Nuclear Power Reactors in the World* (Vienna: IAEA, 2008).

The Coming Nuclear Renaissance

The IAEA expects the global generation of electricity from nuclear energy to rise to 691 GW by 2030 according to the highest estimates, and 447 GW according to the lowest, from its present capacity that had risen to 372 GW by the beginning of 2008, i.e. an increase of 75 to 319 GW.[8]

This expected increase conforms to the higher scenario of the IAEA and is referred to in the relevant literature as the coming "nuclear renaissance." The Agency anticipates that the major reasons for this renaissance will be the demand for energy, fears of climate change resulting from gas emissions and global warming, the anticipated depletion of oil at many of its present sources of export, rising energy prices, and ongoing changes in public opinion that have developed in a number of societies. Such changes in opinion have encouraged these societies to accept methods of radioactive waste DISPOSAL – as in Finland and Sweden – and global cooperation in the field of nuclear energy. The World Nuclear Association (WNA), which is an advocacy group for nuclear technology and includes most of the authorities and companies operating in the field, expects the global electricity generation

[372]

capacity from nuclear energy to rise to 3,500 GWe by 2060, according to the highest estimates, and 1,100 GW according to the lowest, from its present capacity of 372 GW at the beginning of 2008.[9]

Figure 12.3

Expected Increase in Nuclear Reactor Capacity (2007–2030)

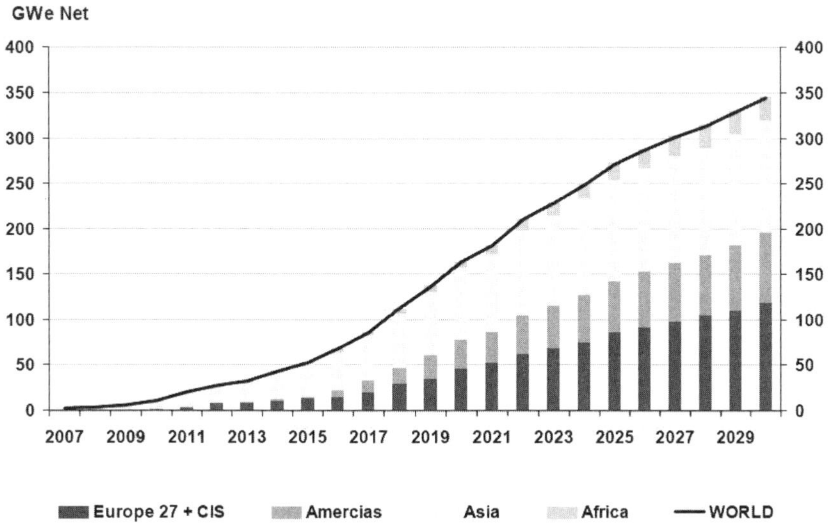

GWe Net

Europe 27 + CIS   Amercias   Asia   Africa   ——WORLD

Source: Philippe Vivien, "Developing Talent in the Energy World: Investing in People and Building Future," ICAPP, 2008.

Generations of Nuclear Reactors

In the field of nuclear technology, reactors can be divided into four generations. Reactors currently in operation represent only a very limited number of the first generation designs. The majority of reactors are from the second and third generations. Third generation reactors will form the nucleus of the nuclear industry of the 21st century and are characterized by their availability with respect to construction. On the other hand, the reactors of the fourth generation represent a paradigm shift in terms of their basic technology, design, safety, multiple uses and broadening applications as a result of global partnerships.

Figure 12.4
Generations of Nuclear Reactors

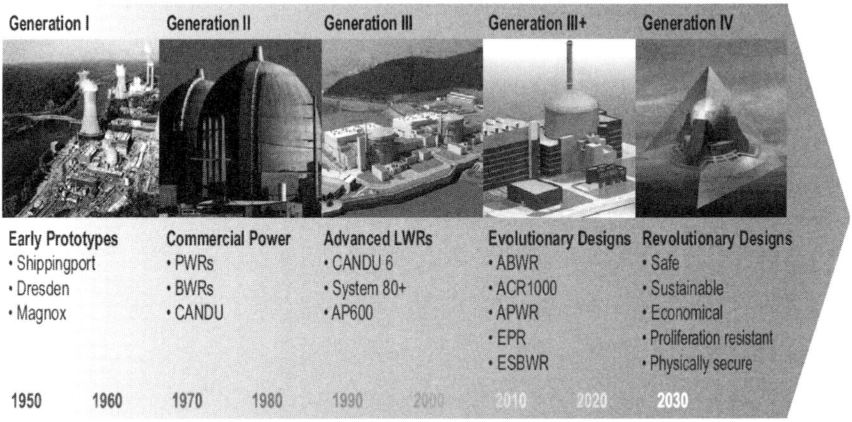

Generation I	Generation II	Generation III	Generation III+	Generation IV
Early Prototypes	**Commercial Power**	**Advanced LWRs**	**Evolutionary Designs**	**Revolutionary Designs**
• Shippingport	• PWRs	• CANDU 6	• ABWR	• Safe
• Dresden	• BWRs	• System 80+	• ACR1000	• Sustainable
• Magnox	• CANDU	• AP600	• APWR	• Economical
			• EPR	• Proliferation resistant
			• ESBWR	• Physically secure
1950 1960	1970 1980	1990 2000	2010 2020	2030

Source: Generation IV International Forum (GIF), *A Technology Roadmap for the Generation IV: Nuclear Energy Systems*, 2002 (http://gif.inel.gov/roadmap/pdfs/gen_iv_roadmap.pdf), accessed November 24, 2008.

The World Energy Council (WEC) estimates that the GCC countries will require an additional 200,000 MW by 2030, compared to 2007, and that 90,000 MW can be provided by nuclear energy plants. This will be the equivalent of one third of estimated total generation capacity. It is also equal to twice the capacity of Japanese nuclear energy plants, which have been developed over half a century. This amount of nuclear capacity can be provided by building 80 large reactors with a capacity of 1,100 MW each, or 55 advanced reactors with a capacity of 1,600 MW each.[10]

If we add to the above the facts that the assumed age of a single reactor is 60 years; that most reactors built in the 1970s and 1980s with an assumed age of 30 years have had this extended to 60 years; and that the period required to build a reactor is 5–10 years, depending on the availability of funding and factories for manufacturing major parts – especially large railway networks – the nuclear energy plan must be both long-term and flexible in order to cope with global economic and political changes.

[374]

The Electricity Grid of the GCC Countries

Nuclear reactors represent significant capital investments, and their economic feasibility is only realized via operation at maximum capacity for their full lifetime. For reactors to secure the basic load of electric power they require a balanced grid connecting all centers of generation to the centers of the load. The GCC countries founded the GCC Interconnection Authority to work on building this grid in three phases, and presently work is continuing on the implementation of the first phase.

Figure 12.5
Components of the First Phase of GCC Interconnection

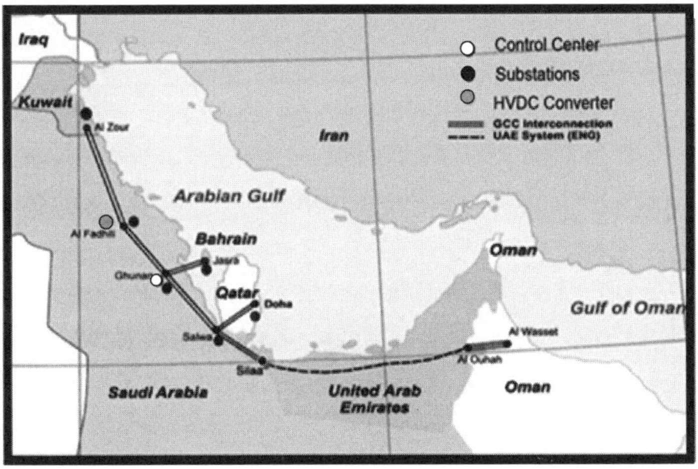

Source: GCC Interconnection Authority (http://www.gccia.com.sa), November 2008.

Conclusion #3: Adherence to Basic Considerations of Nuclear Energy Policies

The option of incorporating nuclear energy into the energy mix in the GCC countries is promising for several reasons. These include: reducing dependence on fossil fuels; reducing greenhouse gas (GHG) emissions that contribute to global warming; expanding the energy mix; and developing and upgrading industrial infrastructure—especially if the plan

[375]

is to pursue gradual participation in terms of producing parts for the nuclear sector, as well as entering into the nuclear fuel cycle.

After preparing the national energy policy of each state and approving a nuclear component, comes the task of nuclear power planning. There are multiple methodologies available for use in preparing this plan which have been laid down by the IAEA, the most important of which are:

- The IAEA brochure, "Considerations to Launch a Nuclear Power Program," March 2007.

- Nuclear Energy series guide NG-G-3.1, "Milestones in the Development of a National Infrastructure for Nuclear Power," IAEA, September 2007.

- TECDOC-1513, "Basic Infrastructure for a Nuclear Power Project," June 2006.

- TECDOC-1522, "Potential for Sharing Nuclear Power Infrastructure between Countries," October 2006.

- TECDOC-1555, "Managing the First Nuclear Power Plant Project," May 2007.

- *Handbook on Nuclear Law*, IAEA, 2003.

- *Interaction of Grid Characteristics with Design and Performance of Nuclear Power Plants: A Guidebook*, Technical Reports Series No. 224 (1983).

- *Policy Planning for Nuclear Power: An Overview of the Main Issues and Requirements* (1993).

- "Nuclear Power Programme Planning: An Integrated Approach," TRS No. 1259 (2001).

There are also other tasks that are required in parallel to the methodological work, including:

- Forming negotiation teams and coordinating with neutral global expertise, working for the national interest and supported by action task forces to process and prepare data and devise methodical mechanisms to make decisions.

- Establishing partnerships and smart alliances based on realistic estimates of expected gains and the obligations of other parties. These include: agreements between countries; joining regional and global organizations; agreements with companies that manufacture systems, components and devices pertinent to the nuclear field; agreements with specialized institutions of education and training; and investment in nuclear technology companies.

- Establishing strong national organizational entities supported by neutral global expertise working for the national interest, followed by new executive entities of a standard that befits the performance and quality required in the nuclear field.

- Dealing with all the added requirements of entering the nuclear field, such as awareness of both local and global public opinion regarding radiation, dealing with the IAEA to obtain maximum benefit from its services and, in exceptional cases, the links between fuel cycle technologies and proliferation. These are fundamental and necessary technologies in the nuclear field and can be restricted to peaceful uses as in the case of Japan, Argentina, Germany and South Africa. Also, their transparency can be guaranteed by the IAEA.

- If building only a limited number of reactors, the GCC countries might not need to enter into the nuclear fuel cycle. But if the goal is to build 50–80 reactors in the region to generate 90 GW, involvement in the nuclear fuel cycle becomes a necessity. Enhancing and reinforcing the security of nuclear fuel is a central issue in energy strategy.

- Planning for a Gulf nuclear program must be conducted on two levels. First, partnerships must be developed and applied as part of a system of international cooperation, guaranteeing nuclear construction and providing fuel. Second, planning must take place on a global level to achieve a true "nuclear renaissance."

[377]

The GCC shares the world's concerns regarding nuclear proliferation. However, there are also other issues facing the nuclear field which the world must come together to combat, especially in light of an anticipated nuclear renaissance. They include:

- Scarcity of cadres in the nuclear field at the global level owing to the neglect of this field over the past three decades.

- The need to raise awareness among students of the coming nuclear renaissance in order to encourage them to pursue professions in this field.

- The limited global manufacturing capacity for major components of nuclear hardware, such as reactor vessel forging facilities.

- Global demand for uranium enrichment, fuel rod manufacturing and fuel transport may surpass present capacities and their anticipated expansion, especially considering the constraints imposed on countries wishing to enter this field.

- Intermediate storage, transport, processing and final storage of irradiated fuel might also be a global problem, especially since most of the irradiated fuel produced over the past decades is still in the intermediate storage stage. Again, there are numerous constraints preventing countries from entering this field.

- Although the present global economic situation is bad, countries with financial surpluses must allocate grants to be invested in companies involved in the nuclear renaissance.

The Option of Nuclear-Powered Desalination

The principle of powering water desalination with nuclear energy was first implemented in the Soviet reactor BN-350 on the coast of the Caspian Sea in present day Kazakhstan. When in operation it generated 135 MW of electricity and 80,000 m^3 of desalinated water daily from 1973. It was officially shut down in 1999, but was actually closed in 2004 after showing increasingly poor performance. The plant also used to produce plutonium and consume some oil and gas to produce a total of 120,000 m^3 of water daily. This plant is considered the only major experiment in this sphere.

In Japan, small amounts of desalinated water have been produced using nuclear energy (1,000–3,000 m^3 daily) for internal use by ten nuclear reactors.[11]

In India, experimental production of desalinated water took place in two small plants: one producing 1,500 m^3 daily, the other 4,500 m^3.[12]

China is now building a nuclear desalination plant in the Yantai region with a daily production capacity of 160,000 m^3. It relies on a reactor generating 200 MWt.[13]

In 2005, the IAEA released a report titled "Optimization of Coupling Nuclear Reactors and Desalination Systems." The report represents the conclusions of studies by interested researchers from Argentina, Canada, China, Egypt, India, Korea, Morocco, Russia, and Tunisia.[14] The principle is simple and co-generation of water and electricity in traditional plants relying on fossil fuels has been a successful technology for decades. However, the economic feasibility of desalinating water by nuclear energy – though promising – has yet to be proven.

Generation IV International Forum

Thus far, the focus has been on using nuclear reactors to generate electric power or naval vehicles like submarines and aircraft carriers. Nuclear reactors have succeeded in these domains and the advanced generation III+ reactors have become the optimal option of the 21st century for generating electricity from nuclear energy in a safe and reliable manner.

However, other potential uses for nuclear reactors require further research. In order to consolidate efforts in this field, the Generation IV International Forum (GIF) was established in 2001 and twelve countries now participate in the forum. These countries are Argentina, Brazil, Canada, France, Japan, Korea, South Africa, the United Kingdom, the

United States, Switzerland, Russia, China and the Euratom grouping. These countries seek to develop the 4th generation of nuclear reactors, targeting new applications such as producing hydrogen, desalination of water, and generating heat for industrial and domestic uses.[15]

Fourth Conclusion

The GCC countries can enter the field of nuclear-powered water desalination by using reactors of the emerging generation IV type after 2030 – or by using generation III+ designs if their economic feasibility is proven – leaving the GCC countries to concentrate on nuclear electricity generation in the short- to medium-term.

The Solar Energy Option

An average of 300 watts per square meter of solar energy falls on the Arabian Gulf region each day, which makes it a promising region for establishing utility-scale solar farms. With the projected development of thin-film and multi-layered nano coatings, which are capable of absorbing all the frequencies of the sun rays, the harvesting of solar energy becomes both more practical and economical. Such developments will enable us to manufacture flexible solar cells which can be easily maintained and operated, and which will be capable of absorbing 96 percent of natural solar energy and converting it directly into electricity. They will also allow us to develop the efficiency of solar complexes. Researchers at the Rensselaer Polytechnic Institute expect such products to be available within two to three years.[16]

Presently there are more than 60 small plants worldwide, each generating more than 10 MW of electricity via solar cells, including the Spanish plant Parque Fotovoltaico Olmedilla de Alacron, which generates 60 MW.[17]

There are also a number of proposed projects that will generate a combined total of 550 MWe when complete, such as the Tobaz Solar Farm[18] and the projects based on the Stirling Engine, which will generate 500–900 MWe.[19]

To generate the equivalent electricity of a large reactor will require less than 4 km^2 in the Arabian Gulf region. This qualifies the sun as one of the largest potential sources of energy in the GCC countries.

Figure 12.6

Average World Sunshine: Round-the-Clock Measurements
(1991–1993, 24 hours/day)

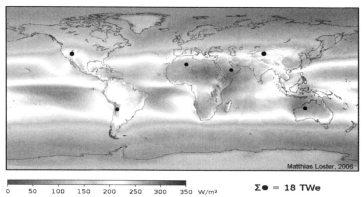

Source: Matthias Loster (http://www.ez2c.de/ml/solar_land_area/), retrieved November 24, 2008.

Conclusion #5

Solar energy must play a part in the energy mix of the GCC countries; there must be investment in R&D in this field, as well as in the companies and factories involved in producing its components.

Conclusion

The sustainable nuclear vision in the GCC countries must be a part of an overarching sustainable vision of energy resource management, and must be based on the rational use of energy, giving priority to renewable sources, and solar energy in particular. The GCC must adopt a program of balanced expansion in terms of different sources of energy – renewables, fossil fuels and nuclear – whilst reducing negative impacts on the environment and taking into consideration the needs of coming generations.

CONTRIBUTORS

HANS BLIX was Associate Professor in International Law at Stockholm University, Sweden, before serving as the Adviser on International Law at the Swedish Ministry of Foreign Affairs from 1963–1976. He served as State Secretary for International Development Cooperation from 1976–78 and then Minister for Foreign Affairs from 1978–79. Dr. Blix took the role of Director General of the International Atomic Energy Agency (IAEA), Vienna, from 1981–1997 and Executive Chairman of the United Nations Monitoring, Verification and Inspection Commission (UNMOVIC) from March 2000–June 2003. Dr. Blix currently chairs the Weapons of Mass Destruction Commission (WMDC) which was formally established in December 2003.

Dr. Blix has written several books on subjects associated with international affairs, constitutional and international law, and has received several honorary doctorates as well as a number of decorations and awards. He is a member of the Institut de Droit International and an honorary member of the American Society of International Law. In November 2006, Dr. Blix was appointed President of the World Federation of United Nations Associations (WFUNA), a position he will hold for a three-year term.

MAHMOUD NASREDDINE obtained his Ph.D in Physical Sciences from the University of Grenoble, France in 1974, and after lecturing at the university was elected to the board of the X-Ray and Crystallography Center based in Grenoble, affiliated to the French National Research Center (CNRS).

In 1974 he was appointed Professor in the Department of Physics at the Lebanese University in Beirut, as well as the Head of the Department of Physics in 1982, and also Director of the Faculty of Science (1983–1985). In addition, he was appointed Dean of the Faculty of Public Health, position he held until 1993. He also chaired the Board of the Institute of Applied Sciences in Lebanon (1983–1986).

In 1993, Dr. Nasreddine joined the National Council for Scientific Research in Lebanon where he was appointed Head of the Department of Energy Sciences and the National Center for Renewable Energy. In 1995 he devoted himself to founding the Lebanese Atomic Energy Commission (LAEC), which started its work that year in close cooperation with the International Atomic Energy Agency (IAEA) and the Arab Atomic Energy Agency (AAEA). He represented Lebanon from 1993–2001 at the conferences of the IAEA and the AAEA, and at a number of international forums relating to nuclear energy and its peaceful applications. During this time he also supervised projects for technological cooperation between Lebanon and the IAEA.

In 2000, Dr. Nasreddine was elected Director General of the AAEA and joined its headquarters in Tunisia in 2001. Since assuming his new tasks, he has worked on developing Arab cooperation in the field of applied research, nuclear safety and protection against radiation. He has also supervised the publication of a number of specialized books in the field of peaceful applications of nuclear energy. In 1998, he was elected a Member of the Advisory Committee on Technical Cooperation at the IAEA, where he is contributing to the formulation of recommendations aimed at developing technological cooperation between the IAEA and its member states, especially those receiving technological assistance.

STEVE THOMAS is Professor of Energy Policy at the Public Services International Research Unit in the Business School of the University of Greenwich, London, where he has led energy research since 2001. He has a BSc in Chemistry (Bristol).

Prof. Thomas has worked as an independent energy policy researcher for more than 30 years. From 1979–2000, he was a member of the Energy Policy Programme at SPRU, University of Sussex and in 2001 he spent 10 months as a visiting researcher in the Energy Planning Programme at the Federal University of Rio de Janeiro.

Prof. Thomas was a member of the team appointed by the European Bank for Reconstruction and Development to carry out the official economic due diligence study for the project to replace the Chernobyl nuclear power plant (1997). He was also a member of an international panel appointed by the South African Department of Minerals and Energy to carry out a study of the technical and economic viability of a new design of nuclear power plant, the Pebble Bed Modular Reactor (2001–02). Furthermore, he was part of an independent team appointed by Eletronuclear (Brazil) to carry out an assessment of the economics of completing the Angra dos Reis 3 nuclear power plant (2002). Prof. Thomas has published extensively on the economics and policy of nuclear power including a major report for Greenpeace, *The Economics of Nuclear Power*, in 2007.

MALCOLM C. GRIMSTON was educated at Scarborough College, UK, and Magdalene College, Cambridge University. He graduated in 1979 having read Natural Sciences and specialized in psychology, after which he worked as a chemistry teacher before joining the Atomic Energy Authority in 1987. In 1995 he joined Imperial College, London University, as a Senior Research Fellow and in 1999 became a Senior Research Fellow at Chatham House (the Royal Institute for International Affairs) in London where he is now an Associate Fellow, conducting an investigation into the future of civil nuclear energy.

Malcolm Grimston is a regular media contributor on energy and nuclear matters. Among his publications are two books co-written with the late Peter Beck: *Double or Quits: The Global Future of Civil Nuclear Energy* and *Civil Nuclear Energy: Fuel of the Future or Relic of the Past?* In 2005 he published his latest study examining the differences between political and technical mindsets and how these impact on major industries such as nuclear energy. He is also taking part in a consortium that includes Manchester, Southampton and City universities to carry out a government-funded project on sustainable nuclear energy. Malcolm Grimston is an elected member of Wandsworth Council, UK, and has executive responsibility for environment and leisure.

YOUSSEF A. SHATILLA (BSc, MSc, nuclear engineering, Alexandria University; DSc nuclear engineering, Massachusetts Institute of Technology) has been a Professor of Nuclear Engineering at the Masdar

Institute of Science and technology since the Spring of 2008. He has also been a Visiting Professor at the Department of Nuclear Science and Engineering at MIT since 2007. From the Fall of 2002 he spent five years as an Associate Professor at the Department of Nuclear Engineering at the King Abdulaziz University, Saudi Arabia. Prof. Shatilla was also a Faculty Fellow at the Idaho National Laboratory for two consecutive years in 2004 and 2005. For the ten years before, Dr. Shatilla was a Principal Engineer in the Core Engineering Department of the Westinghouse Electric Company, where he developed methods and software for nuclear reactor design applications. Prior to joining Westinghouse he was an Assistant Professor at the Department of Nuclear Engineering, Alexandria University.

Prof. Shatilla's research interests include computational reactor physics, in-core nuclear fuel management optimization, advanced reactor design, application of nuclear systems for water desalination and hydrogen production and sustainable energy.

His articles and publications include: "A Pressure-Tube Advanced Burner Test Reactor," *International Journal of Nuclear Engineering and Design*, vol. 238, no. 1, 102, 2008; "A Very High Temperature Reactor Self-Sustainable Oasis Concept," *International Journal of Nuclear Desalination*, vol. 2, no. 2, 193, 2006; Shatilla, Y., P. Hejzlar and M.S. Kazimi, *A PWR Self-Contained Actinide Transmutation System*, MIT-NFC-TR-088, Center for Advanced Nuclear Energy Systems (CANES), Massachusetts Institute of Technology, September 2006; and Gavrilas, M., P. Hejzlar, N. Todreas and Y. Shatilla, *Safety Features of Operating Light Water Reactors of Western Design*, 2nd Edition, Center for Advanced Nuclear Energy Systems (CANES), Department of Nuclear Engineering, Massachusetts Institute of Technology, 2000.

MUJID S. KAZIMI is Professor of Nuclear and Mechanical Engineering at the Massachusetts Institute of Technology (MIT), the faculty of which he joined in 1976. He is the current and founding Director of the Center for Advanced Nuclear Energy Systems (CANES) at MIT. He has extensive experience in the design and safety analysis of nuclear fission reactors, fusion technology devices, and high-level radioactive waste storage facilities. Prof. Kazimi's current interests focus on fuel performance and the fuel cycle issues of nuclear reactors, and the application of nuclear systems for the production of hydrogen and synthetic liquid fuels. He is the author of more than two hundred journal and conference papers and the two-volume textbook *Nuclear Systems* on thermal hydraulics of nuclear reactors.

Prof. Kazimi holds the Tokyo Electric Power Company Professorship in Nuclear Engineering at MIT. He was Head of the Department of Nuclear Engineering at MIT from 1989 to 1997. He is a frequent advisor to US and international institutions on matters involving energy and the environment. He chaired the Department of Energy (DOE) Advisory Panel on High Level Waste Tanks from 1990 to 1995, and was a member of the 2003–2004 National Academy of Engineering (NAE) Committee on the Hydrogen Economy. He is currently a member of the NAE committee on DOE R&D programs for development of nuclear energy. He has been a member of review committees for R&D programs and divisions at the Livermore, Argonne, Los Alamos, Idaho and Brookhaven national laboratories in the United States, as well as laboratories in Holland, Spain, Switzerland and Kuwait. He is a member of the Board of Managers of the Battelle Energy Alliance, charged by the DOE with management of the Idaho National Laboratory. He is also a Fellow of the American Nuclear Society (ANS) and the American Association for the Advancement of Science (AAAS), and has been listed in Who is Who in America since 1995.

Dr. Kazimi obtained a B.Eng degree with Distinction and First Class Honors from the University of Alexandria, Egypt in 1969, and MS and Ph.D degrees from MIT in 1971 and 1973 respectively, all in Nuclear Engineering.

Prof. Kazimi's recent publications include: J.I. Lee, P. Hejzlar, P. Saha and M.S. Kazimi, "Studies Of The Deteriorated Turbulent Heat Transfer Regime for the GAS-COOLED FAST REACTOR Decay Heat Removal System," *Nuclear Engineering and Design* (December 2007); and P. Hejzlar and M.S. Kazimi, "Annular Fuel for High Power Density PWRs: Motivation and Overview."

CHARLES D. FERGUSON is the Philip D. Reed Senior Fellow for Science and Technology at the Council on Foreign Relations (CFR), an Adjunct Assistant Professor in the Security Studies Program at the School of Foreign Service at Georgetown University, and an Adjunct Lecturer in the National Security Studies Program at Johns Hopkins University. His work at CFR focuses on nuclear energy, non-proliferation, and prevention of nuclear terrorism. As part of his work on nuclear energy, Dr. Ferguson wrote the Council Special Report *Nuclear Energy: Balancing Benefits and Risks*, which was published in April 2007. He is currently writing a book on nuclear energy and government decision-making.

Prior to his work at CFR, he was the Scientist-in-Residence in the Monterey Institute's Center for Nonproliferation Studies, where he co-wrote (with William Potter) *The Four Faces of Nuclear Terrorism* (Routledge, 2005). While working at the Monterey Institute he was the lead author of the 2003 report *Commercial Radioactive Sources: Surveying the Security Risks*, which was the first in-depth, post-9/11 study of the "dirty bomb" threat. This report won the 2003 Robert S. Landauer Lecture Award from the Health Physics Society. Dr. Ferguson has acted as a consultant for Sandia National Laboratories and the National Nuclear Security Administration, examining ways to improve the security of radioactive sources. He has also worked as a scientist in the Office of the Senior Coordinator for Nuclear Safety at the US Department of State.

Dr. Ferguson graduated with distinction from the United States Naval Academy and holds a Ph.D in physics from Boston University.

JUNGMIN KANG is a Visiting Scholar at the Stanford Institute for Economic Policy Research (SIEPR), Stanford University, a position he has held since August 2008. Previously, Dr. Kang was a Science Fellow at the Center for International Security and Cooperation (CISAC) at Stanford University from 2006 to 2008, and brought to the study of nuclear policy issues considerable expertise in the technical analysis of nuclear energy issues based on his work in South Korea, Japan and the United States. He has co-authored various articles on areas including: the proliferation-resistance of advanced fuel cycles, spent-fuel storage, plutonium disposition, and South Korea's undeclared uranium enrichment and plutonium experiments. In addition, he has contributed many articles to South Korea's newspapers and magazines on spent-fuel issues and North Korea's nuclear weapons program.

His recent research focuses on technical analyses of dismantlement issues relating to North Korea's nuclear program and issues concerning nuclear fuel cycle and spent-fuel management in Northeast Asia. He is a member of the International Panel on Fissile Materials (IPFM), and has served on South Korea's Presidential Commission on Sustainable Development, where he advised on nuclear energy policy and spent fuel management.

Dr. Kang received his Ph.D in nuclear engineering from the University of Tokyo, and his MSc and BSc degrees in nuclear engineering from Seoul National University.

JUNKO OGAWA is Deputy General Manager and Executive Communicator of the Public Relations Department at the Japan Atomic Power Company (JAPC). She joined JAPC in 1998 and her major responsibility is public awareness regarding the peaceful use of nuclear energy and radiation applications. Junko Ogawa is also a leading member of several non-governmental organizations (NGOs) and is president of Women in Nuclear (WiN) Japan, which is part of the world-wide association WiN Global, consisting of women who are professionally involved in the nuclear industry or nuclear academic field.

Mrs. Ogawa is also a board member of the Atomic Energy Society of Japan (AESJ), an expert committee member of the Japan Atomic Energy Commission (JAEC), and a committee member of the Forum for Nuclear Cooperation in Asia. She regularly gives lectures on nuclear safety and the necessity of nuclear energy in Indonesia and India. As President of WiN Japan, she received the World Nuclear Association's Award for Distinguished Contribution to the Peaceful Use of Nuclear Technology in 2005. She holds a BA degree in Literature from Keio Gijuku University, Tokyo, Japan.

KIRSTEN WESTPHAL is based at the Stiftung Wissenschaft und Politik (German Institute for International and Security Affairs) in Berlin, Germany – an independent academic research center that advises the Bundestag and the

Federal Government on foreign and security policy issues – where she researches international energy relations and global energy security.

Previously, Dr. Westphal was Assistant Professor of International Relations in the Department of Political Science at Justus Liebig University, Giessen, Germany, and also worked at the Peace Research Institute in Frankfurt as well as being a consultant for the energy industry. Her various positions have allowed her to work in several European Union (EU) countries and Organization for Security and Cooperation in Europe (OSCE) missions in Latin America, Eastern Europe, the Commonwealth of Independent States (CIS) and Asia.

Dr. Westphal has published widely on international energy relations and EU external energy relations. Her most recent papers include "The G8's Role in Global Energy Governance since the 2005 Gleneagles Summit" (with Dries Lesage and Thijs van de Graaf) in *Global Governance* (forthcoming, April 2009); "The Relevance of the Wider Black Sea Region to EU and Russian Energy Issues" (with Gerhard Mangott) in D. Hamilton and G. Mangott, *The Wider Black Sea Region in the 21st Century: Strategic, Economic and Energy Perspectives* (Washington, DC: Center for Transatlantic Relations, 2008); and "Germany and the EU–Russia Energy Dialogue" in P. Aalto (ed.), *The EU–Russian Energy Dialogue: Europe's Future Energy Supply* (Aldershot: Ashgate Publishing Ltd., 2007).

RAJAGOPALA CHIDAMBARAM is the Principal Scientific Adviser to the Government of India and the Chairman of the Scientific Advisory Committee to the Cabinet (since 2001). After completing his Ph.D at the Indian Institute of Science, Bangalore, from where he later received his D.Sc. degree, he joined the Bhabha Atomic Research Center (BARC) in 1962 and became its Director in 1990. He was Chairman of the Atomic Energy Commission (AEC) from early 1993 to late 2000. He was also Chairman of the Board of Governors of the IAEA during 1994–95 and a member of the Commission appointed by the Director-General in 2008 to prepare a report on 'The Role of the IAEA to 2020 and Beyond.' He has D.Sc. degrees (honoris causa) from a number of universities in India and abroad. He is currently a member of the Honorary Advisory Board of the international journal *Atoms for Peace*.

RATAN KUMAR SINHA is Director of the Reactor Design and Development Group and the Design, Manufacturing and Automation Group of the Bhabha Atomic Research Centre (BARC). He guides the Indian programs for the design and development of theAdvanced Heavy Water and Compact High Temperature Reactors, which use mainly thorium-based fuel and incorporate several innovative passive safety features. He is currently the Chairman of the Steering Committee of the IAEA's International Project on Innovative Nuclear Reactors and Fuel Cycles (INPRO).

ANSAR PARVEZ gained his Masters degree in Physics and Nuclear Technology in Pakistan, and earned his Ph.D in Nuclear Engineering from the Rensselaer Polytechnic Institute (RPI), United States, in 1977. He began his career as a teacher in the training institutes of the Pakistan Atomic Energy Commission (PAEC), followed by two years at Purdue University, United States, in 1982–84. Later he served as the head of the training center associated with the Karachi Nuclear Power Plant (KANUPP), where he established a Masters program in nuclear power technology in affiliation with the NED University of Engineering and Technology, Karachi.

Dr. Parvez served KANUPP in various capacities for 18 years before taking charge of the construction of the 325 MW Chashma Nuclear Power Project Unit-2. In 2006, he became a member of the PAEC and supervised its research, development and training programs, as well as its nuclear medicine and agriculture programs. Over the years, Dr. Parvez's interests have focused on human resource development, reactor core neutronics and thermohydraulics, nuclear fuel management, nuclear safety, nuclear power planning and economics. Currently, he is serving as Member (Nuclear Power Projects) of the Commission.

GHULAM RASUL ATHAR graduated from Quaid-e-Azam University, Islamabad, with an MSc. in Nuclear Engineering in 1987. Under the IAEA fellowship program, he was a research fellow on the project,

"External Costs Associated with Electricity Generation," at the University of Strathclyde, Glasgow, during 1992–93.

For the last 22 years, he has been participating in research/country studies on various aspects of energy, electricity and nuclear power planning including: energy and electricity demand forecasting, energy/electricity supply system analysis, comparative assesment of environmental impacts of nuclear and other energy options, assessment of external costs of electricity generation, assessment of the role of nuclear power in competitive electricity markets and evaluation of sustainable energy development strategies addressing climate change issues. He is the author/co-author of 35 papers/reports related to energy, electricity and nuclear power.

JOHN STEINBACH is an educator and author, and has written extensively on the environment, nuclear energy, nuclear weapons and social justice. He has a BSc in Environmental Policy and Behavior from the University of Michigan School of Natural Resources and the Environment (SNRE), and has undertaken extensive graduate studies at George Mason University's School of Education and Human Development.

Mr. Steinbach's published works include the exhaustive map and database *Deadly Radiation Hazards USA*, co-authored with his late wife Louise Franklin-Ramirez. Mr. Steinbach's 2002 article in *Covert Action Quarterly (CAQ)*, "Palestine in the Crosshairs: U.S. Policy and the Struggle for Nationhood," received a 2004 Project Censored award. Other published papers include: "Access of Evil: Genocide in Chiapas" in CAQ, and "Israeli Weapons of Mass Destruction: A Threat to Peace" in *Global Outlook.*

Mr. Steinbach lives and works in Prince William County, VA. He is active in numerous community organizations, and was the recipient of the 2007 Prince William Human Rights Award.

ABDELGHANI MOHAMMAD MELAIBARI is Professor of Nuclear Engineering and Chairman of the Department of Nuclear Engineering at King Abdulaziz University (KAU), Jeddah, Kingdom of Saudi Arabia, and is currently supervising the Strategic Planning Department at the University.

[392]

Prof. Melaibari has served as: Chair of Nuclear Engineering Department at KAU from 1999–2005; Member of the Scientific Committee for the First Saudi Symposium on Energy Utilization and Conservation at KAU in 1990; Member of the Fourth Saudi Engineering Conference in 1995; and currently serves as Member of the Scientific Committee for the International Symposium on the Peaceful Applications of Nuclear Technology in the GCC Countries, held in Jeddah in November 2008.

Prof. Melaibari received his Ph.D in Nuclear Engineering from Iowa State University, United States, in 1987. His interests center on nuclear reactor and energy physics.

Chapter 2

1. Terence Creamer, "Eskom Terminates Nuclear 1 Procurement Process, but SA still Committed to Nuclear" *Creamer Media's Engineering News,* December 5, 2008 (http://www.engineeringnews.co.za/article/eskom-terminates-nuclear-1-procurement-process-but-sa-still-committed-to-nuclear-2008-12-05).

2. GE Hitachi, "GE Hitachi Seeks to Renew NRC Certification for ABWR Reactor Design," *Business Wire*, December 15, 2008.

3. Stephen Thomas, Peter Bradford, Antony Froggatt and David Milborrow, *The Economics of Nuclear Power* (Amsterdam: Greenpeace International, 2007); (http://www.greenpeace.org/international/press/reports/the-economics-of-nuclear-power).

4. The Keystone Centre, *Nuclear Power Joint Fact-Finding* (Keystone: Keystone Centre, 2007).

5. "Which Power Generation Technologies Will Take the Lead in Response to Carbon Controls?" *Standard & Poor's Viewpoint*, May 11, 2007.

6. "New Nuclear Generation in the United States: Keeping Options Open vs. Addressing an Inevitable Necessity," Moody's Investors Service, October 10, 2007.

7. Russell Ray, "Nuclear Costs Explode," *Tampa Tribune*, January 15, 2008, p. 1.

8. Robin Pagnamenta, "Reactors Will Cost Twice Estimate, says E.ON Chief," *The Times*, May 5, 2008, p. 32.

9. Housley Carr, "Duke says New Lee Plant to Cost $11 Billion," *Nucleonics Week*, November 6, 2008, p. 1.

10. John Murawski, "Reactors Likely to Cost $9 Billion; Progress Energy Doubles Estimate" *The News & Observer*, October 17, 2008, B1.

11. The volatility of currency exchange rates mean that comparisons between, say, European prices denominated in Euro and US prices in dollars can only be approximate.

12. "Construction Costs to Soar for New US Nuclear Power Plants," Standard & Poor's RatingsDirect, October 15, 2008.

13. Ibid.

14. Ann MacLachlan, "HSE Preparing to Contract for Technical Safety Expertise" *Inside NRC*, November 10, 2008, p. 8.

15. Stephen Thomas, "The Collapse of British Energy: The True Cost of Nuclear Power or a British Failure?" *Economia delle fonti di energia e dell' ambiente*, no. 1–2, 2003: pp. 61–78.

16. For a detailed review of the problems up to March 2007, see Stephen Thomas et al., 2007.

17. Transcript of Finnish Broadcasting Company TV news, 30 January 2007.

18. "Bayern LB Receives 15BN-Euro Bailout," *Frankfurter Allgemeine Zeitung*, December 5, 2008.

19. Ann MacLachlan, "Concrete Pouring at Flamanville-3 Stopped after New Problems Found," *Nucleonics Week*, May 29, 2008, p. 18.

20. Ann MacLachlan, "EDF confirms target of starting up Flamanville-3 in 2012," *Nucleonics Week* November 20, 2008, p. 1.

21. Greg Keller, "EDF to Lead up to Euro 50B in Nuclear Plant Investment," Associated Press Worldstream, December 4, 2008.

22. Patricia Arnold and Rita Hartung Cheng, "The Economic Consequences of Regulatory Accounting in the Nuclear Power

Industry: Market Reaction to Plant Abandonments," *Journal of Accounting and Public Policy*, vol. 19, no. 2, 2000, pp. 161–187.

23. United States Department of Energy (DoE), *A Roadmap to Deploy New Nuclear Power Plants in the United States by 2010* (Washington: USDOE, 2001).

24. United States Department of Energy (DoE), *Analysis of Five Selected Tax Provisions of the Conference Energy Bill of 2003* (Washington: Energy Information Administration, 2004), p. 3 (http://tonto.eia.doe. gov/FTPROOT/service/sroiaf(2004)01.pdf).

25. Congressional Budget Office, *Cost estimate of S.14, Energy Policy Act of 2003* (Washington, Congressional Budget Office); (http://www. cbo.gov/doc. cfm?index=4206).

26. Congressional Research Service (CRS), *Potential Cost of Nuclear Power Plant Subsidies in S.14*, May 7, 2003; requested by Senator Ron Wyden.

27. UK Department of Trade and Industry, *Our Energy Future: Creating a Low Carbon Economy* (The Stationery Office, London, 2003), p. 44 (http://www.berr.gov.uk/files/file10719.pdf).

28. When Tony Blair announced the new nuclear programme in UK in May 2006, he stated, "These facts put the replacement of nuclear power stations, a big push on renewables and a step change on energy efficiency, engaging both business and consumers, back on the agenda with a vengeance." See "Blair Presses the Nuclear Button," *The Guardian*, May 17, 2006 (http://www.guardian.co.uk/environment/ 2006/may/17/energy.business).

29. Department for Business Enterprise and Regulatory Reform, "Meeting the Energy Challenge: A White Paper on Nuclear Power" (London: HM Government Cm 7296, 2003), p. 5 (http://www.berr.gov.uk/files/file 43006.pdf).

30. Select Committee on Trade and Industry, "Minutes of Evidence," October 10, 2006, answer to question 480 (http://www.publications. parliament.uk/pa/ cm200506/cmselect/cmtrdind/1123/6101001.htm).

31. "Brown Expands Nuclear Ambitions," BBC News Online, May 28, 2008 (http://news.bbc.co.uk/1/hi/uk_politics/7424158.stm).

32. Select Committee on Trade and Industry, "Minutes of Evidence," October 10, 2006, answers to questions 483 and 494 (http://www.publications. parliament.uk/pa/cm200506/cmselect/cmtrdind/1123/6101001.htm).

33. Ibid.

34. Department for Business Enterprise and Regulatory Reform, "Consultation on Funded Decommissioning Programme Guidance for New Nuclear Power Stations: Feb 2008" (London: BERR, 2008), p. 15 (http://www.berr.gov.uk/ consultations/page44784.html).

Chapter 3

1. R. Worcester, *Public Opinion in Britain and its Impact on the Future of Britain in Europe* (London: Institute of Directors, 2001).

2. International Energy Agency (IEA), *Nuclear power in the OECD* (Paris: OECD/IEA: 2001).

3. British Nuclear Fuels (BNFL), *MORI UK opinion poll findings* (Warrington, UK: BNFL, 1999).

4. Analysis Group, "Poor Support for the Government's Nuclear Power Phase-out Policy," 2000 (www.analysisgruppen.org/engopin); "Swedish Poll Shows Increasing Support for Nuclear," NucNet, 2008 (http://www.foratom.org/dmdocuments/ne51pdf.pdf).

5. World Nuclear Association, *Nuclear Power in Sweden*, 2008 (http:// www.world-nuclear.org/info/inf42.html). The Conservative-led coalition which came to power was much more positive about nuclear power than

its predecessor. The Centre Party in the coalition had recently changed its view to be more in favour of nuclear power, aligning with the Christian Democrats, Liberals and Conservatives. The phase-out has been abandoned and although planning of new units is not on the agenda during the coalition's first term, several major reactor upgrades will be undertaken. In March 2007 the Christian Democrats changed their policy to allow for new reactors being built after 2010.

6. T.R. Lee, J. Brown, J. Henderson, A. Baillie and J. Fielding, *Psychological Perspectives on Nuclear Energy,* "Report No 2: Results of Public Attitude Surveys towards Nuclear Power Stations Conducted in Five Counties of South West England" (London: Central Electricity Generating Board [CEGB], 1983).

7. D.H. Oughton, *Causing Cancer? Ethical Evaluation of Radiation* (Oslo University, 2001).

8. P. Slovic, B. Fischhoff and S. Lichtenstein, "Facts and Fears; Understanding Perceived Risk" in Shwing, R.C. and W. Al Albers (eds), *Societal Risk Assessment: How Safe Is Safe Enough?* (New York, NY: Plenum, 1980).

9. This is not the case near existing nuclear establishments, where public sentiment is generally much more positive than in the population at large, probably because of the importance of the installation to the local economy and the absence of 'fear of the unknown.'

10. R. Edelman, *Edelman Trust Barometer 2005* (http://www.edelman. com/TRUST/2008/prior/2005/Final_onscreen.ppt#324,1,Slide 1).

11. In *The Downwave: Surviving the Second Depression* (Portsmouth: Milestone Publications, 1983), Robert Beckman argues that a whole range of social factors, such as formality of dress, church-going, tastes in music, etc. can be correlated to perceptions of economic prosperity and comfort.

12. C.P. Snow, *The Two Cultures and the Scientific Revolution* (Cambridge University Press, 1959 [reprinted 1993]).

13. E. Millstone and P. van Zwanenberg, "Politics of Expert Advice: Lessons from the Early History of the BSE Saga," *Science and Public Policy*, vol. 28 no. 2, April 1, 2001 (http://www.ingentaconnect.com/content/beech/spp/2001/00000028/00000002/art00002).

14. A. Schopenhauer and E. Payne, *The World as Will and Representation*, volume 1 (Dover Books, 1819). Schopenhauer notes that each of us tends to live our lives at a fairly constant level of anxiety, independent of what the external environment has on offer to create those anxieties. It seems to the author that this observation has profound implications for our management of risk—particular risks should be regarded as better or worse candidates to justify a general sense of anxiety rather than matters to be 'resolved' in some way, to the greater happiness of mankind. "All willing springs from lack, from deficiency, and thus from suffering. Fulfillment brings this to an end; yet for one wish that is fulfilled there remain at least ten that are denied. Further, desiring lasts a long time, demands and requests go on to infinity, fulfillment is short and meted out sparingly. But even the final satisfaction itself is only apparent; the wish fulfilled at once makes way for a new one. No attained object of willing can give a satisfaction that lasts and no longer declines; but it is always like the alms thrown to a beggar, which reprieves him today so that his misery may be prolonged till tomorrow. Therefore, so long as we are given up to the throng of desires with its constant hopes and fears we never obtain lasting happiness or peace."

15. M. Henderson, "Secret List of Nuclear Dump Sites Revealed," *The Times* (online), June 11, 2005 (http://www.timesonline.co.uk/article/0,,2-1649479,00.html).

16. This is not to denigrate the importance of non-scientific sources of 'knowledge' when it comes to settling our personal values and sense of right and wrong. In the view of this author, though, these sources are less reliable than the output of properly conducted science when it comes to interpreting the behaviour of the physical universe.

17. D. Hume, *A Treatise of Human Nature* (1739); "Should it be asked me whether … I be really one of those skeptics who hold that everything is uncertain, I should reply that neither I nor any other person was ever sincerely and constantly of that opinion. I dine, I play backgammon, I converse and am merry with my friends and when after three or four hours of amusement I would return to these speculations, they appear so cold and strange and ridiculous that I cannot find in my heart to enter into them any further. Thus the skeptic still continues to reason and believe though he asserts he cannot defend his reason by reason."

18. One issue might be how to create institutions and incentives to build interdisciplinary scientific expertise, with a view to establishing reliable bases of 'facts' explaining different standpoints and outlining the advantages and disadvantages of various courses of action, hence aiding individuals who wish to become involved in the debate.

19. International Atomic Energy Agency (IAEA) *PRIS Database* (Vienna: IAEA, 2008); (http://www.iaea.org/programmes/a2/).

Chapter 4

1. International Atomic Energy Agency (IAEA), "Considerations to Launch a Nuclear Power Program" (Vienna: IAEA 2007); (http://www.iaea.org/NuclearPower/Downloads/Launch_NPP/07-11471 _Launch_NPP.pdf).

2. International Atomic Energy Agency (IAEA), "The Management System for Facilities and Activities," IAEA Safety Standards Series GS-R-3 (Vienna: IAEA, 2006).

3. The MIST is also referred to simply as the Masdar Institute.

4. S. Griffiths, et al., "The Masdar Institute Intellectual Platform: Technology, Policy and Systems" (Abu Dhabi: Masdar Institute of Science and Technology, 2008).

5. In the context of the Masdar Institute, "sustainability" refers to the implementation of technologies, policies and systems that can operate in a steady-state manner without adverse environmental consequences. Sustainable solutions include renewable sources of energy such as wind, water, solar, and biomass, as well as nuclear energy (zero CO_2 emission, assuming proper waste management), and fossil fuels subject to emissions remediation technologies such as carbon capture and sequestration (CCS). Sustainability further refers to areas that require advances in energy efficiency, such as water, the environment, buildings, and transportation.

6. REN21, "Renewables 2007 Global Status Report," (REN21: Bali, Indonesia, 2007).

7. J. Makower, R. Pernick, and C. Wilder, *Clean Energy Trends 2008* (San Francisco, CA: Clean Edge Inc., 2008).

8. L. Fleming, "Breakthroughs and the 'Long Tail' of Innovation," *Sloan Management Review* 49(1), 2007, pp. 69–74.

9. F. Moavenzadeh, "The Center for Technology, Policy and Industrial Development: A Working Paper," (Cambridge, MA: MIT, 1999).

10. R. O'Shea, et al., "Delineating the Anatomy of an Entrepreneurial University: The Massachusetts Institute of Technology Experience," *R&D Management*, 37(1), 2007.

11. Example preparatory courses are Mathematical Methods for Engineers, Probability and Statistics in Engineering, Introduction to Numerical Analysis, Principles of Micro- and Macro-economics, Introduction to Hydrology, Thermodynamics and Kinetics with Thermal-Fluid Engineering.

12. S. Griffiths, et al., op. cit.

13. Educational Testing Service, Princeton, NJ, 2008.

14. Masdar Fellow Ph.D students are not included.

15. Affiliates include Imperial College, Tokyo Institute of Technology, the University of Central Florida, Aachen University, and the University of Waterloo.

16. Boston College, Boston University, Bradeis University, Harvard University, MIT, Northeastern University, Tufts University, University of Massachusetts Boston

17. "Engines of Economic Growth: The Economic Impact of Boston's Eight Research Universities on the Metropolitan Boston Area" (New York, NY: Appleseed, 2003).

18. "The Masdar Institute of Science and Technology Business Plan," Abu Dhabi, 2008.

Chapter 5

1. International Atomic Energy Agency (IAEA), "Design Basis Threat," Guidance Information (http://www-ns.iaea.org/security/dbt.htm), accessed on December 29, 2008.

2. Jerrold M. Post, "Prospects for Nuclear Terrorism: Psychological Motivations and Constraints," in Yonah Alexander and Paul A. Leventhal (eds), *Preventing Nuclear Terrorism: The Report and Papers of the International Task Force on Prevention of Nuclear Terrorism* (Lanham, MD: Rowman & Littlefield, 1987), p. 91.

3. Ibid., p. 92.

4. Jerrold M. Post, "Differentiating the Threat of Radiological/Nuclear Terrorism: Motivations and Constraints," Presentation to the IAEA Conference on Nuclear Terrorism Prevention, November 2001.

5. Joseph Lelyveld, "Bombs Damage Atom Plant Site in South Africa," *New York Times*, December 20, 1982, A1; David Beresford, "Man Who Spiked Apartheid's Bomb," *The Guardian* (London), January 2, 1996, p. 9.

6. Center for Nonproliferation Studies (CNS), Terrorism Database (http://cns.miis.edu/research/terror.htm#db).

7. *The 9/11 Commission Report: Final Report of the National Commission on Terrorist Attacks Upon the United States* (New York, NY: W.W. Norton, authorized edition, 2004), p. 245.

8. US Nuclear Regulatory Commission, "NRC Issues Final Rule on New Reactor Aircraft Impact Assessments," Press Release, No. 09-030, February 17, 2009.

9. Scott Sagan, "The Problem of Redundancy Problem: Why More Nuclear Security Forces May Produce Less Nuclear Security," *Risk Analysis*, August 2004, pp. 935–946.

10. Bennett Ramberg, *Nuclear Power Plants as Weapons for the Enemy: An Unrecognized Military Peril* (Berkeley, CA: University of California Press, 1984).

11. US State Department, "Statement of Principles for the Global Initiative to Combat Nuclear Terrorism," Bureau of International Security and Nonproliferation, November 20, 2006 (http://www.state.gov/t/isn/rls/other/76358.htm), accessed on December 29, 2008.

12. International Atomic Energy Agency (IAEA), *IAEA Safeguards: Staying Ahead of the Game*, August 2007.

13. US State Department, "The Nuclear Suppliers Group," Fact Sheet, July 29, 2004 (http://www.state.gov/t/isn/rls/fs/34729.htm), accessed on December 29, 2008.

14. Arms Control Association, "Nuclear-Weapon-Free-Zones at a Glance," Fact Sheet, November 2007 (http://www.armscontrol.org/factsheets/nwfz), accessed on December 29, 2008.

15. Dalia Dassa Kaye, "Time for Arms Talks? Iran, Israel, and Middle East Arms Control," *Arms Control Today*, November 2004; Claudia Baumgart and Harald Müller, "A Nuclear Weapons-Free Zone in the

Middle East: A Pie in the Sky?" *The Washington Quarterly*, Winter 2004–05; Lawrence Scheinman, "Prospects for a Gulf Zone Free of Weapons of Mass Destruction," Comments at SIPRI-GRC Conference, Stockholm, May 30–31, 2005; GRC Project: Weapons of Mass Destruction Free Zone in the Gulf (GWMDFZ), *Security & Terrorism: Regional Security in the Gulf*, November 2006.

Chapter 6

1. Lindsay Windsor and Carol Kessler, *Technical and Political Assessment of Peaceful Nuclear Power Program Prospects in North Africa and the Middle East*, PNNL-16840 (September 2007).

2. Peter Crail and Jessia Lasky-Fink, "Middle Eastern States Seeking Nuclear Power," *Arms Control Today* (May 2008); Raphaeli, Nimrod, "The Middle East Ventures Into Nuclear Energy," The Middle East Media Research Institute, *Inquiry and Analysis*, no. 467, October 7, 2008; and Windsor and Kessler, op. cit.

3. Tariq Rauf and Zoryana Vovchok, "Fuel for Thought," *IAEA Bulletin* 49, 2 (March 2008).

4. Chaim Braun, "Nuclear Fuel Supply Assurance," Draft (October 2007); IAEA, "Communication dated 28 September 2005 from the Permanent Mission of the United States of America to the Agency," INFCIRC/659, September 29, 2005; Meier, Oliver, "News Analysis: The Growing Nuclear Fuel Cycle Debate," *Arms Control Today* (November 2006); Simpson, Fiona, "Reforming the Nuclear Fuel Cycle: Time is Running Out," *Arms Control Today* (September 2008).

5. IAEA, INFCIRC/659, September 29, 2005, op. cit.

6. IAEA, "Communication received from the Resident Representative of the Russian Federation to the Agency transmitting the text of the Statement of the President of the Russian Federation on the Peaceful Use of Nuclear Energy," INFCIRC/667, February 8, 2006.

7. US Department of Energy, *Global Nuclear Energy Partnership Strategic Plan*, GNEP-167312 (January 2007).

8. Braun, op. cit.

9. NPT Preparatory Committee, "Multilateralization of the nuclear fuel cycle/guarantees of access to the peaceful uses of nuclear energy," NPT/CONF.2010/PC.I/WP.61, Working paper submitted on the Multilateralization of the Nuclear Fuel Cycle/Guarantees of Access to the Peaceful Uses of Nuclear Energy by the European Union, Vienna, April 30–May 11, 2007.

10. Braun, op. cit.

11. AEA, "Communication dated 30 May 2007 from the Permanent Mission of the United Kingdom of Great Britain and Northern Ireland to the IAEA concerning Enrichment Bonds: A Voluntary Scheme for Reliable Access to Nuclear Fuel," INFCIRC/707, June 4, 2007.

12. IAEA, "Communication received from the Resident Representative of the Russian Federation to the IAEA on the Establishment, Structure and Operation of the International Uranium Enrichment Centre," INFCIRC/708, June 8, 2007.

13. IAEA, "Communication received from the Resident Representative of Germany to the IAEA with regard to the German proposal on the Multilateralization of the Nuclear Fuel Cycle," INFCIRC/704, May 4, 2007.

14. IAEA, "Communication received from the Federal Minister for European and International Affairs of Austria with regard to the Austrian proposal on the Multilateralization of the Nuclear Fuel Cycle," INFCIRC/706, May 31, 2007.

15. NPT Preparatory Committee, op. cit.

16. Nicole Stracke, "Nuclear Development in the Gulf: A Strategic of Economic Necessity?" *Security and Terrorism* no.7 (December 2007).

17. Frank von Hippel, "National Fuel Stockpiles: An Alternative to a Proliferation of National Enrichment Plants," *Arms Control Today* (September 2008).

18. Massachusetts Institute of Technology (MIT), *The Future of Nuclear Power*, An Interdisciplinary MIT Study (2003), p. 151.

19. Oliver Meier, op. cit.

20. Frank von Hippel, "Managing Spent Fuel in the United States: The Illogic of Reprocessing," Congressional Staff Briefings, March 24, 2008.

21. Jungmin Kang, et al., "Spent Fuel Standard as a Baseline for Proliferation Resistance in Excess Plutonium Disposition Options," *Journal of Nuclear Science and Technology*, vol. 37, no. 8, pp. 691–696 (August 2000); von Hippel, Frank, "Managing Spent Fuel in the United States …" op. cit.

22. D.E. Shropshire, et al., *Advanced Fuel Cycle Cost Basis*, Idaho National Laboratory (INL), INL/EXT-07-12107, pp. E2–16 (April 2007).

23. Harold Feiveson, "Faux Renaissance: Global Worming, Radioactive Waste Disposal, and the Nuclear Future," *Arms Control Today* (May 2007).

24. Frank von Hippel, "Managing Spent Fuel in the United States …" op. cit.

25. Robert Alvarez, et al., "Reducing the Hazards from Stored Spent Power-Reactor Fuel in the United States," *Science and Global Security*, vol. 11, pp. 1–51 (2003); von Hippel, Frank, "Managing Spent Fuel in the United States …" op. cit.

26. von Hippel, Ibid.

27. Matthew Bunn, et al., *Interim Storage of Spent Nuclear Fuel: A Safe, Flexible, and Cost-Effective Near-Term Approach to Spent Fuel Management*, A Joint Report from the Harvard University Project on Managing the Atom and the University of Tokyo Project on Sociotechnics of Nuclear Energy (June 2001).

28. Wolf Hafele, "The Concept of an International Monitored Retrievable Storage System," Uranium Institute Symposium, London, UK (August 1996).

29. Tatsujiro Suzuki, "Nuclear Power in Asia: Issues and Implications of 'ASIATOM' Proposals," United Nations Kanazawa Symposium on Regional Cooperation in Northeast Asia, Kanazawa, Japan, June 2–5, 1997.

30. Wolf Hafele, op. cit.

31. Matthew Bunn, et al., op. cit.

32. Mohamed ElBaradei, "Toward a Safer World," *The Economist*, October 16, 2003; IAEA, *Multilateral Approaches to the Nuclear Fuel Cycle*, Expert Group Report Submitted to the Director General of the International Atomic Energy Agency, INFCIRC/640 (February 2005).

33. S.V. Ruchkin and V. Loginov, "Securing the Nuclear Fuel Cycle: What Next?" IAEA Bulletin 48/1 (September 2006).

34. US Department of Energy (DOE), *GNEP Overview Fact Sheet*, US DOE (July 2007)

35. Matthew Bunn, et al., op. cit.

36. Ibid.

37. IAEA, "The Structure and Content of Agreements Between the Agency and States Required in Connection with the Treaty on the Non-Proliferation of Nuclear Weapon," INFCIRC/153 (June 1972); Shropshire, D.E., et al., op. cit.

38. IAEA, INFCIRC/15, op. cit.; IAEA, "Model Protocol Additional to the Agreement(s) between States(s) and the International Atomic Energy Agency for the Application of Safeguards," INFCIRC/540 (September 1997).

39. David Albright and Andrea Scheel, "Unprecedented Projected Nuclear Growth in the Middle East: Now Is the Time to Create Effective Barriers to Proliferation," ISIS Report, November 12, 2008.

40. United Nations, *Security Council Resolution 1540*, April 28, 2004.

41. Robert Alvarez, et al, op. cit.

42. Wikipedia, "Three Mile Island Accident" (http://en.wikipedia.org/wiki/Three_Mile_Island_accident).

43. MIT, *The Future of Nuclear Power*, op. cit.

44. Robert Alvarez, et al, op. cit.

Chapter 8

1. Mycle Schneider and Anthony Frogatt, "The World Nuclear Industry Status Report 2007," (Brussels, London and Paris: January 2008), p. 6.

2. For more historical details see Radkau, Joachim, Aufstieg und Krise der deutschen Atomwirtschaft 1945-1975 (Hamburg: Rowohlt, 1983), Matthes, Felix Christian, Stromwirtschaft und deutsche Einheit, Eine Fallstudie zur Transformation der Elektrizitätswirtschaft in Ost-Deutschland (Berlin: Books on Demand GmbH, 2000), and Mike Reichert, Kernenergiewirtschaft in der DDR: Entwicklungsbedingugngen, konzptioneller Anspruch und Realisierungsgrad (1955–1990), (St. Katharinen: Scripta Mercaturiae Verlag, 1999).

3. Matthias Corbach, "Atomenergie," in Danyel Reiche, *Grundlagen der Energiepolitik* (Frankufrt am Main: Peter Lang, 2005), pp. 99–117, p. 102.

4. Ibid.

5. See: (http://www.bmwi.de/BMWi/Redaktion/Binaer/energie-daten-ernergiegewinnung,property=blob,bereich=bmwi,sprache=de,rwb=tru e.xls).

6. Hans-Wilhelm Schiffer, "Deutscher Energiemarkt 2007," *Energiewirtschaftliche Tagesfragen*, 58 (2008), pp. 41–42.

7. Deutscher Bundestag (German Parliament), "Kleine Anfrage der Abgeordneten Dr. Reinhard Loske, Hans-Josef Fell, Thilo Hoppe, weiterer Abgeordneter und der Fraktion" BÜNDNIS 90/DIE GRÜNEN – Drucksache 16/1932 – Hermes-Bürgschaften für Atomexporte, Drucksache 16/2185, Wahlperiode 16, July 10, 2006.

8. Alexandre Bredimas and William J. Nuttall, "An International Comparison of Regulatory Organizations and Licensing Procedures for New Nuclear Power Plants," *Energy Policy* 36 (2008) pp. 1344–1354.

9. Ibid.

10. Stefan Deges, "Kein Strom ohne Kohle und Atom, Warum dem Land der Blackout droht," *Die Politische Meinung*, September 2008, p. 20.

11. Silke Linneweber, "Es regnet Geld. Über die zukunft der Energieversorgung," *Internationale Politik und Gesellschaft* no. 466, September 2008, p. 31.

12. Stefan Körner, "Instrumente der Energiepolitik," in Danyel Reiche, *Grundlagen der Energiepolitik* (Frankufrt am Main: Peter Lang, 2005), pp. 219–232.

13. BMU, "Eckpunkte für ein integriertes Energie- und Klimaprogramm" (http://www.bmu.de/files/pdfs/allgemein/application/pdf/klimapaket_ aug2007.pdf).

14. Bundesministerium für Umwelt, Naturschutz und Reaktorsicherheit, "Neues Denken, Neue Energie: Roadmap Energiepolitik 2020" (Berlin: BMU, 2009).

15. See: (www.bfs.de/transport/joint_convention_de_en.pdf).

16. According to the guidelines for the National Report, the second German report on the Joint Convention has been subdivided into 12 sections (http://www.bmu.de/english/nuclear_safety/current/doc/36126.php).

17. See: (http://www.bfs.de/en/transport).

18. See: (http://www.bfs.de/en/transport/zwischenlager/dezentrale_zl.html); and (http://www.bfs.de/en/transport/zwischenlager/einfuehrung.html).

19. See: (http://www.endlager-konrad.de/cln_108/nn_1078232/EN/Home/vorwort.html).

20. For a very detailed discussion of contra see: BMU, "Streitfall Kernenergie, Argumente in Thesen" (Berlin: BMU, 2007).

21. Werner Bartens and Michael Bauchmüller, "Krebsgefahr in der Nähe von Kernkraftwerken." *Süddeutsche Zeitung*, December 17, 2007 (http://www.sueddeutsche.de/wissen/189/426945/text)

22. Simon Koschut, "Demoskopie öffentliche Interesses an Energie und Umweltpolitik," in Josef Braml et al. (eds), *Weltverträgliche Energiesicherheitspolitik: Jahrbuch Internationale Politik 2005–2006* (München, R. Oldenbourg, 2008), pp. 11–15.

23. ZDF poltibarometer, July 11, 2008.

24. IAEA Circular, INFRCIRC/727, communication dated May 30, 2008, received by the Agency from the Permanent Mission of the Federal Republic of Germany with regard to the German proposal for a Multilateral Enrichment Sanctuary Project.

25. Norbert Kriedel and Sebastian Schröer, "Die Auswirkungen des Kernenergieausstiegs auf die Stromerzeugung: Szenarien bis 2020," *Energiewirtschaftliche Tagesfragen*, vol. 58, issue 7, 2007, pp. 18–21.

26. DENA (Deutsche Energie-Agentur GmbH), "Kurzanalyse der Kraftwerks-und Netzplanung in Deutschland bis 2020 (mit Ausblick auf 2030)," Annahmen, Ergebnisse und SchlussfolgerungenBerlin, April 15, 2008 (http://www.dena.de/fileadmin/user_upload/Download/ Dokumente/Meldungen/2008/Kurzanalyse_KuN-Planung_D_2020_2030_ lang_0408. pdf).

Chapter 9

1. V.C. Sahni (ed.), *The Saga of Atomic Energy Programme in India; Volume-I; The Chain Reaction: A Golden Jubilee Commemorative Volume (1954–2004)*, Department of Atomic Energy (DAE), India, 2005.

2. United Nations Development Program (UNDP), *Human Development Report-2007/08* (UNDP, 2008).

3. Homi Bhabha and N.B. Prasad, "A Study on the Contribution of Atomic Energy to a Power Program in India," *Second International Conference on the Peaceful Uses of Atomic Energy*, Geneva, September 1957.

4. Baldev Raj (ed.), *The Saga of Atomic Energy Programme in India; Volume-II ; Atoms with Mission; A Golden Jubilee Commemorative Volume (1954–2004)*; (Department of Atomic Energy [DAE], 2005).

5. H. Bhabha and N.B. Prasad, op. cit.

6. Ibid.

7. Nuclear Power Corporation of India Limited, (http://www.npcil.nic.in).

8. International Atomic Energy Agency (IAEA), Power Reactor Information System (PRIS) homepage (http://www.iaea.org/programs/ a2/).

9. International Atomic Energy Agency (IAEA), *Nuclear Power Reactors in the World*, IAEA-RDS-2/28- Reference Data Series no. 2 (IAEA, 2008)

10. S.B. Bhoje and S. Govindarajan, "The FBR Program in India," *Nu-Power*, vol. 18, no. 2–3 (2004).

11. R. Chidambaram, R. K. Sinha and A. Patwardhan, "Closing the Nuclear Fuel Cycle in the Context of the Global Climate Change Threat," *Nuclear Energy Review*, issue 2, 2007.

12. H. Bhabha and N.B. Prasad, op. cit.

13. Ibid.

14. Ibid.

15. P.K. Dey and N.K. Bansal, "Spent Fuel Reprocessing: A Vital Link in Indian Nuclear Power Program," *Nuclear Engineering and Design*, vol. 236 (2006).

16. H. Bhabha and N.B. Prasad, op. cit.

17. Ibid.

18. World Nuclear Association, "Synroc" (http://www.world-nuclear.org/info/inf58.html).

19. H. Bhabha and N.B. Prasad, op. cit.

20. R. Chidambaram, "Directed Basic Research", *Current Science*, vol. 92(9), 2007, pp. 1229–33.

21. H. Bhabha and N.B. Prasad, op. cit.

22. Bhabha Atomic Research Centre (BARC), Human Resource Development Division (http://bts.barc.gov.in/divisions/hrdd/1HomePg.htm).

23. R. Chidambaram, R. K. Sinha and A. Patwardhan, op. cit.

24. S.K. Jain, "Energy Scenario–Role of Industry in India," *19th Annual Conference of the Indian Nuclear Society*, November 2008.

25. Atomic Energy Regulatory Board (AERB); (http://www.aerb.gov.in).

26. Bhabha Atomic Research Centre (BARC), *Advanced Heavy Water Reactor: The Next Generation Indian HWR*, Brochure.

27. AERB, op. cit.

28. R.K. Sinha and I.V. Dulera, "Indian High Temperature Reactor Program," *International Conference on Non-Electric Applications of Nuclear Power: Seawater Desalination, Hydrogen Production and other Industrial Applications*, Oarai, Japan, April 16–19, 2007.

29. S.S. Kapoor, "Road Map for Development of Accelerator-Driven Sub-Critical Reactor Systems," *BARC/2001/R/004*, Bhabha Atomic Research Centre (BARC), 2001.

30. V.C. Sahni, et al., "Accelerator development in India for ADS program," *PRAMANA Journal of Physics*, vol. 68, no. 2 (February 2007).

31. Institute of Plasma Research (IPR); (http://www.ipr.res.in/).

32. R.K. Sinha and I.V. Dulera, op. cit.

33. A. Kakodkar, "Evolving Indian Nuclear Programme: Rationale and Perspective," *Nuclear India*, vol. 41, no. 11–12 (May–June 2008).

34. G.S.S. Murthy, et al., "Impact of New Trombay Groundnut Varieties," *Nuclear India*, vol. 41, no. 03–04 (September–October 2007), pp. 13–19.

35. Department of Atomic Energy (DAE), *Nuclear Energy and Societal Development* (http://www.dae.gov.in/publ/nucsoc.pdf).

36. K. Shivanna, et al., "Isotope Techniques to Identify Recharge Areas of Springs for Rainwater Harvesting in the Mountainous Region of Gaucher Area, Chamoli District, Uttarakhand," *Current Science*, vol. 94, no. 8 (April 2008), pp. 1003–1011.

37. IPR, op. cit.

38. S.B. Bhoje and S. Govindarajan, op. cit.

39. R.K. Sinha, 2006. op. cit.

Chapter 10

1. According to a census of Electricity Establishments, installed capacity of captive units (self-generation) in 2005/06 was 1,112 MW, which generated 2,841 million kWh of electricity.

2. International Energy Agency (IEA), *Key World Energy Statistics* (Pairs, IEA, 2008).

3. Hydrocarbon Development Institute of Pakistan (HDIP), *Pakistan Energy Yearbook 2008* and earlier issues (Islamabad: HDIP, Ministry of Petroleum and Natural Resources, 2009).

4. Government of Pakistan (GOP), *Pakistan Economic Survey 2007–08* and earlier issues (Islamabad, Economic Adviser's Wing, Finance Division, 2008).

5. HDIP, op. cit.

6. In Pakistan, the financial year starts on July 1 and ends on June 30, i.e., financial year 1980/81 started from July 1, 1980 and ended on June 30, 1981.

7. HDIP, op. cit.

8. Before 1997, the power sector in Pakistan was a state-controlled monopoly consisting of two public utilities; the Water and Power Development Authority (WAPDA) and the Karachi Electric Supply Company (KESC). KESC was responsible for meeting the electricity requirements of Karachi and the surrounding area while WAPDA was responsible for meeting the electricity requirements of the rest of

Pakistan. Both the utilities were engaged in generation (except nuclear power generation, which was responsibility of the PAEC), transmission and distribution of electricity in their respective domains. In 1997, WAPDA was restructured into four generation companies, nine distribution companies and one national transmission and despatch company (NTDC). The thermal power plants have been distributed into four generation companies whereas all the hydro power plants remain with WAPDA. KESC was retained as a single unit responsible for the generation, transmission and distribution in Karachi and the surrounding area. Development and operation of nuclear power plants continues to be the responsibility of the PAEC.

9. Government of Pakistan (GOP), *Medium Term Development Framework 2005–2010* (Islamabad: Planning Commission, 2005); Government of Pakistan (GOP), *Strategic Directions to Achieve Vision 2030: Approach Paper* (Islamabad: Planning Commission, 2006).

10. Asian Development Bank (ADB). *Outlook 2008 Update* (Manila: Asian Development Bank, 2008), 170–176; International Monetary Fund (IMF), *World Economic Outlook October 2008: Financial Stress, Downturns, and Recoveries* (Washington, DC: IMF, 2008).

11. The model for analysis of energy demand (MAED), acquired from the International Atomic Energy Agency, was used for projection of energy and electricity demand in the country. The MAED model uses a bottom-up approach and contains a number of parameters on demography, economy, energy intensities and efficiencies.

12. G.R. Athar, R. Tariq, M. Imtiaz, I. Ahmad, and J. Bashir, *Energy Development Strategies of Pakistan for Addressing Climate Change Issues*. Draft report on country study carried out under IAEA/RCA Project RAS/0/045 (Islamabad: PAEC, 2008).

13. Government of Pakistan, *Medium Term Development Framework 2005–2010*, op. cit.

14. Hydrocarbon Development Institute of Pakistan (HDIP), op. cit.

15. Ibid.

16. Government of Pakistan, *Medium Term Development Framework 2005–2010*, op. cit.

17. Coal resource includes: measured: 3.4 billion tons; indicated: 11.7 billion tons; inferred: 56.6 billion tons; and hypothetical: 114.3 billion tons (HDIP 2009).

18. Hydrocarbon Development Institute of Pakistan (HDIP), op. cit.

19. Ibid.

20. Government of Pakistan, *Medium Term Development Framework 2005–2010*, op. cit.

21. Ibid.

22. Alternative Energy Development Board (AEDB), "Renewable Energy Sector" (http://www.aedb.org), accessed December 4, 2008.

23. International Atomic Energy Agency (IAEA). Design for Safety of Nuclear Power Plants: A Code of Practice, Safety Series No.50-C-D (Vienna, IAEA, 1978).

24. International Atomic Energy Agency (IAEA), Code on the Safety of Nuclear Power Plants: Design, Safety Series No.50-C-D Rev.1 (Vienna: IAEA, 1988).

25. International Atomic Energy Agency (IAEA), Safety of Nuclear Power Plants: Design. IAEA Safety Standard Series No. NS-R-1 (Vienna: IAEA, 2000).

26. Government of Pakistan, *Medium Term Development Framework 2005–2010*, op. cit.; and Government of Pakistan, *Strategic Directions to Achieve Vision 2030: Approach Paper*, op. cit.

27. Pakistan Atomic Energy Commission (PAEC), *PAEC provides equipment and services to CERN, PakAtom* (Islamabad: PAEC, September–October, 2008).

28. The engineering and science graduates coming to PAEC training/education centers have already obtained 16 years of school and university education.

29. Pakistan Atomic Energy Commission (PAEC), *PAEC embarks on the road to promote nuclear desalination as a sustainable solution for water scarcity, PakAtom* (Islamabad: PAEC, March–April 2008).

Chapter 11

1. Walter Pincus, "Push for Nuclear-Free Middle East Resurfaces," *Washington Post*, March 6, 2005 (http://www.washingtonpost.com/wp-dyn/articles/A10418-2005Mar5.html), accessed December 20, 2008.

2. Seymour Hersh, *The Samson Option: Israel's Nuclear Arsenal and American Foreign Policy* (New York, NY: Random House, 1991), p. 319 (A brilliant and prophetic work with much original research).

3. "Former weapons inspector Blix says Olmert's Nuclear Comment Unlikely to Spark Arab Backlash," *International Herald Tribune*, December 13, 2006 (http://www.iht.com/articles/ap/2006/12/13/europe/EU_GEN_Netherlands_Blix_Nuclear.php), accessed December 20, 2008.

4. Ted Flaherty, *Nuclear Database: Israel's Possible Nuclear Delivery Systems*, Center for Defense Information, 1997 (http://www.cdi.org/nuclear/database/isnukes.html), accessed December 20, 2008.

5. Joseph Cirincione, Jon B. Wolfsthal and Miriam Rajkumar, *Deadly Arsenals: Nuclear, Biological and Chemical Threats* (Washington, DC: Carnegie Endowment for International Peace, 2006), p. 265.

6. Aluf Benn, "Israel: Censoring the Past," *Bulletin of the Atomic Scientists*, July/August 2001, pp. 17–19 (http://www.bsos.umd.edu/pgsd/people/staffpubs/Avner-BASreport7-01.htm), accessed December 20, 2008. The Bulletin of the Atomic Scientists was founded in 1945 by dissident Manhattan Project scientists. It is considered an authoritative source on nuclear issues.

7. Israeli Atomic Energy Commission, "Research and publications by IAEC personnel" (http://www.iaec.gov.il/pages_e/card_report_e.asp), accessed March 10, 2009.

8. Israeli Prime Minister's Office, "Atomic Energy Commission" (http://www.pmo.gov.il/PMOEng/PM+Office/Bodies/depenergy.htm), accessed March 10, 2009.

9. Nuclear Threat Initiative (NTI), "Israel Nuclear Facilities: Overview of Organizations and Facilities" (http://www.nti.org/e_research/profiles/Israel/Nuclear/3583.html), accessed March 10, 2009.

10. Global Security, "Israel Atomic Energy Commission" (http://www.globalsecurity.org/wmd/world/israel/iaec.htm), accessed March 10, 2009.

11. David Bedein, "Israel May Expand Its Nuclear Power," *The Philadelphia Bulletin*, November 13, 2008.

12. Amit Baruah, "Now Israel Wants NSG Rules Changed," *The Hindustan Times*, August 28, 2007 (http://www.hindustantimes.com/StoryPage/FullcoverageStoryPage.aspx?sectionName=NLetter&id=9a7f3e9c-db05-4333-beb1-ccece17c6658_Special&Headline=Now%2c+Israel+wants+NSG+rules+changed), accessed March 10, 2009.

13. Dominic Moran, "Middle East Moves," ISN Security Watch (http://www.isn.ethz.ch/isn/Current-Affairs/Security-Watch/Detail/?ots591=4888CAA0-B3DB-1461-98B9-E20E7B9C13D4&lng=en&id=54151), accessed March 10, 2009.

14. Avner Cohen, *Israel and the Bomb* (New York, NY: Columbia University Press, 1998), p. 10.

15. Ibid., pp. 15–16

16. Ibid., p. 16.

17. Ibid., p. 13.

18. Avner Cohen and William Burr, "The Untold Story of Israel's Bomb," *Washington Post*, Sunday, April 30, 2006, B-01 (http://www.washingtonpost.com/wp-dyn/content/article/2006/04/28/AR2006042801326_pf.html), accessed December 20, 2008.

19. Ella Habiba Shohat, "Reflections by an Arab Jew," *Bint Jbeil*, April 17, 1999 (http://www.bintjbeil.com/E/occupation/arab_jew.html), accessed December 20, 2008.

20. "Demographics of Israel," *Wikipedia* (http://en.wikipedia.org/wiki/Demographics_of_Israel), accessed December 20, 2008.

21. Ibid.

22. Uri Avnery, "Sorry, Wrong Continent," *Gush Shalom*, December 23, 2006 (http://zope.gush-shalom.org/home/en/channels/avnery/1166960371), accessed December 20, 2008.

23. Charlotte Halle, "Foreign Ministry Slams Envoy's Comments about 'Yellow Race'," *Haaretz*, October 24, 2006 (http://www.haaretz.com/hasen/spages/774471.html), accessed December 20, 2008.

24. Karen Armstrong, *The Battle for God* (New York, NY: 2000), pp. 278–309.

25. Walter Zander, "Israel and the Anti-Colonial Movement: Orientation Toward the Orient or the Occident?" *Jewish Observer and Middle East Review*, February 24, 1956 (http://walterzander.info/acrobat/Israel%20and%20anti.pdf), accessed December 20, 2008.

26. Shibley Telhami (Principal Investigator), *2008 Annual Arab Public Opinion Poll*, Survey of the Anwar Sadat Chair for Peace and Development at the University of Maryland (with Zogby International), 2008 (http://www.brookings.edu/topics/~/media/Files/events/2008/0414_middle_east/0414_middle_east_telhami.pdf), accessed December 20, 2008 (Egypt, Jordan, Lebanon, Morocco, Saudi Arabia and the United Arab Emirates were surveyed)

27. Geneva Initiative, *New Survey of Israeli Public Opinion, Prospects for Peace*, 2007 (http://www.prospectsforpeace.com/2007/07/new_survey_of_israeli_public_o.html), accessed December 20, 2008.

28. "Israel Nuclear Overview," Nuclear Threat Initiative (NTI), August 2008 (http://www.nti.org/e_research/profiles/Israel/Nuclear/index.html), accessed December 20, 2008.

29. S. Hersh, op. cit., pp. 27–46.

30. Michael Karpin, *The Bomb In The Basement: How Israel Went Nuclear and What It Means for the World* (New York, NY: Simon & Schuster Paperbacks, 2006), pp. 57–95.

31. U.S. Army Lt. Col. Warner D. Farr, "The Third Temple Holy of Holies: Israel's Nuclear Weapons," USAF Counter-proliferation Center, Air War College, September 1999 (www.fas.org/nuke/guide/israel/nuke/farr.htm), accessed December 20, 2008. (This is perhaps the best single condensed history of the Israeli nuclear program.)

32. Peter Hounam, *Woman from Mossad: The Torment of Mordechai Vanunu* (London: Vision Paperbacks, 1999), pp. 34–35. (The most complete and up to date account of the Vanunu story by the London *Sunday Times* reporter who broke the story.)

33. S. Hersh, op. cit., p. 30.

34. "Israel Nuclear Overview," NTI, op. cit.

35. P. Hounam, op. cit., p. 36.

36. A. Cohen, 1998, p. 187.

37. W.D. Farr, op. cit.

38. M. Karpin, op. cit., pp. 268–269.

39. Benjamin Pincus, "Atomic Power to Israel's Rescue: French-Israeli Nuclear Cooperation 1949–1957," *Israel Studies*, vol. 7, no. 1 (Spring 2002), pp. 104–138 (http://muse.jhu.edu/login?uri=/journals/israel_studies/v007/7.1pinkus.html; http://www.nti.org/e_research/profiles/Israel/Nuclear/index.html), accessed December 20, 2008.

40. M. Karpin, op. cit., p. 276.

41. A. Cohen, 1998, pp. 273–274.

42. S. Hersh, op. cit., p. 131.

43. Mark Gaffney, *Dimona the Third Temple: The Story Behind the Vanunu Revelation* (Brattleboro, VT: Amana Books, 1989), pp. 68–69.

44. "Uranium: The Israeli Connection," *Time Magazine*, May 30, 1977 (http://www.time.com/time/magazine/article/0,9171,914952,00.html), accessed December 20, 2008.

45. S. Hersh, op. cit., pp. 242–257.

46. "Israel: Uranium Processing and Enrichment," Wisconsin Project on Nuclear Arms Control, *The Risk Report*, vol. 2, no. 4, 1996, (http://www.wisconsinproject.org/countries/israel/uranium.html), accessed December 20, 2008.

47. Gary Milholland, "A Heavy Water Whitewash," Arbeiderbladet (Oslo), April 20, 1988 (http://www.wisconsinproject.org/pubs/editorials/1988/heavywaterwhitewash.htm), accessed December 20, 2008.

48. W.D. Farr, op. cit., p. 8.

49. S. Hersh, op. cit., pp. 271–283.

50. W.D. Farr, op. cit., p. 12.

51. "South Africa's Nuclear Autopsy," Wisconsin Project On Nuclear Arms Control, *The Risk Report*, vol. 2, no. 1, 1996, p. 4, 5, 10 (http://www.wisconsinproject.org/countries/safrica/autopsy.html), accessed December 20, 2008; Farr, W.D., op. cit., p. 8.

52. Mark Wade Jericho, Encyclopedia Astronautica (http://www.astronautix.com/lvfam/jericho.htm), accessed December 20, 2008.

53. Barbara Rogers and Zdenek Cervenka, *The Nuclear Axis: The Secret Collaboration Between West Germany and South Africa* (New York, NY: Times Books, 1978), pp. 325–328 (the definitive history of the Apartheid Bomb).

54. M. Gaffney, op. cit., p. 34.

55. "Israel Profile: Nuclear Overview," Nuclear Threat Initiative (http://www.nti.org/e_research/profiles/Israel/Nuclear/index.html), accessed December 20, 2008.

56. A. Cohen, op. cit., pp. 336–338.

57. Ibid., p. 323.

58. Stephen Zunes, "The Release of Mordechai Vanunu and U.S. Complicity in the Development of Israel's Nuclear Arsenal," *Foreign Policy in Focus*, April 21, 2004 (http://www.fpif.org/fpiftxt/1134), accessed December 20, 2008.

59. P. Hounan, op. cit., pp. 155–168.

60. S. Hersh, op. cit., p. 213.

61. Ibid., pp. 3–17.

62. "Israel Gets US Supercomputer for Secret Military Site," Wisconsin Project on Nuclear Arms Control, *The Risk Report*, vol. 2, no. 3, 1996

(http://www.wisconsinproject.org/countries/israel/israelgets.htm),
accessed December 20, 2008.

63. Gerard C. Smith and Helen Cobban, "A Blind Eye to Nuclear
Proliferation," *Foreign Affairs* (Summer 1989), pp. 59–63 (http://
www.foreignaffairs.org/19890601faessay5960/gerard-c-smith-helena-
cobban/a-blind-eye-to-nuclear-proliferation.html), accessed December
20, 2008.

64. W.D. Farr, op. cit., p. 11.

65. S. Hersh, op. cit., pp. 290–291.

66. Peter Hounam, "Mordechai Vanunu: The *Sunday Times* Articles,"
April 21, 2004 (http://www.timesonline.co.uk/tol/news/article830147.
ece), accessed December 20, 2008.

67. Ibid., pp. 35–45.

68. Ibid, p. 45

69. Hersh, S., op. cit., p. 197.

70. M. Gaffney, op. cit., pp. 7–8.

71. Eileen Fleming, "The Vanunu Saga 2008," *Counter Currents*, January
22, 2008 (http://www.countercurrents.org/fleming220108.htm), accessed
December 20, 2008.

72. P. Hounam, pp. 189–203.

73. US Strategic Bombing Survey, "The Effects of the Atomic Bombs on
Hiroshima and Nagasaki" (Washington, DC: US Government Printing
Office, 1946), pp. 9–14.

74. Fredrick Solomon and Robert Q. Marston, *The Medical Implications
of Nuclear War* (Washington, DC: National Academy of Science
Press, 1986), pp. 211–213.

75. S. Hersh, op. cit., pp. 199–200.

76. Ibid., 312.

77. John Pike, Israel Special Weapons Guide Website, Federation of American Scientists (http://www.fas.org/nuke/guide/israel/index.html), accessed December 20, 2008. (An invaluable internet resource.)

78. Usi Mahnaimi and Peter Conradi, "Fears of New Arms Race as Israel Tests Cruise Missiles," *Sunday Times*, June 18, 2000 (available at: http://www.fas.org/news/israel/e20000619fearsof.htm), accessed December 20, 2008.

79. Center for Nonproliferation Studies, "Weapons of Mass Destruction In the Middle East," 2006 (http://cns.miis.edu/research/wmdme/israel.htm).

80. Ibid.

81. John Pike, "Weapons of Mass Destruction: Jericho 1," Global Security, 2008 (http://www.globalsecurity.org/wmd/world/israel/jericho-1.htm), accessed December 20, 2008. (Global Security is the most up-to-date, authoritative source for information about weapons of mass destruction worldwide.)

82. Anthony Cordesman, "Israeli Weapons of Mass Destruction: An Overview" (working draft), Center for Strategic and International Studies (CSIS), 2008 (http://www.csis.org/media/csis/pubs/080602_israeliwmd.pdf), accessed December 20, 2008.

83. "Israel Profile: Missile Overview," Nuclear Threat Initiative (http://www.nti.org/e_research/profiles/Israel/Missile/index.html), accessed December 20, 2008.

84. Serge Schmemann, "Israel Clings to its 'Nuclear Ambiguity'," *New York Times*, June 21, 1998 (http://www.google.com/search?hl=en&q=Israel+Clings+to+its+'Nuclear+Ambiguity&btnG=Google+Search), accessed December 20, 2008.

85. Mark Wade, "Jericho, 2008" Encyclopedia Astronautica (http://www.astronautix.com/lvfam/jericho.htm), accessed December 20, 2008. (According to the Encyclopedia Astronautica, much of what we know about Israel's ballistic missile program has been gleaned from examining the official South African records of its missile program.)

86. A. Cordesman, op. cit., p. 5 (Cordesman is referencing the authoritative *Jane's Intelligence Review*).

87. Yaakov Katz, "Israel Test-Fires Long-Range Ballistic Missile," *Jerusalem Post*, July 17, 2008 (http://www.jpost.com/servlet/Satellite?cid=1200475902683&pagename=JPost%2FJPArticle%2FShowFull), accessed December 20, 2008.

88. Michael Barletta and Erik Jorgensen, "Weapons of Mass Destruction in the Middle East," Center for Non-Proliferation Studies (http://cns.miis.edu/research/wmdme/), accessed December 20, 2008.

89. STRATFOR, "Military: A New Ballistic Missile for Israel?" January 17, 2008 (http://www.stratfor.com/analysis/military_new_ballistic_missile_israel), accessed December 20, 2008.

90. Barletta, M. and Jorgensen, E., op. cit.

91. John Pike, "Military: F-4E Phantom II," Global Security, (http://www.globalsecurity.org/military/systems/aircraft/f-4e.htm), accessed December 20, 2008.

92. Ibid.

93. T. Flaherty, op. cit.

94. "McDonnell Douglas F4 Phantom," Israeli Airforce: The Official Website (http://www.iaf.org.il/Templates/Aircraft/Aircraft.IN.aspx?lang=EN&lobbyID=69&folderID=78&subfolderID=184&docfolderID=184&docID=18176), accessed December 20, 2008.

95. T. Flaherty, op. cit.

96. Peter Beaumont, "Was Israeli Raid a Dry Run for an Attack on Iran?" *The Observer*, September 16, 2007 (http://www.guardian.co.uk/world/ 2007/sep/16/iran.israel), accessed December 20, 2008.

97. John Pike, "F-15I Ra'am (Thunder)," Global Security (http://www.globalsecurity.org/military/world/israel/f-15i.htm), accessed December 20, 2008.

98. Ibid.

99. George Friedman, "Mediterranean Flyover: Telegraphing an Israeli Punch," *Jewish World Review*, June 25, 2008 (http://www.stratfor.com/weekly/mediterranean_flyover_telegraphing_israeli_punch), accessed December 20, 2008.

100. John Pike, "Weapons of Mass Destruction: Dolphin," Global Security (http://www.globalsecurity.org/wmd/world/israel/sub.htm), accessed December 20, 2008.

101. Nuclear Threat Initiative, "Submarine Proliferation: Israel Current Capabilities, 2008" (http://www.nti.org/db/submarines/israel/index.html), accessed December 20, 2008).

102. Uzi Mahnaimi and Matthew Campbell, "Israel Makes Waves With Submarine Missile Test," *Sunday Times*, June 18, 2000 (available at: http://www.fas.org/news/israel/e20000619israelmakes.htm), accessed December 20, 2008).

103. J. Pike, op. cit.

104. Marineforce International, "Submarines: Dolphin" (http://www.marineforce.net/dolphin.html), accessed December 20, 2008.

105. Jim Dunnigan, "Israel Gets Super Subs," Strategy Page, August 27, 2006 (http://www.strategypage.com/htmw/htsub/articles/20060827.aspx), accessed December 20, 2008.

106. "SSK Dolphin Class Submarine, Israel, 2008," Naval-Technology. com (http://www.naval-technology.com/projects/dolphin/), accessed December 20, 2008.

107. J. Pike, op. cit.

108. Cirincione, J., et al., op. cit., p. 261.

109. Usi Mahnaimi, "Israeli Jets Equipped for Chemical Warfare," *Sunday Times*, October 4, 1998 (article posted on the MER website at: http://www.middleeast.org/archives/1998_11_08.htm), accessed December 20, 2008.

110. Avner Cohen, "Israel and Chemical/Biological Weapons: History, Deterrence, and Arms Control," *The Nonproliferation Review*, Fall–Winter 2001 (http://www.vho.org/aaargh/fran/livres3/Cohen.pdf), accessed December 20, 2008.

111. "Israel Profile: Chemical Overview," Nuclear Threat Initiative (NTI) 2008 (http://www.nti.org/e_research/profiles/Israel/Chemical/index. html).

112. "Israel Profile: Biological Overview," Nuclear Threat Initiative (NTI) 2008 (http://www.nti.org/e_research/profiles/Israel/Biological/ index.html), accessed December 20, 2008.

113. "Strategic Israel: The Secret Arsenal of the Jewish State," MSNBC (http://www.sweetliberty.org/issues/israel/strategic/index.shtml), accessed December 20, 2008. (A remarkable interactive map of Israel's nuclear facilities. No longer stored at the MSNBC website, it is archived at the Sweet Liberty site.)

114. John Pike, "Special weapons Facilities: Israel," Global Security (http://www.globalsecurity.org/wmd/world/israel/facility.htm), accessed December 20, 2008.

115. Nuclear Threat Initiative (NTI), "Israel Nuclear Facilities" (http://www.nti.org/e_research/profiles/Israel/Nuclear/3583.html), accessed December 20, 2008.

116. Greenpeace International, "An Overview of Nuclear Facilities In Iran, Israel and Turkey," February, 2007 (http://www.greenpeace. org/raw/content/mediterranean/reports/an-overview-of-nuclear-facilit. pdf), accessed December 20, 2008. (Greenpeace issues the following disclaimer: "Due to the highly secretive nature of the Israeli nuclear programme and the complete lack of official information, the chapter on Israeli nuclear facilities was written based upon the best available informative yet unofficial sources" — this disclaimer doesn't apply to Turkey or Iran.)

117. J. Pike, Special weapons Facilities-Israel, op. cit.

118. Ibid.

119. Carnegie Endowment for International Peace, "Israel: Nuclear Infrastructure," 1998 (http://www.carnegieendowment.org/files/ Tracking_israelmap.pdf), accessed December 20, 2008.

120. Greenpeace International, op. cit.

121. Michael Adler, "Israel's Soreq Nuclear Reactor: The One They Show to Journalists," Agence France Press, July 18, 2004 (available at: http://www.globalsecurity.org/org/news/2004/040718-israel-soreq.htm), accessed December 20, 2008.

122. John Pike, "Soreq Nuclear Research Center," Global Security (http://www.globalsecurity.org/wmd/world/israel/soreq.htm), accessed December 20, 2008.

123. Yael Ivri-Darel, "Soreq Nuclear Research Center's Reactor to Shut Down in 8 Years," Ynetnews, April 3, 2008 (http://www.ynet. co.il/english/articles/0,7340,L-3514334,00.html), accessed December 20, 2008.

124. Peter Hounam, "Revealed: The Secrets of Israel's Nuclear Arsenal / Atomic Technician Mordechai Vanunu Reveals Secret Weapons Production," *Sunday Times*, October 5, 1986 (http://www.timesonline.co.uk/tol/news/article830147.ece), accessed December 20, 2008.

125. John Pike, "Dimona Negev Nuclear Research Center," Global Security, 2008 (http://www.globalsecurity.org/wmd/world/israel/dimona.htm), accessed December 20, 2008.

126. John Pike, "Negev Nuclear Research Center," Global Security (http://www.globalsecurity.org/wmd/world/israel/dimona.htm), accessed December 20, 2008.

127. "Israel Reviews its Nuclear Deterrent," *Jane's Intelligence Review*, November 1, 1998 (http://www.jane's.com/extract/jir98/jir00807.html), accessed December 20, 2008.

128. Brian Whitaker and Richard Norton-Taylor, "Revealed: Israel's Nuclear Site," *The Guardian*, August 23, 2000 (http://www.guardian.co.uk/technology/2000/aug/23/news.israel), accessed December 20, 2008.

129. John Pike, "Plutonium Production," Global Security, 2008 (http://www.globalsecurity.org/wmd/intro/pu-prod.htm), accessed December 20, 2008.

130. Arjun Makhijani, Howard Hu and Katherine Yih (eds), *Nuclear Wastelands: A Global Guide to Nuclear Weapons Production and Its Health & Environmental Effects* (Cambridge, MA: MIT Press, 2000). To date, the most comprehensive and authoritative overview of the environmental and health consequences of nuclear weapons production.

131. Jonathon Broder, "A Challenge to Israel's Nuclear Blind Spot," *The Washington Post*, Sunday, March 11, 2001; Page B02 (http://www.bsos.umd.edu/pgsd/people/staffpubs/Avner-WashPost3-11-01.htm), accessed December 20, 2008.

132. Bennett Ramberg, "Should Israel Close Dimona? The Radiological Consequences of a Military Strike on Israel's Plutonium-Production Reactor," *Arms Control Today*, May 2008 (http://www.armscontrol.org/act/2008_05/Dimona), accessed December 20, 2008.

133. "Israel Reviews Its Nuclear Deterrent," *Jane's Intelligence Review*, November 1, 1998 (http://www.jane's.com/extract/jir98/jir00807.html), accessed December 20, 2008.

134. Arabs Against Discrimination, "Employees at Nuclear Reactors Demand Compensation," April 17, 2005 (http://www.aad-online.org/2005/English/8-August/13-18/13-8/aad20/1.htm), accessed December 20, 2008.

135. Liat Collins, "Family of Nuclear Cancer Victim Awarded NIS 2.5m," *Jerusalem Post*, October 13, 1997 (http://nuclearweaponarchive.org/Israel/Dimonanews.txt), accessed December 20, 2008.

136. Rayna Moss, "Israel: Dimona Death Factory Exposed," WISE Nuclear Monitor, February 1, 2002, (http://www10.antenna.nl/wise/index.html?http://www10.antenna.nl/wise/beyondbomb/3-3.html), accessed December 20, 2008.

137. Palestinian National Authority (PNA), "Dimona Reactor … a Mystery Threatening Middle East," PNA Press Release, Thursday, September 18, 2003 (http://www.scoop.co.nz/stories/WO0309/S00228.htm), accessed December 20, 2008.

138. "Jordanian Fears over Israeli Dimona Nuclear Reactor," *Arabic News* April 28, 2005 (http://www.arabicnews.com/ansub/Daily/Day/050428/2005042832.html), accessed December 20, 2008.

139. Ibid.

140. Yuval Azoulay, "Deputy CEO of Negev Nuclear Center: Dimona Reactor is Safe," *Haaretz*, August 23, 2007 (http://www.haaretz.co.il/hasen/spages/896676.html), accessed December 20, 2008.

141. "Israel Reviews Its Nuclear Deterrent," op. cit.

142. *Sunday Times*: Close Dimona Nuclear Facility," IsraelWire, February 7, 2000 (http://www.fas.org/news/israel/000207-israel1. htm), accessed December 20, 2008.

143. Aluf Benn, "Israel: Censoring the Past," *Bulletin of the Atomic Scientists*, July/August 2001 (http://www.bsos.umd.edu/pgsd/people/ staffpubs/Avner-BASreport7-01.htm), accessed December 20, 2008.

144. BBC World Service, BBC Transcript of "Israel's Secret Weapon, Electric Intifada," March 17, 2003 (http://electronicintifada.net/v2/ article1665.shtml), accessed December 20, 2008.

145. Ibid.

146. Ibid.

147. J. Cirincione, et al., p. 265.

148. A. Cohen, 1998, op. cit., pp. 12–14.

149. Israel Shahak, *Open Secrets: Israeli Nuclear and Foreign Policies* (London: Pluto Press, 1997), p. 40. (For many years Shahak translated news from the Hebrew language press. This volume contains many such examples.)

150. S. Hersh, op. cit., p. 319.

151. M. Gaffney, op. cit., p. 163.

152. C.R. Jayachandran, "Pak Fears Joint Indo-Israel Strikes," *Times of India*, October 14, 2003 (http://timesofindia.indiatimes.com/ articleshow/232733.cms), accessed December 20, 2008.

153. "History of Lebanon," Wikipedia (http://en.wikipedia.org/wiki/History _of_Lebanon#Israeli_invasion_and_international_intervention:_1982. E2.80.9384), accessed December 20, 2008.

154. Chris Hedges, "Israel's 'Crime Against Humanity'," TruthDig.com, December 15, 2008 (http://www.truthdig.com/report/item/20081215_israels_crime_against_humanity/), accessed December 20, 2008. (In December, 2008, Israel denied entry to the Occupied Territories by Professor Falk.)

155. James A. Russell, "Nuclear Strategy and the Modern Middle East," *Middle East Council Journal* (Fall 2004); (http://www.mepc.org/journal_vol11/0409_russell.asp), accessed December 20, 2008.

156. Ibid.

157. M. Gaffney, op. cit., p. 131.

158. Avner Cohen, "Did Nukes Nudge the PLO?" *Bulletin of the Atomic Scientists* (December 1993), pp. 11–13.

159. Robert W. Tucker, "Israel & the US: From Dependence to Nuclear Weapons?" *Commentary* (November 1975), pp. 41–42 (http://www.Commentarymagazine.com/viewarticle.cfm/israel-and-the-united-states-from-dependence-to-nuclear-weapons-5582), accessed December 20, 2008.

160. Ibid.

161. Wisconsin Project On Nuclear Arms Control, "Israel's Nuclear Weapon Capability: An Overview," *The Risk Report* (July–August 1996); (http://www.wisconsinproject.org/countries/israel/nuke.html), accessed December 20, 2008.

162. M. Gaffney, op. cit., p. 147.

163. Ibid., p. 153.

164. "Israel–United States Military Relations," Wikipedia, December 10, 2008 (http://en.wikipedia.org/wiki/Israel-United_States_military_relations), accessed December 20, 2008.

165. I. Shahak, op. cit., pp. 39–40.

166. Ido Kanter, "Changing Nuclear Equation: Should Nuclear Arms be used in Response to Powerful Conventional Attacks?" *Israel Opinion*, September 4, 2006 (http://www.ynetnews.com/articles/0,7340,L-3299440,00.html), accessed December 20, 2008.

167. S. Hersh, op. cit., p. 19.

168. Aronson, Geoffrey, "Hidden Agenda: US-Israeli Relations and the Nuclear Question," *Middle East Journal* (Autumn 1992), pp. 619–630 (http://www.mideasti.org/middle-east-journal/volume-46/4/hidden-agenda-us-israeli-relations-and-nuclear-question), accessed December 20, 2008.

169. Jonathan Schell and Martin Sherwin, "Israel, Iran and the Bomb," *The Nation*, August 8, 2008 (http://www.thenation.com/doc/20080818/schell_sherwin), accessed December 20, 2008.

170. I. Shahak, op. cit., p. 150.

171. Norman Soloman, "Israel's Future Leader? (Benjamin Netanyahu)" Alternet, January 6, 2006 (http://www.thirdworldtraveler.com/Israel/Benjamin_Netanyahu.html), accessed December 20, 2008.

172. S. Hersh, op. cit., p. 319.

173. Ibid., p. 153.

174. Ibid., pp. 285–305.

175. John Pike, "Israel: Strategic Doctrine," Global Security (http://www.globalsecurity.org/wmd/world/israel/doctrine.htm), accessed December 20, 2008.

176. M. Gaffney, op. cit., p. 194.

177. Shai Feldman, *Nuclear Weapons and Arms Control in the Middle East* (Cambridge, MA: MIT Press, 1997), pp. 243–244; emphasis in original.

178. Avner Cohen and Marvin Miller, "Country Perspectives on the Challenges to a Fissile Material (Cutoff) Treaty," International Panel on Fissile Materials, Israel, 2008, pp. 27–34 (http://www.ipfmlibrary.org/gfmr08cv.pdf), accessed December 20, 2008.

179. Yossi Melman, "Report suggests Obama Press Israel Over Nuke Program," *Haaretz*, November 17, 2008 (http://www.haaretz.com/hasen/spages/1037558.html), accessed December 20, 2008.

180. Ibid., p. 149.

181. S. Feldman, pp. 245–248.

182. Greenpeace Briefing, "Conditions for a Nuclear Free Middle East," February, 2007 (http://www.greenpeace.org/raw/content/mediterranean/reports/conditions-for-NFME.pdf), accessed December 20, 2008.

183. Zeev Maoz, "The Mixed Blessing of Israel's Nuclear Policy," *International Security*, vol. 28, no. 2 (Fall 2003), pp. 44–77 (http://belfercenter.ksg.harvard.edu/files/maoz.pdf), accessed December 20, 2008.

184. Michael Donovan, "Iran, Israel and Nuclear Weapons in the Middle East," Center for Defense Information Terrorism Project, February 14, 2002 (http://www.cdi.org/terrorism/menukes.cfm), accessed December 20, 2008.

185. I. Shahak, op. cit., p. 34.

186. "United Nations Security Resolution 487," The Avalon Project, Yale Law School, (http://www.yale.edu/lawweb/avalon/un/un487.htm), accessed December 20, 2008.

187. Richard Beeston, "Six Arab States Join Rush to Go Nuclear," TimesOnline, November 6, 2006 (http://www.timesonline.co.uk/tol/news/world/middle_east/article624855.ece), accessed December 20, 2008.

188. Dominic Moran, "Israel Mulls Open Nuke Program," *Security Watch*, August 3, 2007 (http://www.isn.ethz.ch/isn/Current-Affairs/Security-Watch/Detail/?ots591=4888CAA0-B3DB-1461-98B9-E20E7B9C13 D4&lng=en&id=53614), accessed March 10, 2009.

189. "Arab League Vows to Drop Out of NPT if Israel Admits It has Nuclear Weapons," AP Wire, March 5, 2008 (http://www.2020 visioncampaign.org/pages/351/Arab_League_vows_to_drop_out_of_ NPT_if_Israel_admits_it_has_nuclear_weapons), accessed December 20, 2008.

190. International Pugwash, "A Middle East Weapons of Mass Destruction Free Zone," Pugwash Online, June, 2008 (http://www.pugwash.org/reports/rc/me/middle-east-WMDFZ.htm), accessed December 20, 2008.

191. Haider Rizvi, "Israeli Arsenal Vexes Nuclear Negotiators," Inter Press Service, Inter Press Service, May 20, 2005 (http://www.antiwar.com/ips/rizvi.php?articleid=6033), accessed December 20, 2008.

192. Z. Maoz, op. cit.

193. Ibid.

194. Glen Kessler, "In 2003, U.S. Spurned Iran's Offer of Dialogue: Some Officials Lament Lost Opportunity," *Washington Post*, June 18, 2006, A-16 (http://www.washingtonpost.com/wpdyn/content/article/2006/06/17/AR2006061700727.html), accessed December 20, 2008.

195. Gareth Porter, "Iran Proposal to US Offered Peace with Israel," Inter Press Service, May 24, 2006 (http://ipsnews.net/news.asp?idnews=33348), accessed December 20, 2008.

196. Gershom Gorenberg and Hiam Watzman, "Is Hamas Looking For a Two-State Solution? Should We Listen?" *South Jerusalem*, April 10, 2008 (http://southjerusalem.wordpress.com/2008/04/10/is-hamas-looking-for-a-two-state-solution-should-we-listen/).

197. Yossi Verter, "Poll: Most Israelis Back Direct Talks with Hamas on Shalit," *Haaretz*, February 27, 2008 (http://haaretz.com/hasen/spages/958473.html), accessed December 20, 2008.

198. Barak Ravid, Aluf Benn and Assaf Uni, "Israel, PA Agree to Strive for Deal by End of 2008," *Haaretz*, November 27, 2007 (http://www.haaretz.com/hasen/pages/ShArt.jhtml?itemNo=928637).

199. Tom Allard, "Israeli Nukes A Key to Peace," *The Sydney Morning Herald*, November 9, 2004 (http://www.smh.com.au/articles/2004/11/08/1099781324113.html?from=storylhs).

200. Andrew Tulley, "Radio Free Europe US: Bush Signals New Interpretation of Nonproliferation Treaty," Radio Free Europe US, March 16, 2005 (http://www.globalsecurity.org/wmd/library/news/usa/2005/usa-050316-rferl01.htm), accessed December 20, 2008.

201. Bernd Debusmann, "Can Obama Avert an Arab-Israeli Disaster?" Reuters, December 11, 2008 (http://blogs.reuters.com/great-debate/2008/12/11/can-obama-avert-an-arab-israeli-disaster/), December 20, 2008.

202. Lily Galili, "Netanyahu, Yisrael Beiteinu Strike Deal, Lieberman to get Foreign Ministry," Haaretz.com, March 9, 2009 (http://www.haaretz.com/hasen/spages/1069317.html), accessed March 9, 2009.

203. "Arab Stocks Volatile on Falling Oil Prices, Global Uncertainty," Business News, Deutsche Presse-Agenteur, November 28, 2008 (http://www.monstersandcritics.com/news/business/news/article_1445534.php).

204. Rebecca Johnson, "Is the NPT Being Overtaken by Events?" *Disarmament Diplomacy* (Spring 2008); (http://www.acronym.org.uk/dd/dd87/87npt.htm), accessed December 20, 2008.

205. W. Pincus, op. cit.

206. "UAE Calls for a Nuclear-Free World," Bahrain News Agency, May 4, 2005 (http://english.bna.bh/?ID=31216), accessed December 20, 2008.

207. Amy Goodman and Juan Gonzalez, "Former Chief UN Weapons Inspector Hans Blix on the US Rush to War in Iraq, the Threat of an Attack on Iran, and the Need for a Global Nuclear Ban to Avoid Further Catastrophe," *Democracy Now*, May 21, 2008 (http://www.democracynow.org/2008/5/21/former_chief_un_weapons_inspector_hans), accessed December 20, 2008.

Chapter 12

1. Mohammed El-Genk, "On the Introduction of Nuclear Power in Middle East Countries: Promise, Strategies, Vision and Challenges," *Energy Conversion and Management*, vol. 49, Issue 10 (October 2008).

2. IAEA, "Nuclear Power Reactors in the World," *Reference Data Series No.2* (Vienna: IAEA, 2008).

3. Hamad Al Ka'abi, "UAE Policy on the Evaluation and Potential Development of Peaceful Nuclear Energy," ECSSR 14th Annual Energy Conference, Abu Dhabi, November 2008.

4. Ghassan Ejjeh, "Desalination and Water Reuse: A Technology for the Future"(www.pub.iaea.org/MTCD/Meetings/PDFplus/2007/cn152/cn152p/Ejjeh%20IAEA%20Conference%20%20presentation.pdf), 2008.

5. International Desalination Association (IDA), *Desalination Yearbook 2008–2009* (Global Water Intelligence, 2008).

6. Farrah Imam Andejani, "Review of WEC IWPPs in Saudi Arabia," *IFAT* (July 2008).

7. IAEA, "Nuclear Power Reactors in the World," op. cit.

8. IAEA, *Annual Report 2007* (Vienna: IAEA, September 2007); (http://www.iaea.org/Publications/Reports/Anrep2007/anrep2007_full.pdf).

9. World Nuclear Association (WNA), "The Nuclear Renaissance" (http://www.world-nuclear.org/info/inf104.html), September 2007.

10. Mohammed El-Genk, op. cit.

11. WNA, "The Nuclear Renaissance," op. cit.

12. Ibid.

13. Ibid.

14. IAEA, "Optimization of Coupling of Nuclear Reactors and Desalination Systems," IAEA-TECDOC-1444 (Vienna: IAEA, September 2005).

15. Photovoltaic Resources, "Large-scale Photovoltaic Power Plants," (http://www.pvresources.com/en/top50pv.php), 2007.

16. Mei-Ling Kuo, et al., "Realization of a Near-Perfect Anti-Reflection Coating for Silicon Solar Energy Utilization," *Optics Letters*, vol. 33, Issue 21 (2008).

17. Generation IV International Forum (GIF), "A Technology Roadmap for the Generation IV Nuclear Energy Systems," GIF, 2002 (http://gif.inel.gov/roadmap/pdfs/gen_iv_roadmap.pdf), accessed November 24, 2008.

18. OptiSolar, "Topaz Solar Farm 550 Megawatt PG&E Solar Power Agreement" (http://www.optisolar.com/081408_opti_pge_sp.htm).

19. Stirling Energy Systems, "Solar One" (http://www.stirlingenergy.com/projects/solar-one.asp).

[439]

BIBLIOGRAPHY

"Arab League Vows to Drop out of NPT if Israel Admits it has Nuclear Weapons." AP Wire, March 5, 2008 (http://www.2020visioncampaign. org/pages/351/Arab_League_vows_to_drop_out_of_NPT_if_Israel_admits _it_has_nuclear_weapons).

"Arab Stocks Volatile on Falling Oil Prices, Global Uncertainty." Business News, Deutsche Presse-Agenteur, November 28, 2008 (http://www.monstersandcritics.com/news/business/news/article_1445 534.php).

"Bayern LB Receives 15BN-Euro Bailout." *Frankfurter Allgemeine Zeitung*, December 5, 2008.

"Blair Presses the Nuclear Button." *The Guardian*, May 17, 2006 (http://www.guardian.co.uk/environment/2006/may/17/energy.business.

"Brown Expands Nuclear Ambitions." BBC News Online, May 28, 2008 (http://news.bbc.co.uk/1/hi/uk_politics/7424158.stm).

"Business Chance for Advanced Technology under Global Warming and Oil Soaring." *Nihon Keizai Sinbun*, October 27, 2008.

"Construction Costs to Soar for New US Nuclear Power Plants." Standard & Poor's RatingsDirect, October 15, 2008.

"Demographics of Israel." Wikipedia (http://en.wikipedia.org/wiki/ Demographics_of_Israel).

"Engines of Economic Growth: The Economic Impact of Boston's Eight Research Universities on the Metropolitan Boston Area" (New York, NY: Appleseed, 2003).

"Former weapons inspector Blix says Olmert's Nuclear Comment Unlikely to Spark Arab Backlash." *International Herald Tribune*, December 13, 2006 (http://www.iht.com/articles/ap/2006/12/13/ europe/EU_GEN_Netherlands_Blix_Nuclear.php).

[441]

"GE Hitachi Seeks to Renew NRC Certification for ABWR Reactor Design." *Business Wire*, December 15, 2008.

"Graduate Record Examinations: Guide to the Use of Scores 2007–2008." Educational Testing Service, 2008.

"History of Lebanon." Wikipedia (http://en.wikipedia.org/wiki/History_of_Lebanon#Israeli_invasion_and_international_intervention:_1982.E2.80.9384).

"Israel Atomic Energy Commission." GlobalSecurity.org (http://www.globalsecurity.org/wmd/world/israel/iaec.htm), accessed March 10, 2009.

"Israel Gets US Supercomputer for Secret Military Site." Wisconsin Project on Nuclear Arms Control, *The Risk Report*, vol. 2, no. 3 (1996); (http://www.wisconsinproject.org/countries/israel/israelgets.htm).

"Israel Reviews its Nuclear Deterrent." *Jane's Intelligence Review*, November 1, 1998 (http://www.jane's.com/extract/jir98/jir00807.html).

"Israel: Uranium Processing and Enrichment." Wisconsin Project on Nuclear Arms Control, *The Risk Report*, vol. 2, no. 4 (1996); (http://www.wisconsinproject.org/countries/israel/uranium.html).

"Israel–United States Military Relations." Wikipedia, December 10, 2008 (http://en.wikipedia.org/wiki/Israel-United_States_military_relations).

"Jordanian Fears over Israeli Dimona Nuclear Reactor." *Arabic News,* April 28, 2005 (http://www.arabicnews.com/ansub/Daily/Day/050428/2005042832.html).

"McDonnell Douglas F4 Phantom." Israeli Airforce: The Official Website (http://www.iaf.org.il/Templates/Aircraft/Aircraft.IN.aspx?lang=EN&lobbyID=69&folderID=78&subfolderID=184&docfolderID=184&docID=18176).

"New Nuclear Generation in the United States: Keeping Options Open vs. Addressing an Inevitable Necessity." Moody's Investors Service, October 10, 2007.

"South Africa's Nuclear Autopsy." Wisconsin Project on Nuclear Arms Control, *The Risk Report*, vol. 2, no. 1 (1996); (http://www. wisconsinproject.org/countries/safrica/autopsy.html).

"SSK Dolphin Class Submarine, Israel, 2008," Naval-Technology.com (http://www.naval-technology.com/projects/dolphin/).

"Strategic Israel: The Secret Arsenal of the Jewish State," MSNBC (http://www.sweetliberty.org/issues/israel/strategic/index.shtml).

"*Sunday Times*: Close Dimona Nuclear Facility." IsraelWire, February 7, 2000 (http://www.fas.org/news/israel/000207-israel1.htm).

"Swedish Poll Shows Increasing Support for Nuclear," NucNet, 2008 (http://www.foratom.org/dmdocuments/ne51pdf.pdf).

"UAE Calls for a Nuclear-Free World." Bahrain News Agency, May 4, 2005 (http://english.bna.bh/?ID=31216).

"United Nations Security Resolution 487." The Avalon Project, Yale Law School (http://www.yale.edu/lawweb/avalon/un/un487.htm).

"Uranium: The Israeli Connection." *Time Magazine*, May 30, 1977 (http://www.time.com/time/magazine/article/0,9171,914952,00.html).

"Weapons of Mass Destruction In the Middle East." Center for Nonproliferation Studies, 2006 (http://cns.miis.edu/research/wmdme/ israel.htm).

"Which Power Generation Technologies Will Take the Lead in Response to Carbon Controls?" *Standard & Poor's Viewpoint*, May 11, 2007.

Adler, Michael. "Israel's Soreq Nuclear Reactor: The One They Show to Journalists." Agence France Press, July 18, 2004 (available at: http://www.globalsecurity.org/org/news/2004/040718-israel-soreq.htm).

Akimoto, Y. and Y. Fujiie. "50th Anniversary of the IAEA's Founding," *Energy Review* (June 2007).

Al Ka'abi, Hamad. "UAE Policy on the Evaluation and Potential Development of Peaceful Nuclear Energy." ECSSR 14th Annual Energy Conference, Abu Dhabi, November 2008.

Albright, David and A. Scheel. *Unprecedented Projected Nuclear Growth in the Middle East: Now Is the Time to Create Effective Barriers to Proliferation*. ISIS Report, November 12, 2008.

Allard, Tom. "Israeli Nukes a Key to Peace." *The Sydney Morning Herald*, November 9, 2004 (http://www.smh.com.au/articles/2004/11/08/109978 1324113.html?from=storylhs).

Alternative Energy Development Board (AEDB). *Renewable Energy Sector* (http://www.aedb.org), accessed December 4, 2008.

Alvarez, Robert, et al. "Reducing the Hazards from Stored Spent Power-Reactor Fuel in the United States." *Science and Global Security* vol. 11 (2003).

American Nuclear Society. "Use of Nuclear Energy for Desalination." Position Statement, March 2005.

Analysis Group. "Poor Support for the Government's Nuclear Power Phase-out Policy," 2000 (www.analysisgruppen.org/engopin).

Andejani, Farrah Imam. "Review of WEC IWPPs in Saudi Arabia," *IFAT* (July 2008).

Arab Atomic Energy Agency (AAEA). *Research Reactors* [in Arabic] (Tunisia: AAEA, 2008).

Arab Atomic Energy Agency (AAEA). *The Future of Generating Electricity by Nuclear Energy* [in Arabic] (Tunisia: AAEA, 2006).

Arabs Against Discrimination. "Employees at Nuclear Reactors Demand Compensation," April 17, 2005 (http://www.aad-online.org/2005/ English/8-August/13-18/13-8/aad20/1.htm).

Argonne National Laboratory. "Fast Reactor Development." International Nuclear Safety Center (INSC); (http://www.insc.anl.gov/dbfiles/ plant_category/FBR/bn.2.html), accessed November 24, 2008.

Arms Control Association. "Nuclear-Weapon-Free-Zones at a Glance." Fact Sheet, November 2007 (http://www.armscontrol.org/factsheets/nwfz).

Armstrong, Karen. *The Battle for God* (New York, NY: Ballantine, 2000).

Arnold, Patricia and Rita Hartung Cheng. "The Economic Consequences of Regulatory Accounting in the Nuclear Power Industry: Market Reaction to Plant Abandonments." *Journal of Accounting and Public Policy*, vol. 19, no. 2 (2000).

Aronson, Geoffrey. "Hidden Agenda: US-Israeli Relations and the Nuclear Question." *Middle East Journal* (Autumn 1992); (http://www.mideasti.org/middle-east-journal/volume-46/4/hidden-agenda-us-israeli-relations-and-nuclear-question).

Asian Development Bank (ADB). *Outlook 2008 Update* (Manila: Asian Development Bank, 2008).

Athar, G.R., R. Tariq, M. Imtiaz, I. Ahmad, and J. Bashir. *Energy Development Strategies of Pakistan for Addressing Climate Change Issues*. Draft report on country study carried out under IAEA/RCA Project RAS/0/045 (Islamabad: Pakistan Atomic Energy Commission, 2008).

Atomic Energy Society of Japan. "Code of Ethics," November 11, 2005 (http://wwwsoc.nii.ac.jp/aesj/rinri/committee/kensho-e.html).

Atomic Energy Society of Japan. "The Era Brought by Nuclear Power," July 6, 1998.

Avnery, Uri. "Sorry, Wrong Continent." *Gush Shalom*, December 23, 2006 (http://zope.gush-shalom.org/home/en/channels/avnery/1166960371).

Azoulay, Yuval. "Deputy CEO of Negev Nuclear Center: Dimona Reactor is Safe." *Haaretz*, August 23, 2007 (http://www.haaretz.co.il/hasen/spages/896676.html).

Barletta, Michael and Erik Jorgensen. "Weapons of Mass Destruction in the Middle East." Center for Non-Proliferation Studies (http://cns.miis.edu/research/wmdme/).

Baruah, Amit. "Now Israel Wants NSG Rules Changed." *The Hindustan Times*, August 28, 2007 (http://www.hindustantimes.com/StoryPage/FullcoverageStoryPage.aspx?sectionName=NLetter&id=9a7f3e9c-db05-4333-beb1-ccece17c6658_Special&Headline=Now%2c+Israel+wants+NSG+rules+changed).

Baumgart, Claudia and Harald Müller. "A Nuclear Weapons-Free Zone in the Middle East: A Pie in the Sky?" *The Washington Quarterly* (Winter 2004–05).

Beaumont, Peter. "Was Israeli Raid a Dry Run for an Attack on Iran?" *The Observer*, September 16, 2007 (http://www.guardian.co.uk/world/2007/sep/16/iran.israel).

Beckman, R. *The Downwave: Surviving the Second Depression* (Portsmouth: Milestone Publications, 1983).

Bedein, David. "Israel May Expand Its Nuclear Power." *The Philadelphia Bulletin*, November 13, 2008.

Beeston, Richard. "Six Arab States Join Rush to Go Nuclear." TimesOnline, November 6, 2006 (http://www.timesonline.co.uk/tol/news/world/middle_east/article624855.ece).

Benn, Aluf. "Israel: Censoring the Past." *Bulletin of the Atomic Scientists*, (July/August 2001); (http://www.bsos.umd.edu/pgsd/people/staffpubs/Avner-BASreport7-01.htm).

Beresford, David. "Man Who Spiked Apartheid's Bomb." *The Guardian* (London), January 2, 1996.

Braun, Chaim. "Nuclear Fuel Supply Assurance," Draft (October 2007).

Bredimas, Alexandre, and William J. Nuttall. "An International Comparison of Regulatory Organizations and Licensing Procedures for New Nuclear Power Plants." *Energy Policy* 36 (2008).

British Broadcasting Corporation (BBC). BBC World Service, BBC Transcript of "Israel's Secret Weapon, Electric Intifada," March 17, 2003 (http://electronicintifada.net/v2/article1665.shtml).

British Nuclear Fuels Ltd. (BNFL). *MORI UK opinion poll findings* (Warrington, UK: BNFL, 1999).

Broder, Jonathon. "A Challenge To Israel's Nuclear Blind Spot." *The Washington Post*, Sunday, March 11, 2001; Page B02 (http://www.bsos.umd.edu/pgsd/people/staffpubs/Avner-WashPost3-11-01.htm).

Bundesamt für Strahlenschutz (http://www.bfs.de), 2008.

Bunn, Matthew, et al. *Interim Storage of Spent Nuclear Fuel: A Safe, Flexible, and Cost-Effective Near-Term Approach to Spent Fuel Management.* A Joint Report from the Harvard University Project on Managing the Atom and the University of Tokyo Project on Sociotechnics of Nuclear Energy (June 2001).

Carr, Housley. "Duke says New Lee Plant to Cost $11 Billion." *Nucleonics Week*, November 6, 2008.

Center for Nonproliferation Studies (CNS). Terrorism Database (http://cns.miis.edu/research/terror.htm#db).

Central Intelligence Agency (CIA). *The 2008 World Fact Book* (https://www.cia.gov/library/publications/the-world-factbook/), November 2008.

Cirincione, Joseph, Jon B. Wolfsthal and Miriam Rajkumar. *Deadly Arsenals: Nuclear, Biological and Chemical Threats* (Washington, DC: Carnegie Endowment for International Peace, 2006).

Cohen, Avner and Marvin Miller. "Country Perspectives on the Challenges to a Fissile Material (Cutoff) Treaty." International Panel on Fissile Materials, Israel, 2008 (http://www.ipfmlibrary.org/gfmr08cv.pdf).

Cohen, Avner and William Burr. "The Untold Story of Israel's Bomb." *Washington Post*, Sunday, April 30, 2006, B-01 (http://www.washingtonpost.com/wp-dyn/content/article/2006/04/28/AR2006042801326_pf.html).

Cohen, Avner. "Did Nukes Nudge the PLO?" *Bulletin of the Atomic Scientists*, December, 1993.

Cohen, Avner. "Israel and Chemical/Biological Weapons: History, Deterrence, and Arms Control." *The Nonproliferation Review* (Fall–Winter 2001); (http://www.vho.org/aaargh/fran/livres3/Cohen.pdf).

Cohen, Avner. *Israel and the Bomb* (New York, NY: Columbia University Press, 1998).

Collins, Liat. "Family of Nuclear Cancer Victim Awarded NIS 2.5m." *Jerusalem Post*, October 13, 1997 (http://nuclearweaponarchive. org/Israel/Dimonanews.txt).

Congressional Budget Office. *Cost estimate of S.14, Energy Policy Act of 2003* (Washington, Congressional Budget Office); (http://www.cbo. gov/doc.cfm?index=4206).

Congressional Research Service (CRS). *Potential Cost of Nuclear Power Plant Subsidies in S.14*, May 7, 2003.

Corbach, Matthias. "Atomenergie," in Danyel Reiche, *Grundlagen der Energiepolitik* (Frankfurt am Main: Peter Lang, 2005).

Cordesman, Anthony. "Israeli Weapons of Mass Destruction: An Overview" (working draft). Center for Strategic and International Studies (CSIS), 2008 (http://www.csis.org/media/csis/pubs/080602_ sraeliwmd.pdf).

Crail, Peter and Jessia Lasky-Fink. "Middle Eastern States Seeking Nuclear Power." *Arms Control Today* (May 2008)

Creamer, Terence. "Eskom Terminates Nuclear 1 Procurement Process, but SA still Committed to Nuclear." *Creamer Media's Engineering News,* December 5, 2008 (http://www.engineeringnews.co.za/article/eskom-terminates-nuclear-1-procurement-process-but-sa-still-committed-to-nuclear-2008-12-05).

Debusmann, Bernd. "Can Obama Avert an Arab-Israeli Disaster?" Reuters, December 11, 2008 (http://blogs.reuters.com/great-debate/ 2008/12/11/can-obama-avert-an-arab-israeli-disaster/).

Deges, Stefan. "Kein Strom ohne Kohle und Atom, Warum dem Land der Blackout droht." *Die Politische Meinung*, no. 466/2008, September 2008.

DENA (Deutsche Energie-Agentur GmbH). "Kurzanalyse der Kraftwerks-und Netzplanung in Deutschland bis 2020 (mit Ausblick auf 2030)." Annahmen, Ergebnisse und SchlussfolgerungenBerlin, April 15, 2008 (http://www.dena.de/fileadmin/user_upload/Download/ Dokumente/Meldungen/2008/Kurzanalyse_KuNPlanung_D_2020_2030_ lang_0408.pdf.pdf).

Department for Business Enterprise and Regulatory Reform. "Consultation on Funded Decommissioning Programme Guidance for New Nuclear Power Stations: Feb 2008" (London: BERR, 2008); (http://www.berr.gov.uk/consultations/page44784.html).

Department for Business Enterprise and Regulatory Reform. "Meeting the Energy Challenge: A White Paper on Nuclear Power" (London: HM Government Cm 7296, 2003); (http://www.berr.gov.uk/files/file4 3006.pdf).

Desalination as a Sustainable Solution for Water Scarcity, PakAtom (Islamabad: PAEC, March–April 2008).

Donovan, Michael. "Iran, Israel and Nuclear Weapons in the Middle East." Center for Defense Information Terrorism Project, February 14, 2002 (http://www.cdi.org/terrorism/menukes.cfm).

Dunnigan, Jim. "Israel Gets Super Subs." Strategy Page, August 27, 2006 (http://www.strategypage.com/htmw/htsub/articles/20060827.aspx).

Edelman, R., *Edelman Trust Barometer 2005* (http://www.edelman.com/ TRUST/2008/prior/2005/Final_onscreen.ppt#324,1,Slide 1).

Ejjeh, Ghassan. "Desalination and Water Reuse: A Technology for the Future" (www.pub.iaea.org/MTCD/Meetings/PDFplus/2007/cn152/cn 152p/Ejjeh%20IAEA%20Conference%20%20presentation.pdf), accessed November 24, 2008.

ElBaradei, Mohamed. "Toward a Safer World." *The Economist*, October 16, 2003.

[449]

El-Genk, Mohammed. "On the Introduction of Nuclear Power in Middle East Countries: Promise, Strategies, Vision and Challenges." *Energy Conversion and Management*, vol. 49, Issue 10 (October 2008).

European Commission. "Second Strategic Energy Review: An EU Energy Security and Solidarity Action Plan," Commission Staff Working Document COM/2008/0781, Brussels, 2008.

European Commission. *Eurobarometer Special 271: Europeans and Nuclear Safety* (Brussels, February 2007).

Farr, Lt. Col. Warner D. "The Third Temple Holy of Holies: Israel's Nuclear Weapons." USAF Counter-proliferation Center, Air War College, September 1999 (www.fas.org/nuke/guide/israel/nuke/farr.htm).

Federal Ministry for the Environment, Nature Conservation and Nuclear Safety. "Convention on Nuclear Safety." Report by the Government of the Federal Republic of Germany for the Fourth Review Meeting in April 2008 (Berlin: BMU, 2007).

Federation of Electric Power Companies of Japan. "Nuclear Power Generation Seasonal Report No. 43," August 2008.

Feiveson, Harold. "Faux Renaissance: Global Worming, Radioactive Waste Disposal, and the Nuclear Future." *Arms Control Today* (May 2007).

Feldman, Shai. *Nuclear Weapons and Arms Control in the Middle East* (Cambridge, MA: MIT Press, 1997).

Flaherty, Ted. *Nuclear Database: Israel's Possible Nuclear Delivery Systems*, Center for Defense Information, 1997 (http://www.cdi.org/nuclear/database/isnukes.html).

Fleming, Eileen. "The Vanunu Saga 2008." *Counter Currents*, January 22, 2008 (http://www.countercurrents.org/fleming220108.htm).

Fleming, L. "Breakthroughs and the 'Long Tail' of Innovation." *Sloan Management Review* 49(1); (2007).

Friedman, George. "Mediterranean Flyover: Telegraphing an Israeli Punch." *Jewish World Review*, June 25, 2008 (http://www.stratfor. com/weekly/mediterranean_flyover_telegraphing_israeli_punch).

Gaffney, Mark. *Dimona the Third Temple: The Story Behind the Vanunu Revelation* (Brattleboro, VT: Amana Books, 1989).

Galili, Lily. "Netanyahu, Yisrael Beiteinu Strike Deal, Lieberman to get Foreign Ministry." Haaretz.com, March 9, 2009 (http://www.haaretz. com/hasen/spages/1069317.html).

Generation IV International Forum (GIF). "A Technology Roadmap for the Generation IV Nuclear Energy Systems," 2002 (http://gif.inel. gov/roadmap/pdfs/gen_iv_roadmap.pdf), accessed November 24, 2008.

Geneva Initiative. *New Survey of Israeli Public Opinion, Prospects for Peace*, 2007 (http://www.prospectsforpeace.com/2007/07/new_survey_ of_israeli_public_o.html).

Goodman, Amy and Juan Gonzalez. "Former Chief UN Weapons Inspector Hans Blix on the US Rush to War in Iraq, the Threat of an Attack on Iran, and the Need for a Global Nuclear Ban to Avoid Further Catastrophe." *Democracy Now*, May 21, 2008 (http://www.democracynow.org/ 2008/5/21/former_chief_un_weapons_inspector_hans).

Gorenberg, Gershom and Hiam Watzman. "Is Hamas Looking For a Two-State Solution? Should We Listen?" *South Jerusalem*, April 10, 2008 (http://southjerusalem.wordpress.com/2008/04/10/is-hamas-looking-for-a-two-state-solution-should-we-listen/).

Government of Pakistan (GOP), *Pakistan Economic Survey 2007–08* and earlier issues (Islamabad, Economic Adviser's Wing, Finance Division, 2008).

Government of Pakistan (GOP). *Medium Term Development Framework 2005–2010* (Islamabad: Planning Commission, 2005).

Government of Pakistan (GOP). *Pakistan Economic Survey 2007–08 and Earlier Issues* (Islamabad: Finance Division, 2008).

Government of Pakistan (GOP). *Strategic Directions to Achieve Vision 2030: Approach Paper* (Islamabad: Planning Commission, 2006).

Greenpeace International. "An Overview of Nuclear Facilities In Iran, Israel and Turkey," February, 2007 (http://www.greenpeace.org/raw/content/mediterranean/reports/an-overview-of-nuclear-facilit.pdf). Carnegie Endowment for International Peace. "Israel: Nuclear Infrastructure," 1998 (http://www.carnegieendowment.org/files/Tracking_israelmap.pdf).

Greenpeace. "Conditions for a Nuclear Free Middle East." Briefing, February 2007 (http://www.greenpeace.org/raw/content/mediterranean/reports/conditions-for-NFME.pdf).

Griffiths, S., et al. *The Masdar Institute Intellectual Platform: Technology, Policy and Systems* (Abu Dhabi: The Masdar Institute of Science and Technology, 2008).

Gulf Research Center (GRC). "Weapons of Mass Destruction Free Zone in the Gulf (GWMDFZ)." *Security & Terrorism: Regional Security in the Gulf*, November 2006.

Hafele, Wolf. "The Concept of an International Monitored Retrievable Storage System." Uranium Institute Symposium, London, UK (August 1996).

Halle, Charlotte. "Foreign Ministry Slams Envoy's Comments about 'Yellow Race'." *Haaretz*, October 24, 2006 (http://www.haaretz.com/hasen/spages/774471.html).

Hedges, Chris. "Israel's 'Crime Against Humanity'." TruthDig.com, December 15, 2008 (http://www.truthdig.com/report/item/20081215_israels_crime_against_humanity/).

Heller, Jeffrey. "Analysis: Netanyahu on Course for Israeli Election Win-Polls." Reuters, December 10, 2008 (http://www.reuters.com/article/vcCandidateFeed2/idUSLA371329).

[452]

Henderson M. "Secret List of Nuclear Dump Sites Revealed." *The Times* (online), June 11, 2005 (http://www.timesonline.co.uk/article/0,,2-1649479,00.html).

Hennicke, Peter, and Michael Müller. *Weltmacht Energie, Herausforderung für Demokratie und Wohlstand*, Wuppertal Institute for Climate, Environment and Energy, 2006.

Hersh, Seymour. *The Samson Option: Israel's Nuclear Arsenal and American Foreign Policy* (New York, NY: Random House, 1991).

Hobohm, Jens. "Energy in the Baltic: The German Case." Presentation in Riga, November 7, 2008.

Hounam, Peter. "Mordechai Vanunu: The *Sunday Times* Articles," April 21, 2004 (http://www.timesonline.co.uk/tol/news/article830147.ece).

Hounam, Peter. "Revealed: The Secrets of Israel's Nuclear Arsenal/Atomic Technician Mordechai Vanunu Reveals Secret Weapons Production." *Sunday Times*, October 5, 1986 (http://www.timesonline.co.uk/tol/news/article830147.ece).

Hounam, Peter. *Woman from Mossad: The Torment of Mordechai Vanunu* (London: Vision Paperbacks, 1999).

Hume, D., *A Treatise of Human Nature* (1739).

Hydrocarbon Development Institute of Pakistan (HDIP), *Pakistan Energy Yearbook 2008* and earlier issues (Islamabad: HDIP, Ministry of Petroleum and Natural Resources, 2009).

Hydrocarbon Development Institute of Pakistan (HDIP). *Pakistan Energy Yearbook 2008 and earlier issues* (Islamabad: HDIP, Ministry of Petroleum and Natural Resources, 2009).

International Atomic Energy Agency (IAEA). "Advanced Applications of Water Cooled Nuclear Power Plants." TECDOC-158-IAEA (Vienna: IAEA, 2007).

International Atomic Energy Agency (IAEA). *Annual Report 2007* (Vienna: IAEA, September 2007); (http://www.iaea.org/Publications/ Reports/Anrep2007/anrep2007_full.pdf).

International Atomic Energy Agency (IAEA). "Basic Infrastructure for a Nuclear Power Project." IAEA-TEDOC-1513 (Vienna: IAEA, 2006).

International Atomic Energy Agency (IAEA). "Climate Change and the NPT" (Vienna: IAEA, 2008).

International Atomic Energy Agency (IAEA). *Code on the Safety of Nuclear Power Plants: Design, Safety Series No.50-C-D Rev.1* (Vienna: IAEA, 1988).

International Atomic Energy Agency (IAEA). "Communication dated 28 September 2005 from the Permanent Mission of the United States of America to the Agency." INFCIRC/659, September 29, 2005.

International Atomic Energy Agency (IAEA). "Communication dated 30 May 2007 from the Permanent Mission of the United Kingdom of Great Britain and Northern Ireland to the IAEA concerning Enrichment Bonds: A Voluntary Scheme for Reliable Access to Nuclear Fuel." INFCIRC/707, June 4, 2007.

International Atomic Energy Agency (IAEA). "Communication received from the Resident Representative of the Russian Federation to the Agency transmitting the text of the Statement of the President of the Russian Federation on the Peaceful Use of Nuclear Energy." INFCIRC/667, February 8, 2006.

International Atomic Energy Agency (IAEA). "Communication received on 12 September 2006 from the Permanent Mission of Japan to the Agency concerning arrangements for the assurance of nuclear fuel supply." INFCIRC/683, September 15, 2006.

International Atomic Energy Agency (IAEA). "Communication received from the Resident Representative of Germany to the IAEA with regard to the German proposal on the Multilateralization of the Nuclear Fuel Cycle." INFCIRC/704, May 4, 2007.

International Atomic Energy Agency (IAEA). "Communication received from the Federal Minister for European and International Affairs of Austria with regard to the Austrian proposal on the Multilateralization of the Nuclear Fuel Cycle." INFCIRC/706, May 31, 2007.

International Atomic Energy Agency (IAEA). "Communication received from the Resident Representative of the Russian Federation to the IAEA on the Establishment, Structure and Operation of the International Uranium Enrichment Centre." INFCIRC/708, June 8, 2007.

International Atomic Energy Agency (IAEA). "Considerations to Launch a Nuclear Power Program" (Vienna: IAEA, 2007); (http://www.iaea.org/NuclearPower/Downloads/Launch_NPP/07-11471_Launch_NPP.pdf).

International Atomic Energy Agency (IAEA). "Design Basis Threat," Guidance Information (http://www-ns.iaea.org/security/dbt.htm).

International Atomic Energy Agency (IAEA). *Design for Safety of Nuclear Power Plants: A Code of Practice, Safety Series No.50-C-D* (Vienna: IAEA, 1978).

International Atomic Energy Agency (IAEA). *IAEA Safeguards: Staying Ahead of the Game* (August 2007).

International Atomic Energy Agency (IAEA). "International Status and Prospects of Nuclear Power" (Vienna: IAEA, 2008).

International Atomic Energy Agency (IAEA). "Milestones in the Development of a National Infrastructure for Nuclear Power." NE series guide NG-G-3.1 (Vienna: IAEA, September 2007).

International Atomic Energy Agency (IAEA). *Model Protocol Additional to the Agreement(s) between States(s) and the International Atomic Energy Agency for the Application of Safeguards.* INFCIRC/540 (September 1997).

International Atomic Energy Agency (IAEA). *Multilateral Approaches to the Nuclear Fuel Cycle.* Expert Group Report Submitted to the Director General of the International Atomic Energy Agency. INFCIRC/640 (February 2005).

International Atomic Energy Agency (IAEA). "Nuclear Power Reactors in the World." Reference Data Series No.2 (Vienna: IAEA, 2008).

International Atomic Energy Agency (IAEA). "Nuclear Power Reactors in the World" (Vienna: IAEA, 2004).

International Atomic Energy Agency (IAEA). "Optimization of Coupling of Nuclear Reactors and Desalination Systems," IAEA-TECDOC-1444 (Vienna: IAEA, September 2005).

International Atomic Energy Agency (IAEA). *PRIS Database* (Vienna: IAEA, 2008); (http://www.iaea.org/programmes/a2/).

International Atomic Energy Agency (IAEA). "Radiation, People and the Environment" (Vienna: IAEA, 2004).

International Atomic Energy Agency (IAEA). "Safeguards Current Status" (http://www.iaea.org/OurWork/SV/Safeguards/sir_table.pdf).

International Atomic Energy Agency (IAEA). *Safety of Nuclear Power Plants: Design. IAEA Safety Standard Series No. NS-R-1* (Vienna: IAEA, 2000).

International Atomic Energy Agency (IAEA). "The Management System for Facilities and Activities." IAEA Safety Standards Series GS-R-3 (Vienna: IAEA, 2006).

International Atomic Energy Agency (IAEA). *The Structure and Content of Agreements Between the Agency and States Required in Connection with the Treaty on the Non-Proliferation of Nuclear Weapon.* INFCIRC/153 (June 1972).

International Desalination Association (IDA). *Desalination Yearbook 2008–2009* (Global Water Intelligence, 2008).

International Energy Agency (IEA). *Key World Energy Statistics* (Pairs: IEA, 2008).

International Energy Agency (IEA). *Nuclear power in the OECD* (Paris: OECD/IEA: 2001).

International Monetary Fund (IMF). *World Economic Outlook October 2008: Financial Stress, Downturns, and Recoveries* (Washington, DC: IMF, 2008).

International Pugwash. "A Middle East Weapons of Mass Destruction Free Zone." Pugwash Online, June, 2008 (http://www.pugwash.org/reports/rc/me/middle-east-WMDFZ.htm).

Israeli Atomic Energy Commission. "Research and publications by IAEC personnel" (http://www.iaec.gov.il/pages_e/card_report_e.asp), accessed March 10, 2009.

Israeli Prime Minister's Office. "Atomic Energy Commission" (http://www.pmo.gov.il/PMOEng/PM+Office/Bodies/depenergy.htm), accessed March 10, 2009.

Ivri-Darel, Yael. "Soreq Nuclear Research Center's Reactor to Shut Down in 8 Years." Ynetnews, April 3, 2008 (http://www.ynet.co.il/english/articles/0,7340,L-3514334,00.html).

Japan Atomic Energy Commission. "Framework for Nuclear Energy Policy in Japan," October 11, 2005 (http://www.aec.go.jp/jicst/NC/iinkai/teirei/siryo2005/kettei/speech051019.pdf).

Japan Atomic Energy Commission. "White Paper on Nuclear Energy, 2007," March 2008 (http://www.aec.go.jp/jicst/NC/about/hakusho/hakusho2007/wp_e.pdf).

Japan Atomic Energy Commission. Newsletter, October 3, 2008.

Japan Atomic Industrial Forum. "Energy Awareness Fact-Finding Survey," December 2008.

Jayachandran, C.R. "Pak Fears Joint Indo-Israel Strikes." *Times of India*, October 14, 2003 (http://timesofindia.indiatimes.com/articleshow/232733.cms).

Jericho, Mark Wade. "Encyclopedia Astronautica" (http://www.astronautix.com/lvfam/jericho.htm).

Johnson, Rebecca. "Is the NPT Being Overtaken by Events?" *Disarmament Diplomacy* (Spring 2008); (http://www.acronym.org.uk/dd/dd87/87npt.htm).

Kang, Jungmin, et al. "Spent Fuel Standard as a Baseline for Proliferation Resistance in Excess Plutonium Disposition Options." *Journal of Nuclear Science and Technology*, vol. 37, no. 8 (August 2000).

Kanter, Ido. "Changing Nuclear Equation: Should Nuclear Arms be used in Response to Powerful Conventional Attacks?" *Israel Opinion*, September 4, 2006 (http://www.ynetnews.com/articles/0,7340,L-3299440,00.html).

Karpin, Michael. *The Bomb In The Basement: How Israel Went Nuclear and What It Means for the World* (New York, NY: Simon & Schuster Paperbacks, 2006).

Katz, Yaakov. "Israel Test-Fires Long-Range Ballistic Missile." *Jerusalem Post*, July 17, 2008 (http://www.jpost.com/servlet/Satellite?cid=1200475902683&pagename=JPost%2FJPArticle%2FShowFull).

Kaye, Dalia Dassa. "Time for Arms Talks? Iran, Israel, and Middle East Arms Control." *Arms Control Today*, November 2004.

Kelle, Alexander, and Annette Schaper. "Bio- und Nuklearterrorimus, Eine kritische Analyse der Risiken nach dem 11 September 2001." *HSFK-Report 10/2001*.

Keller, Greg. "EDF to Lead up to Euro 50B in Nuclear Plant Investment." Associated Press Worldstream, December 4, 2008.

Kessler, Glen. "In 2003, U.S. Spurned Iran's Offer of Dialogue: Some Officials Lament Lost Opportunity." *Washington Post*, June 18, 2006, A-16 (http://www.washingtonpost.com/wpdyn/content/article/2006/06/17/AR2006061700727.html).

Körner, Stefan. "Instrumente der Energiepolitik," in Danyel Reiche, *Grundlagen der Energiepolitik* (Frankfurt am Main: Peter Lang, 2005).

Koschut, Simon. "Demoskopie öffentliche Interesses an Energie und Umweltpolitik," in Josef Braml, et al. (eds). *Weltverträgliche Energiesicherheitspolitik: Jahrbuch Internationale Politik 2005/2006* (München: R. Oldenbourg, 2008).

Kuo, Mei-Ling, et al. "Realization of a Near-Perfect Anti-Reflection Coating for Silicon Solar Energy Utilization." *Optics Letters*, vol. 33, Issue 21 (2008).

Lee, T.R., J. Brown, J. Henderson, A. Baillie and J. Fielding. *Psychological Perspectives on Nuclear Energy,* "Report No 2: Results of Public Attitude Surveys towards Nuclear Power Stations Conducted in Five Counties of South West England" (London: Central Electricity Generating Board [CEGB], 1983).

Lelyveld, Joseph. "Bombs Damage Atom Plant Site in South Africa." *New York Times*, December 20, 1982.

Linneweber, Silke. "Es regnet Geld." *Die Politische Meinung* no. 466 (September 2008).

Loster, Matthias. "Total Primary Energy Supply: Required Land Area" (http://www.ez2c.de/ml/solar_land_area/) accessed November 24, 2008.

MacLachlan, Ann. "Concrete Pouring at Flamanville-3 Stopped after New Problems Found." *Nucleonics Week*, May 29, 2008.

MacLachlan, Ann. "EDF confirms target of starting up Flamanville-3 in 2012." *Nucleonics Week* November 20, 2008.

MacLachlan, Ann. "HSE Preparing to Contract for Technical Safety Expertise." *Inside NRC*, November 10, 2008.

Mahnaimi, Usi and Matthew Campbell. "Israel Makes Waves with Submarine Missile Test." *Sunday Times*, June 18, 2000 (available at: http://www.fas.org/news/israel/e20000619israelmakes.htm).

Mahnaimi, Usi and Peter Conradi. "Fears of New Arms Race as Israel Tests Cruise Missiles." *Sunday Times*, June 18, 2000 (http://www.fas.org/news/israel/e20000619fearsof.htm).

Mahnaimi, Usi. "Israeli Jets Equipped for Chemical Warfare." *Sunday Times*, October 4, 1998 (article posted on the MER website at: http://www.middleeast.org/archives/1998_11_08.htm).

Makhijani, Arjun, Howard Hu and Katherine Yih (eds). *Nuclear Wastelands: A Global Guide to Nuclear Weapons Production and Its Health & Environmental Effects* (Cambridge, MA: MIT Press, 2000).

Makower, J., R. Pernick and C. Wilder. "Clean Energy Trends 2008" (Clean Edge, Inc., 2008).

Maoz, Zeev. "The Mixed Blessing of Israel's Nuclear Policy." *International Security*, vol. 28, no. 2 (Fall 2003); (http://belfercenter.ksg.harvard.edu/files/maoz.pdf).

Marineforce International. "Submarines: Dolphin" (http://www.marineforce.net/dolphin.html).

Masdar Institute of Science and Technology (MIST). "Masdar Institute Research Plan" (Abu Dhabi, UAE: MIST, 2008).

Massachusetts Institute of Technology (MIT). *The Future of Nuclear Power*, An Interdisciplinary MIT Study (2003).

Matthes, Felix Christian. *Stromwirtschaft und deutsche Einheit, Eine Fallstudie zur Transformation der Elektrizitätswirtschaft in Ost-Deutschland* (Berlin: Books on Demand GmbH, 2000).

Meier, Oliver. "News Analysis: The Growing Nuclear Fuel Cycle Debate." *Arms Control Today* (November 2006).

Melman, Yossi. "Report suggests Obama Press Israel Over Nuke Program." *Haaretz*, November 17, 2008 (http://www.haaretz.com/hasen/spages/1037558.html).

Milholland, Gary. "A Heavy Water Whitewash," Arbeiderbladet (Oslo), April 20, 1988 (http://www.wisconsinproject.org/pubs/editorials/1988/heavywaterwhitewash.htm).

Millstone E. and P. van Zwanenberg. "Politics of Expert Advice: Lessons from the Early History of the BSE Saga." *Science and Public Policy*, vol. 28 no. 2, April 1, 2001 (http://www.ingentaconnect.com/content/beech/spp/2001/00000028/00000002/art00002).

Ministry of Economy, Trade and Industry (Japan). "The Challenges and Directions for Nuclear Energy Policy in Japan: Japan's Nuclear Energy National Plan." Agency for Natural Resources and Energy, December 1, 2006 (http://www.enecho.meti.go.jp/english/report/rikkoku. pdf).

Moavenzadeh, F. "The Center for Technology, Policy and Industrial Development: A Working Paper." (Cambridge, MA: MIT, 1999).

Moran, Dominic. "Israel Mulls Open Nuke Program," *Security Watch*, August 3, 2007 (http://www.isn.ethz.ch/isn/Current-Affairs/Security-Watch/Detail/?ots591=4888CAA0-B3DB-1461-98B9-E20E7B9C13D4&lng=en&id=53614).

Moran, Dominic. "Middle East Moves." ISN Security Watch (http://www.isn.ethz.ch/isn/Current-Affairs/Security-Watch/Detail/?ots591=4888CAA0-B3DB-1461-98B9-E20E7B9C13D4&lng=en&id=54151), accessed March 10, 2009.

Moss, Rayna. "Israel: Dimona Death Factory Exposed." WISE Nuclear Monitor, February 1, 2002 (http://www10.antenna.nl/wise/index.html?http://www10.antenna.nl/wise/beyondbomb/3-3.html).

Murawski, John. "Reactors Likely to Cost $9 Billion; Progress Energy Doubles Estimate." *The News & Observer*, October 17, 2008.

NPT Preparatory Committee. "Multilateralization of the nuclear fuel cycle/guarantees of access to the peaceful uses of nuclear energy." NPT/CONF.2010/PC.I/WP.61, Working paper submitted on the Multilateralisation of the Nuclear Fuel Cycle/Guarantees of Access to the Peaceful Uses of Nuclear Energy by the European Union, Vienna, April 30–May 11, 2007.

Nuclear Threat Initiative (NTI). "Israel Nuclear Facilities: Overview of Organizations and Facilities." (http://www.nti.org/e_research/profiles/Israel/Nuclear/3583.html), accessed March 10, 2009.

Nuclear Threat Initiative (NTI). "Israel Nuclear Facilities." (http://www.nti.org/e_research/profiles/Israel/Nuclear/3583.html).

Nuclear Threat Initiative (NTI). "Israel Nuclear Overview." August 2008 (http://www.nti.org/e_research/profiles/Israel/Nuclear/index.html).

Nuclear Threat Initiative (NTI). "Israel Profile: Biological Overview." 2008 (http://www.nti.org/e_research/profiles/Israel/Biological/index.html).

Nuclear Threat Initiative (NTI). "Israel Profile: Chemical Overview," 2008 (http://www.nti.org/e_research/profiles/Israel/Chemical/index.html).

Nuclear Threat Initiative (NTI). "Israel Profile: Nuclear Overview" (http://www.nti.org/e_research/profiles/Israel/Nuclear/index.html).

Nuclear Threat Initiative (NTI). "NTI in Action: Creating an International Nuclear Fuel Bank" (http://www.nti.org/b_aboutnti/b7_fuel_bank. html).

Nuclear Threat Initiative (NTI). "Submarine Proliferation: Israel Current Capabilities, 2008" (http://www.nti.org/db/submarines/israel/index.html).

O'Shea, R., et al. "Delineating the Anatomy of an Entrepreneurial University: The Massachusetts Institute of Technology Experience." *R&D Management* 37(1), 2007.

OptiSolar. "Topaz Solar Farm 550 Megawatt PG&E Solar Power Agreement" (http://www.optisolar.com/081408_opti_pge_sp.htm).

Oughton, D.H. *Causing Cancer? Ethical Evaluation of Radiation* (Oslo University, 2001).

Pagnamenta, Robin. "Reactors Will Cost Twice Estimate, says E.ON Chief." *The Times*, May 5, 2008.

Pakistan Atomic Energy Commission (PAEC). "PAEC Embarks on the Road to Promote Nuclear Desalination as a Sustainable Solution for Water Scarcity." PAKATOM Newsletter, March–April, 2008.

Pakistan Atomic Energy Commission (PAEC). "PAEC Provides Equipment and Services to CERN, PakAtom" (Islamabad: PAEC, September–October 2008).

Palestinian National Authority (PNA). "Dimona Reactor … a Mystery Threatening Middle East." PNA Press Release, Thursday, September 18, 2003 (http://www.scoop.co.nz/stories/WO0309/S00228.htm).

Photovoltaic Resources. "Large-scale Photovoltaic Power Plants" (http://www.pvresources.com/en/top50pv.php), accessed November 24, 2008.

Pike, John. "Dimona Negev Nuclear Research Center." Global Security, 2008 (http://www.globalsecurity.org/wmd/world/israel/dimona.htm).

Pike, John. "F-15I Ra'am (Thunder)." Global Security (http://www.globalsecurity.org/military/world/israel/f-15i.htm).

Pike, John. "Israel: Strategic Doctrine." Global Security (http://www.globalsecurity.org/wmd/world/israel/doctrine.htm).

Pike, John. "Military: F-4E Phantom II." Global Security, (http://www.globalsecurity.org/military/systems/aircraft/f-4e.htm).

Pike, John. "Negev Nuclear Research Center." Global Security (http://www.globalsecurity.org/wmd/world/israel/dimona.htm).

Pike, John. "Plutonium Production," Global Security, 2008 (http://www.globalsecurity.org/wmd/intro/pu-prod.htm).

Pike, John. "Soreq Nuclear Research Center." Global Security (http://www.globalsecurity.org/wmd/world/israel/soreq.htm).

Pike, John. "Special weapons Facilities: Israel." Global Security (http://www.globalsecurity.org/wmd/world/israel/facility.htm).

Pike, John. "Weapons of Mass Destruction: Dolphin." Global Security (http://www.globalsecurity.org/wmd/world/israel/sub.htm).

Pike, John. "Weapons of Mass Destruction: Jericho 1." Global Security, 2008 (http://www.globalsecurity.org/wmd/world/israel/jericho-1.htm).

Pike, John. Israel Special Weapons Guide Website, Federation of American Scientists (http://www.fas.org/nuke/guide/israel/index.html).

Pincus, Benjamin. "Atomic Power to Israel's Rescue: French-Israeli Nuclear Cooperation 1949–1957." *Israel Studies*, vol. 7, no. 1 (Spring 2002); (http://muse.jhu.edu/login?uri=/journals/israel_studies/v007/7.1pinkus.html).

Pincus, Walter. "Push for Nuclear-Free Middle East Resurfaces," *Washington Post*, March 6, 2005 (http://www.washingtonpost.com/wp -dyn/articles/A10418-2005Mar5.html), accessed December 20, 2008.

Porter, Gareth. "Iran Proposal to US Offered Peace with Israel." Inter Press Service, May 24, 2006 (http://ipsnews.net/news.asp?idnews =33348).

Post, Jerrold M. "Differentiating the Threat of Radiological/Nuclear Terrorism: Motivations and Constraints." Presentation to the IAEA Conference on Nuclear Terrorism Prevention, November 2001.

Post, Jerrold M. "Prospects for Nuclear Terrorism: Psychological Motivations and Constraints," in Yonah Alexander, Y. and Leventhal, P.A. (eds), *Preventing Nuclear Terrorism: The Report and Papers of the International Task Force on Prevention of Nuclear Terrorism* (Lanham, MD: Rowman & Littlefield, 1987).

Radkau, Joachim. *Aufstieg und Krise der deutschen Atomwirtschaft 1945– 1975* (Hamburg: Rowohlt, 1983).

Raj, Baldev (ed.). *The Saga of Atomic Energy Programme in India; Volume-II; Atoms with Mission: A Golden Jubilee Commemorative Volume (1954–2004)*. Department of Atomic Energy (DAE), 2005.

Ramberg, Bennett. "Should Israel Close Dimona? The Radiological Consequences of a Military Strike on Israel's Plutonium-Production Reactor." *Arms Control Today* (May 2008); (http://www.armscontrol. org/act/2008_05/Dimona).

Ramberg, Bennett. *Nuclear Power Plants as Weapons for the Enemy: An Unrecognized Military Peril* (Berkeley, CA: University of California Press, 1984).

[464]

Raphaeli, Nimrod. "The Middle East Ventures Into Nuclear Energy." The Middle East Media Research Institute, Inquiry and Analysis, no. 467, October 7, 2008.

Rauf, Tariq and Zoryana Vovchok. "Fuel for Thought." *IAEA Bulletin* 49/2 (March 2008).

Ravid, Barak, Aluf Benn and Assaf Uni. "Israel, PA Agree to Strive for Deal by End of 2008." *Haaretz*, November 27, 2007 (http://www.haaretz.com/hasen/pages/ShArt.jhtml?itemNo=928637).

Ray, Russell. "Nuclear Costs Explode." *Tampa Tribune*, January 15, 2008.

Reiche, Danyel. *Grundlagen der Energiepolitik* (Frankufrt am Main: Peter Lang, 2005).

Reichert, Mike. *Kernenergiewirtschaft in der DDR. Entwicklungsbedingugngen, konzptioneller Anspruch und Realisierungsgrad (1955 – 1990)*; (St. Katharinen, Scripta Mercaturiae Verlag, 1999).

Renewable Energy Policy Network for the 21st Century (REN21). "REN21 Renewables 2007 Global Status Report" (Bali, Indonesia: REN21, 2007).

Rizvi, Haider. "Israeli Arsenal Vexes Nuclear Negotiators." Inter Press Service, Inter Press Service, May 20, 2005 (http://www.antiwar.com/ips/rizvi.php?articleid=6033).

Rogers, Barbara and Zdenek Cervenka. *The Nuclear Axis: The Secret Collaboration Between West Germany and South Africa* (New York, NY: Times Books, 1978).

Ruchkin, S.V. and V. Loginov. "Securing the Nuclear Fuel Cycle: What Next?" IAEA Bulletin 48/1 (September 2006).

Russell, James A. "Nuclear Strategy and the Modern Middle East." *Middle East Council Journal* (Fall 2004); (http://www.mepc.org/journal_vol11/0409_russell.asp).

Sagan, Scott. "The Problem of Redundancy Problem: Why More Nuclear Security Forces May Produce Less Nuclear Security." *Risk Analysis* (August 2004).

Sahni, V.C. (ed.). *The Saga of Atomic Energy Programme in India; Volume-I; The Chain Reaction: A Golden Jubilee Commemorative Volume (1954–2004)*. Department of Atomic Energy (DAE), 2005.

Scheinman, Lawrence. "Prospects for a Gulf Zone Free of Weapons of Mass Destruction." Comments at SIPRI-GRC Conference, Stockholm, May 30–31, 2005.

Schell, Jonathan and Martin Sherwin. "Israel, Iran and the Bomb." *The Nation*, August 8, 2008 (http://www.thenation.com/doc/20080818/schell_sherwin).

Schiffer, Hans-Wilhelm. "Deutscher Energiemarkt 2007." *Energiewirtschaftliche Tagesfragen*, 58, 2008.

Schmemann, Serge. "Israel Clings to its 'Nuclear Ambiguity'." *New York Times*, June 21, 1998.

Schneider, Mycle and Anthony Frogatt. *The World Nuclear Industry Status Report 2007* (Brussels, London Paris, January 2008).

Schopenhauer A. and E. Payne. *The World as Will and Representation*, volume 1 (Dover Books, 1819).

Select Committee on Trade and Industry. "Minutes of Evidence," October 10, 2006, answer to question 480 (http://www.publications.parliament.uk/pa/cm200506/cmselect/cmtrdind/1123/6101001.htm).

Select Committee on Trade and Industry. "Minutes of Evidence," October 10, 2006, answers to questions 483 and 494 (http://www.publications.parliament.uk/pa/cm200506/cmselect/cmtrdind/1123/6101001.htm).

Shahak, Israel. *Open Secrets: Israeli Nuclear and Foreign Policies* (London: Pluto Press, 1997).

Shohat, Ella Habiba. "Reflections by an Arab Jew." *Bint Jbeil*, April 17, 1999 (http://www.bintjbeil.com/E/occupation/arab_jew.html).

Shropshire, D.E., et al. "Advanced Fuel Cycle Cost Basis." Idaho National Laboratory (INL), INL/EXT-07-12107 (April 2007).

Simpson, Fiona. "Reforming the Nuclear Fuel Cycle: Time Is Running Out." *Arms Control Today* (September 2008).

Slovic, P., B. Fischhoff and S. Lichtenstein. "Facts and Fears; Understanding Perceived Risk" in Shwing, R.C. and W. Al Albers (eds). *Societal Risk Assessment: How Safe Is Safe Enough?* (New York, NY: Plenum, 1980).

Smith, Gerard C. and Helen Cobban. "A Blind Eye to Nuclear Proliferation." *Foreign Affairs* (Summer 1989); (http://www.foreignaffairs.org/19890601faessay5960/gerard-c-smith-helena-cobban/a-blind-eye-to-nuclear-proliferation.html).

Snow, C.P. *The Two Cultures and the Scientific Revolution* (Cambridge University Press, 1959 [reprinted 1993]).

Soloman, Norman. "Israel's Future Leader? (Benjamin Netanyahu)." Alternet, January 6, 2006 (http://www.thirdworldtraveler.com/Israel/Benjamin_Netanyahu.html).

Solomon, Fredrick and Robert Q. Marston. *The Medical Implications of Nuclear War* (Washington, DC: National Academy of Science Press, 1986).

Stirling Energy Systems. "Solar One" (http://www.stirlingenergy.com/projects/solar-one.asp).

Stracke, Nicole. "Nuclear Development in the Gulf: A Strategic of Economic Necessity?" *Security and Terrorism* no.7 (December 2007).

STRATFOR. "Military: A New Ballistic Missile for Israel?" January 17, 2008 (http://www.stratfor.com/analysis/military_new_ballistic_missile_israel).

Sundaram, C.V., L.V. Krishnan, and T.S. Iyengar. *Atomic Energy in India:50 years*. Department of Atomic Energy (DAE), India, 1998.

Suzuki, Tatsujiro. "Nuclear Power in Asia: Issues and Implications of 'ASIATOM' Proposals." United Nations Kanazawa Symposium on Regional Cooperation in Northeast Asia, Kanazawa, Japan, June 2–5, 1997.

Telhami, Shibley. "2008 Annual Arab Public Opinion Poll." Survey of the Anwar Sadat Chair for Peace and Development at the University of Maryland (with Zogby International), 2008 (http://www. brookings.edu/topics/~/media/Files/events/2008/0414_middle_east/04 14_middle_east_telhami.pdf).

The 9/11 Commission Report: Final Report of the National Commission on Terrorist Attacks Upon the United States (New York, NY: W.W. Norton, authorized edition, 2004).

The Keystone Centre. *Nuclear Power Joint Fact-Finding* (Keystone: Keystone Centre, 2007).

Thomas, Stephen, Peter Bradford, Antony Froggatt and David Milborrow. *The Economics of Nuclear Power* (Amsterdam: Greenpeace International, 2007); (http://www.greenpeace.org/international/press/ reports/the-economics-of-nuclear-power).

Thomas, Stephen. "The Collapse of British Energy: The True Cost of Nuclear Power or a British Failure?" *Economia delle fonti di energia e dell' ambiente*, no. 1–2, 2003.

Tucker, Robert W. "Israel & the US: From Dependence to Nuclear Weapons?" *Commentary* (November 1975); (http://www.commentary magazine.com/viewarticle.cfm/israel-and-the-united-states-from-dependence-to-nuclear-weapons-5582).

Tulley, Andrew. "Radio Free Europe US: Bush Signals New Interpretation of Nonproliferation Treaty." Radio Free Europe US, March 16, 2005 (http://www.globalsecurity.org/wmd/library/ news/usa/2005/usa-050316-rferl01.htm).

UK Department of Trade and Industry. *Our Energy Future: Creating a Low Carbon Economy* (The Stationery Office, London, 2003); (http://www.berr.gov.uk/files/file10719.pdf).

United Nations (UN). "Security Council resolution 1540," April 28, 2004.

United States Department of Energy (DoE). *A Roadmap to Deploy New Nuclear Power Plants in the United States by 2010* (Washington: USDOE, 2001).

United States Department of Energy (DoE). *Analysis of Five Selected Tax Provisions of the Conference Energy Bill of 2003* (Washington: Energy Information Administration, 2004); (http://tonto.eia.doe.gov /FTPROOT/service/sroiaf(2004)01.pdf).

US Department of Energy (DOE). *Global Nuclear Energy Partnership Strategic Plan.* GNEP-167312 (January 2007).

US Department of Energy (DOE). *GNEP Overview Fact Sheet*, US DOE (July 2007)

US State Department. "Statement of Principles for the Global Initiative to Combat Nuclear Terrorism." Bureau of International Security and Nonproliferation, November 20, 2006 (http://www.state.gov/t/isn/rls/ other/76358.htm).

US State Department. "The Nuclear Suppliers Group." Fact Sheet, July 29, 2004 (http://www.state.gov/t/isn/rls/fs/34729.htm).

US Strategic Bombing Survey. "The Effects of the Atomic Bombs on Hiroshima and Nagasaki" (Washington, DC: US Government Printing Office, 1946).

Venugopal, V. (ed.). *The Saga of Atomic Energy Programme in India; Volume-III; Atoms for Health and Prosperity: A Golden Jubilee Commemorative Volume (1954–2004)*. Department of Atomic Energy (DAE), 2005.

Verter, Yossi. "Poll: Most Israelis Back Direct Talks with Hamas on Shalit." *Haaretz*, February 27, 2008 (http://haaretz.com/hasen/spages/958473.html).

Vivien, Philippe. "Developing Talent in the Energy World: Investing in People and Building Future." International Congress on Advances in Nuclear Power Plants (ICAPP), 2008.

von Hippel, Frank. "Managing Spent Fuel in the United States: The Illogic of Reprocessing." Congressional Staff Briefings, March 24, 2008.

von Hippel, Frank. "National Fuel Stockpiles: An Alternative to a Proliferation of National Enrichment Plants." *Arms Control Today* (September 2008).

von Hippel, Frank. *Managing Spent Fuel in the United States: The Illogic of Reprocessing*. A research report of the International Panel on Fissile Materials (January 2007).

Whitaker, Brian and Richard Norton-Taylor. "Revealed: Israel's Nuclear Site." *The Guardian*, August 23, 2000 (http://www.guardian.co.uk/technology/2000/aug/23/news.israel).

Wikipedia. "Three Mile Island Accident" (http://en.wikipedia.org/wiki/Three_Mile_Island_accident).

Windsor, Lindsay and Carol Kessler. *Technical and Political Assessment of Peaceful Nuclear Power Program Prospects in North Africa and the Middle East*. PNNL-16840 (September 2007).

Wisconsin Project on Nuclear Arms Control. "Israel's Nuclear Weapon Capability: An Overview." *The Risk Report* (July–August 1996); (http://www.wisconsinproject.org/countries/israel/nuke.html).

Worcester R. *Public Opinion in Britain and its Impact on the Future of Britain in Europe* (London: Institute of Directors, 2001).

World Energy Council (WEC). "2007 Survey of Energy Resources" (London: WEC, 2007).

World Nuclear Association (WNA). "Nuclear Desalination" (http://www.world-nuclear.org/info/inf71.html), accessed November 24, 2008.

World Nuclear Association (WNA). "Nuclear Power in Sweden." 2008 (http://www.world-nuclear.org/info/inf42.html).

World Nuclear Association (WNA). "The Nuclear Renaissance" (http://www.world-nuclear.org/info/inf104.html), accessed November 24, 2008.

World Nuclear Association (WNA). "World Nuclear Power Reactors 2007-09 and Uranium Requirements." April 1, 2009 (http://www.nei.org/resourcesandstats/nuclear_statistics/worldstatistics/).

Zander, Walter. "Israel and the Anti-Colonial Movement: Orientation Toward the Orient or the Occident?" *Jewish Observer and Middle East Review*, February 24, 1956 (http://walterzander.info/acrobat/Israel%20and%20anti.pdf).

Zunes, Stephen. "The Release of Mordechai Vanunu and U.S. Complicity in the Development of Israel's Nuclear Arsenal." *Foreign Policy in Focus*, April 21, 2004 (http://www.fpif.org/fpiftxt/1134).